ARID LANDS OF AUSTRALIA

Proceedings of a Symposium held in the
Academy of Science, Canberra
May 1969

ARID LANDS OF AUSTRALIA

Edited by

R. O. SLATYER & R. A. PERRY

AUSTRALIAN NATIONAL UNIVERSITY PRESS

CANBERRA 1969

Registered at the General Post Office, Sydney, for transmission by post as a book.

SBN 7081 0693 5

Library of Congress Catalog card no. 73-93688

National Library of Australia reg. no. AUS 69-2289

CONTENTS

PREFACE

HISTORY HAS SHOWN that man's utilisation of the world's arid lands has, in general, caused progressive deterioration of the natural vegetation, of the animal resources, and finally of the landscape itself.

In Australia, although little arid land settlement extends for much longer than 100 years, limited areas have already been seriously damaged, and signs of increasing degradation are apparent in most regions currently being grazed. There seems little doubt that progressive deterioration will occur if present practices are continued.

In order to arrest this tendency and preserve the arid lands for the future in a manner which is consistent with non-destructive utilisation, it is clearly desirable that the national character of the problems should be recognised and steps taken to solve them.

The problems are complex. They call for an assessment of the resources of the arid lands, the changes that have resulted from utilisation and the significance of those changes, and planning to assure long-term conservation of the arid lands.

Differences of opinion as to the best means of arid land utilisation are inevitable. In particular, there is a conflict of interest between those whose primary concern is to conserve the arid land resource and those who wish to obtain an economic return from it.

In this regard it is important to realise that, while utilisation can be consistent with ecological principles and economic realities, the introduction of non-indigenous animals to any ecosystem must bring change to the ecosystem. Preservation, in the sense of avoiding other than natural changes, is therefore quite incompatible with utilisation. This is particularly true in arid ecosystems where the plant, animal, and environmental factors associated with ecosystem stability are so delicately balanced. The introduced animals have different grazing habits from the indigenous species, different plants may be grazed and, in any case, the plants are likely to react differently to the altered character and pressure of grazing. The key factor on which stable, long-term utilisation depends is whether the change so induced represents progressive deterioration or is a shift towards a different but still balanced ecosystem.

The criteria for this assessment are, as yet, poorly defined. However, change is acceptable, in ecological terms, as long as it is associated with a succession in which different species, better adapted to the altered herbivore component, become the dominant vegetation component; as long as the energy exported through the animal does not exceed that flowing into the ecosystem, so that there is not a continuous run down in biomass, and as long as the stability of the landscape itself is not impaired.

If criteria such as these are established, most contemporary grazing practices, in truly arid lands (where there is no well defined growth period at the same time each year), do not appear to meet them. Not only is there an absence of any definite succession of the type just mentioned, but the deliberate attempt, consciously or unconsciously, of the grazier to maximise production, at the expense of the other components of the ecosystem, clearly tends to cause progressive deterioration.

If this is to be avoided, there is a challenge for research to develop management standards which maximise for long-term ecosystem stability rather than for short-term animal productivity. In some of the major communities at present being grazed this objective may be inconsistent with any utilisation at all. The present species may be so vulnerable to grazing that, instead of natural succession, landscape deterioration occurs, and the deliberate introduction of new species, better able to withstand grazing, is impossible either because they are unavailable, or because of the cost of introduction, or both. In some other communities, it seems highly probable that the natural species can be utilised as long as the grazing pressure is carefully regulated.

Before any such changes in property management could be introduced, it seems that many other changes would have to occur. Some of these are linked with the ecological principles of management already mentioned, which would require, for effective implementation, a larger and more suitably trained extension service than now exists. Others are linked with the whole economic climate in which management decisions must be taken. In turn, this has several facets associated with the flexibility of the grazier's financial arrangements, and the number of alternative management practices he can pursue. Ecologically, the land tenure provisions, the ease with which the grazier can transport stock on to or away from his property, and the arrangements by which he can borrow money from financial institutions are of primary importance.

The interplay between these factors can be seen in a hypothetical situation where a region may be enduring a drought. Plant growth is at a standstill. Ecologically it is desirable that the grazier should reduce grazing pressure by reducing his stock numbers; therefore it is desirable that he should be positively motivated to do so, not only because of the condition of his range but also by other factors. Thus the transportation system should be such that he can quickly and economically shift his animals and the financial system should enable him to borrow money on a long-term basis. In addition, it would seem desirable that the land tenure system should be such that it is stock number and range condition, rather than property size, that should determine land rental.

This volume deals primarily with the Australian arid land resources, with the patterns of utilisation and management which have been developed since the date of initial stocking by domesticated animals, and with a number of the factors which affect these practices. Although some aspects have not been covered, the volume provides a comprehensive picture of the Australian arid zone both in historical and ecological perspective, and in terms of the socio-economic environment in which utilisation is now carried out.

Although this picture is not one of uniform optimism, it is clear that many of the problems which tend to mitigate against stable, long-term utilisation have been recognised and some are under study. However, much of this activity is superficial and geographically fragmented. Before it can be really effective the problems need to be looked at nationally rather than within the artificial confines of State boundaries. In this context the success of bodies such as the Australian Water Resources Council suggests that a similar organisation might fulfil a valuable role in arid land utilisation. If established, it could consider a wide range of matters from research through extension to management, although the development of the Australian Arid Lands Research Association may mean that it need not cover research liaison, except in a complementary way. A body such as this could possibly play a major role in ensuring that the arid lands are used wisely and conserved for future generations.

The chapters of this book represent the papers presented at a symposium organised in Canberra in May 1969, by the Australian Academy of Science. The costs of the conference were substantially defrayed by financial support from Elder Smith Goldsbrough Mort, the Bank of New South Wales, Commonwealth Development Bank, Dalgety and New Zealand Loan, Colonial Sugar Refining Co. Ltd., Ian Potter Foundation, Conzinc Riotinto of Australia and the Reserve Bank of Australia.

R. O. SLATYER

Canberra
July 1969

NOTES ON CONTRIBUTORS

DR ALAN BARNARD is Senior Fellow in Economic History at the Australian National University. His research over the past decade has been closely associated with the history of Australia's pastoral industries, which he has approached through the history of wool marketing and through business biography and company history. He has published *The Australian Wool Market 1840-1900, Visions and Profits: studies in the business career of T. S. Mort*, and (ed.) *The Simple Fleece*.

DR T. W. BOX is the Director of the International Center for Arid and Semi-Arid Land Studies and Professor of Range Management at Texas Technological University in Lubbock, Texas. He worked with the Rangelands Research Unit of CSIRO in 1969. He is the recipient of the E. Harris Harbison Award for Distinguished Teaching and has developed curricula for training scientists both in America and overseas.

R. L. CASKEY is grazier of Byrnedale Station in the Broken Hill area. He is Past President of the Pastoralists Association of West Darling, Chairman of the Broken Hill Pastures Protection Board, and member of the N.S.W. Pastures Protection Boards Council of Advice.

R. W. CONDON, Assistant Commissioner for Western Lands, Western Lands Commission of N.S.W., graduated in Agriculture at the University of Sydney. Until 1968 he was Special Soil Conservationist, Soil Conservation Service of N.S.W., and worked on erosion survey and land assessments. He led a consultant team from the Soil Conservation Service of N.S.W., which studied drought and erosion problems in the Northern Territory. In 1968 he joined Western Lands Commission where he is concerned with arid and semi-arid land administration and management.

DR A. B. COSTIN, Assistant Chief, Division of Plant Industry, CSIRO, was formerly Principal Research Officer, Soil Conservation Authority of Victoria and Soil Conservation Service of New South Wales. He has worked extensively on ecological, hydrological, and conservation problems of mountain environments. His present duties include general responsibility for the Division of Plant Industry's Riverina Laboratory at Deniliquin.

DR N. H. FISHER, Director of the Bureau of Mineral Resources, is a graduate of the University of Queensland. He worked at Mt Isa and in New Guinea before joining the Commonwealth Government and was Chief Geologist of the Bureau of Mineral Resources from its formation in 1946 until 1969. He is convener/deputy chairman of the Technical Committee on Underground Water of the Australian Water Resources Council.

DR W. J. GIBBS is Director of Meteorology in Australia and First Vice-President of the World Meteorological Organisation. He is a science graduate of Sydney University and the Massachusetts Institute of Technology. Melbourne University recently awarded him a degree of D.Sc. (Honoris Causa). His published papers include works on tropical and antarctic meteorology and more recently on drought occurrence in Australia.

DR JAMES R. GRAY is a Professor of Agricultural Economics and Agricultural Business at New Mexico State University in Las Cruces, New Mexico. His teaching and research for the past twenty years have concentrated on range and ranch economics with the United States Department of Agriculture and New Mexico State University. He has published a textbook entitled *Ranch Economics* (1968).

DR R. L. HEATHCOTE, Senior Lecturer in Geography at the Flinders University of South Australia, graduated from University College, London, the University of Nebraska, and the Australian National University. He has worked on pastoral land appraisal in semi-arid Australia, on which he has published *Back of Bourke* (1965), and on agricultural land appraisal on the Great Plains of the U.S.A. His current research is concerned with the problems of land settlement in arid Australia.

Professor E. S. HILLS is Research Professor of Geology and Deputy Vice-Chancellor in the University of Melbourne. His work in geomorphology and regional geology led to his involvement with Army research and he travelled widely in central and north Australia during the war. He has subsequently been Chairman of the National Committee for Hydrology of the Academy of Science, Chairman of the UNESCO Advisory Committee on Arid Zone Research, and Advisor to the Government of Egypt on the development of the Desert Research Institute, Cairo.

DR J. H. LEIGH, Officer-in-Charge of the Riverina Laboratory, CSIRO Division of Plant Industry at Deniliquin, N.S.W., graduated in 1960 from the University of the Witwatersrand, Johannesburg, South Africa. He later studied the significance of plants and grazing animals in the nitrogen cycle of one ecosystem at the Grasslands Research Institute at Hurley, U.K. He joined the staff of the Riverina Laboratory in 1961 where his main research interests have been the ecology of semi-arid plant communities including diet selection by sheep, and the effects of grazing on the dominant species.

Professor J. A. MABBUTT, Head of the School of Geography at the University of New South Wales, and formerly Principal Research Scientist, Division of Land Research, CSIRO, is a graduate of Cambridge University. He studied desert landscapes in southern Africa whilst on the staff of the University of Cape Town, and since coming to Australia has also worked mainly in the arid zone.

A. R. MAIN, Professor of Zoology, University of Western Australia, is a graduate of the University of Western Australia, who, since 1952, has studied the adaptations to aridity of a wide variety of native animals. His work on kangaroos and wallabies has proceeded under both field and laboratory situations and has concentrated on the role played by physiology in the survival of arid zone populations.

J. C. NEWMAN, District Soil Conservationist, Soil Conservation Service of N.S.W., graduated in Agriculture at the University of Sydney. Since 1958 he has been working on reclamation techniques and management of arid and semi-arid lands in western N.S.W. In 1965 he was a member of a consultant team of the Soil Conservation Service of N.S.W., which studied drought and erosion problems in the Northern Territory.

MR J. C. NOBLE was employed as an agronomist by the N.S.W. Department of Agriculture from 1959 to 1965 during which time he was concerned primarily with irrigated crops and pastures in the southern irrigation districts of N.S.W. In 1965 he joined the staff of the CSIRO Riverina Laboratory at Deniliquin and has since become increasingly involved in dryland agronomy, particularly the establishment of sown pastures in the winter rainfall semi-arid zone.

R. A. PERRY is Leader of the CSIRO Rangelands Research Unit established in 1969. Previously he was leader of the Natural Resource Survey Section of the CSIRO Division of Land Research where he gained an extensive experience of the resources of arid Australia and the grazing industries based on them. In 1960-1 he spent a year in U.S.A. during which he studied range science at the University of Arizona and visited most of the range research stations in the western states. He has consulted on range research in India, Pakistan, and the Middle East. Since 1967 he has been Australian Arid Zone Research Liaison Officer.

W. J. RICKETTS is Director of the Northern Division of the Commonwealth Department of National Development. The Northern Division is concerned with the development of the northern half of Australia in which the pastoral industry provides the dominant land use. The Division has as a particular task the investigation and evaluation of the program of beef road development in northern Australia.

Professor R. O. SLATYER is Professor of Biology, Research School of Biological Sciences, Australian National University, Canberra. Previously he was

Associate Chief and Head of the Environmental Biology program at CSIRO Division of Land Research, Canberra, and was Arid Zone Research Liaison Officer, 1960-7. He has been interested in the effect of environmental factors on the physiological responses and ecological distribution of plants, with special interest in stress physiology and plant-water relationships.

E. J. WARING is an Assistant Director of the Bureau of Agricultural Economics and Head of its Wool and Beef Research Branch. He graduated in Agriculture from Sydney in 1944 and after working for some years in agronomy joined the Faculty of Agricultural Economics at the University of New England as a Senior Lecturer in Farm Management, being awarded the M.Ag.Ec. degree in 1962. He is interested in resource allocation problems under conditions of uncertainty.

DR C. WAYNE-COOK is Head of the Range Science Department at Colorado State University, U.S.A. His research and teaching over twenty-seven years is concerned with range livestock nutrition, range rehabilitation, and range ecosystem ecology, on which he has published many papers.

Professor MARTYN J. WEBB is Foundation Professor of Geography at the University of Western Australia and was formerly of Balliol College and the School of Geography, Oxford. His broad field of interest lies in the location of economic activity, land use, and economic development with special reference to Australia and South Asia. He is presently concerned with problems of northern Western Australia and the Cauvery Delta, India. He has recently completed a study based upon the mapping and the reconstruction of Greco-Roman agriculture in the semi-arid to sub-humid area in Cyrenaica, North Africa. His studies in Australia, India, and North Africa are all directed towards the problem of resource utilisation and physical planning.

1

HISTORY OF THE WORLD'S ARID LANDS

EDWIN S. HILLS

I WOULD LIKE to commence this history of the world's arid lands by recalling that it is about twenty years since UNESCO, which itself was almost brand new at the time, first set up a panel to advise its Director General on what might be done to support the development of arid lands, particularly through the application of scientific knowledge and the encouragement of research.

This move followed a discussion at the United Nations itself, where it was pointed out that over a third of the land area of the earth is arid: that this third holds about one-eighth of the world's people, and that apart from certain favourable areas such as the great river valleys, living conditions in the truly arid regions are harsh and productivity low. So, with the aim of improving the lot of the millions who live out their lives in the scorched deserts and semi-deserts, UNESCO set up an Arid Zone Program in 1951.

Australian scientists played a major role in the preliminary planning and in the resulting program itself, which fired so much enthusiasm that it was given more funds and was transformed into a major project in 1957, covering North Africa and the Near and Middle East as far as Pakistan and India. The whole effort was certainly one of UNESCO's most successful undertakings; but it was not long after the major project had been launched that a representative of India at the General Conference of UNESCO, a lady, rose in the gathering of delegates of Member States and questioned the practical results of the UNESCO program, which had been based essentially on a long-term plan for scientific study and research. Her questioning illustrated two things about arid land development: first, that most people expect quick action and quick results; and second, that any major operations are very much entangled with governments and politics, and the lady's real aim may well have been to induce UNESCO to spend more of its internationally derived funds in India rather than elsewhere.

I mention such matters because I must, in dealing with the whole range of arid climates and the complex eco-systems of land, vegetation, animals and man, also deal especially with the things that only man can do to modify his environment, and also give some attention to his motivation.

One should recognise, to start with, the enormous complexity of cultures at levels that we call primitive—of men who live as the lower animals do in deserts; but if their cultures are complex and their motives hidden, they do have their ingenious practical devices. The Bushman of the Kalahari induces grandmother to sit all day at a sip-well, sucking up mouthfuls of water from the damp sand of some nearly dry soak and squirting them into a container for communal use. The Bushmen store their little supplies of water and food in hollow gourds or ostrich eggshells hidden in the sand and under the rocks, very like the ant. For men who live like this, naked or near-naked, without permanent dwellings or agriculture, the natural desert is their *Heimat*. They belong with it—they react directly to it—and any thought of attempting to change what the Great Spirits have provided might well be cast out as sacrilegious, or at least anti-social. It is surely quite remarkable that in this age when some men are about to land on the moon, others, like the Bindibu, still live so intimately entwined with Mother Earth.

The step from the nomadic hunter and food gatherer to the nomadic herdsman is often regarded as not a great one, and historically it may not have been. But as of today, the nomads of North Africa and Asia are culturally remote indeed from the primitive hunters, not only in the level of their material existence but also in that they live symbiotically with agricultural and urban communities. Their ancient habits of warlike pillage have now been replaced by more peaceful penetration. It appears that they not only enter the caravanserai for short stop-overs to purchase or barter goods and food for themselves and their animals but some may alternate between urban and nomadic existence.

Detailed studies of nomadism have clearly shown, however, that the rhythms and patterns of movement and rest of the herdsmen are as intimately connected with the regional geography—the distribution and growth-patterns of herbage, seasonal weather, and trade routes—as are the lives of primitive hunters and food gatherers. To change such cultural patterns, tied apparently irrevocably to the environment—or to change the environment and thus to affect the cultures—required of old a combination of physical power and psychological persuasion—if that is a strong enough word for the priestly support granted to royalty to maintain the ruling hierarchy. And let me emphasise that each of the major arid areas of the world is unique in its climate and micro-climates, its rocks and soils, vegetation and animals, and races and cultures of mankind.

One cannot, in a short address, hope to describe the complex interplay of such factors for all the world's arid areas, but one example may suffice to demonstrate the magnitude of the exercise. This is the history of Mesopotamia and the whole Tigrus-Euphrates basin as told by R. O. Whyte in UNESCO's *History of Land Use in Arid Regions*. From archaeology, carbon-dating, archaeo-magnetism, and pollen analysis, and from bibliographic references starting with cuneiform tablets and going on to biblical texts and later documents, it is possible to get a remarkably long history and pre-history of Mesopotamia dating back at least to 6000 years B.C. and, with less certainty, to 8000 B.C., that is back to Neolithic times.

The decline of this cradle of civilisation, which is so often quoted as an example of the ephemeral in human history, cannot be basically ascribed to climatic change, since although there may have been a very arid episode sometime before 2000 B.C. following a previous longer period of rather pluvial conditions, it is agreed that since Roman and Byzantine times the climate as a whole has been more or less constant, with minor variations having a period of perhaps a few hundred years. This is not to say, however, that micro-climates have remained the same. The areas that are still covered by anything in the way of vegetation have suffered ecological regression for thousands of years, and micro-climates have been seriously affected by progressive desiccation, especially of rocky soils. The trees in such areas were being burnt over ten thousand years ago, and timber-getting for fuel and for wood to make tools and houses, increasing as communities became larger, clearly would have had a devastating effect. It is said that 'the bald mountains and foothills of the Anatolian Plateau and Iran stand as stark witnesses of millennia of uncontrolled utilization'. What results this had on run-off and flooding, siltation and soil deterioration can only be guessed at, but they must have been of major proportions.

It is, indeed, remarkable that despite the poetic references to shade in biblical and other writings, most of the people in the region care nothing for trees and will cut them down without thought of the future, so that experimental regeneration plots have required protection by armed troops.

In ancient Mesopotamia the earliest development of small settlements of an agricultural kind, with cultivation of wheat and barley by the digging stick and hoe, the domestication of sheep, goats and cattle, and the beginnings of a ceramics technology, was in the uplands, the catchment areas of the great rivers. It was only later that people shifted to the lowlands where plow irrigation agriculture and expansion of technology with the appearance of craft specialisations, went on. The potter's wheel, the cart, the chariot, sail, copper metallurgy, and early phases in the development of writing all then made their appearance on the plains, where rapid growth of food surpluses permitted the concentration of people in cities, with their attendant priestly hierarchies and political groupings. Thus emerged the concepts of city states, of warfare and fortifications, organised trade and, in fact, life very much as we know it today. It was in the course of the Bronze Age that the agricultural economy of this part of the ancient world assumed the form which it was to retain with very little change until mediaeval times. By 1200 B.C. the best adapted types of cereals had been selected and developed. The methods and the tools then used changed little over another millennium except that iron took the place of bronze. All the domestic animals of the present day were already serving man, although the horse was limited to the chariot. The fig, apple, pomegranate, peach, and mulberry were cultivated and dates were an important product of Southern Mesopotamia. A considerable variety of vegetables was produced and beer was brewed.

How far back did all this go? The origins of agriculture are placed before the beginning of Neolithic time, because the earliest archaeological records of the Neolithic present a picture of an already accomplished domestication of plants and animals, and of peasant and pastoral life resembling the basic primitive conditions that may still be found in some parts of Asia; but at that time Mesopotamia was not considered suitable for human occupation and was apparently used chiefly for grazing in spring and early summer, and for hunting. Southern Iraq, or Lower Mesopotamia, was first settled in the Al Ubaïd period (about 4200 B.C.) by the Sumerians. A well-developed system of land ownership evolved, with money-lenders, bankers, contracts between landlords and cultivators, and indeed all the complexities of civilised life. The land-tenure system was included in the Laws of Hammurabi. Alexander the Great, who conquered the Persian kings in 331 B.C., undertook to urbanise the country. Greek and Macedonian colonists were brought in in this deliberate decentralisation and the Hellenistic influence was strong especially in the higher classes of society. The colonists were given land, a farm, a house, seeds, and cattle, and the land was allotted to the cities and was cultivated by the peasants attached to these. It was about this time that rice first appeared in Iraq, probably as a result of the contacts made with the East through Alexander's expeditions to India.

This period of colonialism probably marks the acme of the Mesopotamian civilisation, for although the Romans, who conquered the north soon after, built temples and arenas, the country was already divided in Roman times and subsequently several different ruling races occupied parts of it for various periods. But before continuing, we should note that in the Tigrus-Euphrates valley there are two great natural problems—flooding and siltation on the one hand and salinisation on the other—and it was with these that the peasants had to cope, bewildered as they must have been by the Greco-Roman and later cultural domination to which they were exposed.

Siltation was very rapid and it is possible to delineate the historical changes in the complex delta system entering the head of the Persian Gulf in a series of maps. Siltation also affected the irrigation canals, which needed to be continually freed from deposits. Salinisation was the other great problem, and a special study has been made of the history of salinity in the region. The present soils in the irrigated plain are alluvial and largely man-made as a result of irrigation. The first soil has been dated at 4000 B.C. and is now some 30 feet below the surface at a site called Tell Asmar, and up to 6 feet at Ur.

The ground water which moves through the old settlement soil layers is extremely saline and is the immediate source of salts today.

A rough outline of the distributional picture of former salinity has been worked out from carbonised plant remains, imprints in pottery and bricks, and references in ancient documents; even the types of crops grown have been discovered, as well as whether they had any relation to increasing salinity.

Types of wheat and barley cultivated in Northern Iraq from about 6000 B.C. to 4500 B.C. were very close to the wild species, while Southern Iraq, which was deficient in rain, still lay uninhabited and visited only by nomads. When Southern Iraq was finally settled, about 4200 B.C., settlers brought Emmer wheat and barley—not the two-row variety of the north, but a six-row variety of unknown origin. The only other field crop was flax. The cultivation of Emmer wheat and barley with the new techniques of irrigation continued through to 3500 B.C. and the crops appeared to have been of high quality judging from imprints of grain on ceramics. New crops began to appear about 3300 B.C.—naked wheat and millet. Evidence shows that a fall in the acreage of the less salt-tolerant wheat in favour of the more salt-tolerant barley coincides with the early references to salinity in Southern Iraq. Wheat as a proportion of total crop acreage in Southern Babylonia was 16 per cent in 2400 B.C., 1·8 per cent in 2100 B.C., and zero in 2000-1700 B.C. By way of comparison, in the rain-fed areas wheat represented 32·9 per cent of the cultivated land in 2300 B.C.

There was a particularly close relation between irrigation agriculture and the presence of stable and vigorous central governments. When the central governments weakened, internal conditions became disturbed and disastrous abandonments of land took place accompanied by incursions of nomadic tribes and damage to the irrigation systems. The evidence shows that after some 1,000-1,500 years of successful irrigation which produced the brilliant Sumerian civilisation, a very serious salinity problem developed from about 2400 to 2000 B.C., with detrimental effects on crops and yields. When the country became progressively weakened economically, the long political and cultural supremacy of the south could not be maintained and the last dynasty, the third dynasty of Ur, fell. By 1700 B.C., when wheat had completely disappeared in the south, the centre of civilisation shifted to the north, where Babylon under Hammurabi began to emerge. From this disastrous first salting up of its lands, the south never fully recovered, although human skill and tenacity made it possible to maintain a certain amount of production. With deteriorating dykes and canals the system began to fail, and in the seventh century A.D. a great flood such as had never before been seen wrecked everything. Subsequent invasions consolidated the ruin; but according to Toynbee these invasions were not the cause but the consequence of a decline in population and prosperity, which was in itself due to internal social causes, especially in the seventh and thirteenth centuries A.D. when civilisation was at such a low ebb and insecurity was so extreme that nobody had either the means of investing capital nor the motive to employ energy in river conservancy and irrigation work. In the seventh century it was the great Roman-Persian war, followed by the overrunning of Iraq by the primitive Muslim Arabs, and in the thirteenth century the Mongol (Mogul) invasion, which dealt the decaying society its *coup de grâce*.

It is not possible to deal in any detail here with other ancient civilisations in arid lands; but let us note that despite the tremendous differences of terrain, as for example between the Nile Valley and the high deserts of the Andes, there are remarkable similarities in the history of occupancy. An early period when these regions were not permanently settled was followed by the development of well-controlled irrigation going along with the rise of highly organised civilisations, some of which lasted for thousands of years. In Mesopotamia, progressive changes in crops and farm practices were introduced with obvious understanding of the soil conditions; one can readily imagine the frustration of knowing, as I am sure they did know, that the end had finally come and, and again I would guess, knowing how to deal with it if only they had had adequate resources of power. But they were before their time. Even today, when we know what should be done in the Indus Valley, the lack of cheap power for pumping has been one of the main hindrances to soundly-based attempts at lowering the saline groundwater-table which was resulting, a few years ago, in the loss of more land due to salinisation than was being brought into use by new major irrigation schemes.

But I cannot play down the results of war, invasion and conquest, slaughter, destruction, and pillage, which reduced what resources were left to a quite inadequate residuum, or, as with the overthrow of the Inca civilisation, to zero. But why go back to the Conquistadores? It was similar rapacity that led, only two or three generations ago, to the utter despoliation of the Algerian oak forests by European landlords anxious to make their fortunes in their own lifetimes, but causing thereby complete deterioration of the semi-arid hill country and thus changing the whole delicately balanced system of run-off, stream relations, erosion, deposition of detritus and, of course, the long-established and stable land-use practices of the indigenous people.

So on the one hand we seem to see stable and adjusted systems of land use, with people attempting to readjust to changing conditions over thousands of years, and with considerable success; and on the other hand, plain exploitation of some resource immediately available.

With a policy of conservancy, a lot of capital and labour must go into the development of arid lands. To take an example or two from the Old World, I may quote the *qanats* as they are known in the Middle East—*foggeras* in Africa—by which underground water in alluvial cones is collected in tunnels and gradually led out to the surface downslope. These are of ancient origin, and it is estimated that there are some 40,000 of them in Iran alone, aggregating 100,000 miles in length and yielding 20,000 cubic yards of water a second. They are in use from western China to the Atlas Mountains, and were introduced into Chile and Peru by the Spaniards. Dixey has pointed out that the work, effort, and cost invested in *qanats* since the first was constructed about 500 B.C. at Persepolis is stupendous, but their operation is much cheaper than any water delivery system requiring power-operated pumps.

One sees too in northern Egypt the *kharms*—ridges built up by scraping rock and soil together to create a rectangular field onto which run-off from the ridges collects, so concentrating the low rainfall on a restricted area. These were established by the Romans when north Africa was the granary of Rome. It is a simple proposition and was apparently relatively successful.

The Nabatean minor irrigation works in the Negev, where small check dams in the lateral valleys collected enough water to supply vegetable and crop growth on the alluvium of the larger valley, are now well known. These systems silted up long ago, but are being re-established by the Israelis. Each one of these examples represents an ingenious kind of nibbling at the desert, but is also in the tradition of conservancy, and as R. O. Whyte in his article in the UNESCO book comments, a system based on conservation of natural resources and their improvement costs much more than an exploitative system.

Sir Samuel Wadham has written that 'the courage of new investment of money and sweat and enterprise which has the micro-climatic odds against it may be magnificent, like the charge of the Light Brigade at Balaclava, but it does little in forwarding man's progress in the war against the desert'. Now these remarks may be true; but at the same time what they really highlight is another angle and that is the actual searching for micro-climatically suitable areas for development in broad regions, or in a similar vein what one might call micro-agricultural practices and changes in practices, which, in fact, may permit more sympathetic use of land. By this I imply the totality of the researches that are being undertaken in plant-soil-water relations, reduction of evaporation, plant and animal breeding, and a host of others. I call them 'micro' in the same sense that Sir Samuel used the word in referring to micro-climates. He meant that it was the short period of adverse climate, whether of high temperature or low, that did most of the damage. I imply that much of our research is directed also at particular details of relationships which become critical in a naturally adverse environment. With major irrigation, although this has its problems and they are not small, the answers are often clearer because ample water is provided; but they are clearer with one big question, and that is with regard to the social side—education of people, including politicians as well as farmers, in proper land-use practices, and also the requirement for large funds, in general from the taxpayer, for major works.

It is at this level that we are entering the realm of controversy, for it is not everyone who believes that this expenditure of energy and capital in arid lands is the best use of these resources, especially in a country like Australia where we have a lot of better-watered land which it is still possible to develop further. It is always a source of wonderment to me that the total agricultural and pastoral productivity of Great Britain should equal that of Australia and New Zealand put together, so I look forward to some lively controversy on the subject of cost-benefit studies.

Now I would like to venture one or two conclusions. It seems that there is a primitive stage or condition in which men live with the land in a state of nature as hunters and food gatherers, and that this condition can and does persist for thousands of years. Nomadic herdsmen I put aside as too hard to comprehend, but what seems to be the chief characteristic of a more advanced civilisation, with cities and architecture and so forth, is perhaps the ability to conceive and execute the organisation of labour and the accumulation of capital goods.

Perhaps the temples and monumental structures that go along with this stage are incidental products of this major new factor that seems to be involved. At least one can say that every civilisation of any consequence in arid and semi-arid lands showed some manifestation in this style, in Mesopotamia, Egypt, and Greece, in the realms of the Incas and the Aztecs, and in the stupas and cave temples of India.

It would be an interesting point to argue whether this ability to organise a labour force within a stratified community evolved with indigenous peoples or whether it was, as it certainly was with the Incas, something additive and brought in by a conquering race. One is in either case tempted to hypothecate that the origination of concepts such as mass organisation and the accumulation of capital that can go along with it were as much inventions, arising *ad hoc* in the mind of some man and acceptable to his colleagues, as was any invention in the physical world. Having reached a stage of development based on conservancy and capital, civilisation seems to demand continuing expenditure on experimentation, including experiments in the development of marginal lands. Consolidation and expansion are perhaps deeply ingrained along with the organising skills; but whatever the natural propensities of people may be, it does also appear that there is a strong conditioning process. Conquerors ravish and destroy; but their descendants, left to cohabit with the indigenes, themselves often start anew the development and the toil that goes with it. Broadly speaking and but for a few races of mankind, this seems to be a deep and fundamental thing. And after all it may be very simple. Men have energies; they find themselves, after some migratory impulse, in a region; they can't help but develop it as best they may, like the peasant who, after a volcanic eruption, returns to the site of his old farm, now covered with volcanic ashes and lava. He has no other home to which he may go.

So the rationalists, to judge from the history of Australian arid land development as well as from that of other countries, fight against a tide which is strongly emotional. And at the same time they have to cope with the exploiters, and I don't use the word with any derogatory implications, but to indicate the attitude of 'take it as and where you find it'. This applies to land and, of course, to mineral deposits. As to these latter, we need note only that the large-scale exploitation in mines belongs to that same stage of organisation with which the irrigation civilisations were concerned. The oldest mine map is carved on a rock in China; but probably the next oldest is Egyptian, referring to a gold mine in the Eastern Desert at the time of Seti I, some 1300 years B.C. I don't propose to discuss the effects of mining, but merely to indicate that they are often like those of the invader and conqueror of old, in relation to the indigenous people of a region newly to be developed. This we have seen with oil in the Middle East, and now in Australia with oil and minerals. It affects the local sophisticated people generally advantageously, but with some reservations; but as to the true indigenes—in Australia the Aborigines, in the Middle East and North Africa the nomadic tribes—great adjustments are called for. I will only say in this context that we may in the end come to regard favourably the preservation of large areas of arid land in a state of nature, or near nature. We might use those areas for large-scale exercises, say for rocketry; for recreation, and for regeneration; and, who knows, as true reserves for our Aboriginal people where they might live without too much disturbance of their traditional ways of life while time brings its slow changes and greater understanding.

Now as to those optimistic people, the developers: they do get a lot of sympathy and practical help when the luck is against them, either because of what they call drought (but which one might, in the environment we are discussing, call expectable low rainfall periods) or because of so-called unseasonable rains—also expectable by the way—that may ruin the crops.

I freely admit to being emotionally involved with government subsidies and assistance for arid land development; but I must say, and I take myself to be fairly normal, that my emotions also urge me to support conservation and planned scientific development. To what extent I am conditioned by being an Australian I don't know; but if I am rational I think I can see in the history of arid lands a record of achievement that is impressive indeed, even where failure eventually occurred, but only after millennia of intensive land use. I believe too that with the aid of present scientific knowledge and technology, development such as took place in Mesopotamia, Persia, the African oases and elsewhere could have been continued indefinitely.

The land is there forever and in the end it will certainly be used for purposes that we cannot at present foresee. So we must tend it and modify it but not destroy it or ravish it. And as to our researchers, the lesson of history is clear. They are the scions of innumerable researchers of the past, and their motivation derives both from the intellect and from the heart. I trust that their advices pro or con this or that scheme may be widely judged by an informed community, for if there is one thing which seems to distinguish Australia from other homoclimes I would say that it is the feeling that the land belongs to us all and is to be nurtured as a common heritage.

If I interpret the history of land use in arid Australia aright I would say that with one major exception, and that is our treatment of the Aborigines, our efforts have been reasonably honest. The present climate of opinion is clearly veering towards even greater rationality and more sophisticated judgments, and he would indeed be rash who attempted, rather than to follow the logical precepts of the ancient Sumerians, to promulgate the ideas of the suburban land-developer, whether in the Great Sandy Desert, or the Big Desert, or even in the Little Desert!

REFERENCES

Hills, E. S. (ed.) (1966). *Arid lands—a geographical appraisal.* Methuen, London; UNESCO, Paris.

Stamp, Dudley (ed.) (1961). *A history of land use in arid regions.* UNESCO, Paris. Arid Zone Research Series, No. XVII.

2

LANDFORMS OF ARID AUSTRALIA

J. A. MABBUTT*

SINCE the processes which fashion the earth's surface are so largely controlled by climate, it is predictable that each major climatic zone will have characteristic landform assemblages. This is particularly true of the extreme, physically dominated régimes, and climatic geomorphologists recognise a desert morphogenetic system in which, as a result of inadequate moisture, the vegetation and soils are too scanty to afford effective protection to rock surfaces against the atmospheric agents (Tricart and Cailleux, 1960). The gross outlines of desert relief have their origin in the geologic build of the continent and in a geomorphic history which may reach back beyond the present pattern of climate: the imprint of present aridity shows most clearly in surface detail, in the dynamic status of landforms, and in the balance of landforming processes. In this account of Australian arid landscapes, consideration will first be given to landscape dynamics, a matter of importance to land use with its implication of man-induced changes, and then the major relief and physiographic desert types of Australia will be described within the broad tectonic framework.

DYNAMICS OF DESERT LANDSCAPES

An arid landscape is one of marked and abrupt contrasts: between rugged uplands and smooth plains; between vegetated drainage tracts and bare wastes of rock outcrop or dune sand; between sites which receive and store water and those which shed or otherwise lose it. It is convenient to describe the dynamics of such a compartmented environment under distinctive physical settings, each illustrative of a set of landforming processes.

Desert Hillslopes

Hillslopes in deserts generally show an alternation of rock outcrop and débris, with crusted and pitted surfaces commonly bearing the dark ferruginous coatings known as desert varnish. Chemical weathering is clearly important, but restricted in many ways through lack of moisture. It tends to be superficial because of the limited penetration and capillary return of weathering solutions with strong evaporation; it is localised in hollows and undersurfaces where water may be trapped and protected from rapid evaporation; and it is also

* School of Geography, University of New South Wales

partial, in that the formation of weathering crusts tends to disrupt the rock superficially and to release the mineral grains before decomposition takes place. This is reflected in coarse textures and general lack of clay in weathering products.

Since rock breakdown involves physical disruption before advanced chemical alteration, all parting planes in the rock are exploited, from bedding surfaces to the boundaries of mineral grains. On a large scale, this can be seen in the mural outlines of a well-jointed sandstone, but minor cleavages are also developed, particularly where each joint block remains intact to the end, for instance the spalling and blocky disintegration of quartzite. Processes invoked to explain such small-scale break-down include heating by insolation or grass fires, the expansion of clay fillings on wetting, and the release of connate stresses (offloading) (Ollier, 1963, 1967). Such tendencies are emphasised by the selectiveness of desert weathering; a shaded cleft in which waste is trapped, water collected and held, and vegetation established becomes a weathering oasis, and irregularities are enlarged to the point of instability. Many forms attest to the superficiality of arid weathering, for instance the weathered shells and flakes shed from granite surfaces. Conversely, case-hardening may result by cementation or coating from evaporating solutions at rock faces, and this forms a protective shell behind which cavernous weathering can occur through the detachment of grains in the moist environment of a shaded hollow; such are tafoni and weathering pits in granite and honeycombing in sandstones.

In contrast with the slow processes of preparatory weathering, erosion by rainwash, sheetwash, and rilling is remarkably effective on desert hillslopes, as reflected in the sediment yields of upland catchments. In a study of sediment yield in relation to effective rainfall, that is the amount of precipitation required to produce a given amount of run-off, Langbein and Schumm (1958) found that maximum erodibility occurs in semi-arid regions with between 10 and 14 inches annual rainfall, where desert shrubs fail to protect and bind the ground. Sediment yields with between 4 and 10 inches rainfall, a range which applies to most of the Australian arid zone, are but little lower.

Since erosion tends to wait upon weathering, colluvial mantles are generally thin and discontinuous and true soils are lacking on slopes greater than 5 per cent. Save where the rock mass itself becomes saturated and mobilised, profiles tend not to be smoothed by the mass movement of colluvium but to remain rugged, with each rock stratum expressed in relief according to its resistance; hence the clarity with which geologic structure is expressed in the forms of desert uplands.

Desert Piedmonts

In contrast to the steep, rugged hillslopes, desert piedmonts characteristically sweep with smooth and gentle concavity to meet the hill base with an abruptness which can locally amount to an angular junction. These piedmont plains are in part depositional, where fans and aprons have been built up by streams and slope rills from waste washed clear from the hillslides. In arid areas of marked earth movement, like Chile, piedmont deposits of great thickness may drown the lower landscape, but in the more stable Australian setting the cover is generally thin and may locally be absent, and the bedrock surface is here seen to share the smooth profile of the mantle.

Such cut surfaces are known as pediments and the plains formed by their coalescence as pediplains; their origins are uncertain, undoubtedly complex, and probably various—geomorphologists by and large have yet to accept the idea of convergence in landforms—but some general factors clearly operate. One is the exploitation of lithologic contrasts in deserts, such that the junction between the resistant relief-builder and the weaker rock of the piedmont is drawn sharply in relief. Another is the tendency for enhanced rock weathering in the moister environment beneath a shallow cover, whereby the slope foot is trimmed back and the smoothness of a depositional profile transmitted to the rock base (Mabbutt, 1966a). And there is the effectiveness of rills and sheet-floods in regrading the piedmont, particularly by the removal of débris from slowly weathering rock surfaces, a process termed 'rock-floor robbing' (Davis, 1938).

By and large, cut or thinly mantled pediments are typical of granitic uplands whilst ranges of sedimentary rocks are flanked by aprons and fans. There appear to be many contributory causes. Under desert conditions, the contrast in resistance between the exposed and buried rock is greatest in granite, whereby thinly mantled lowland surfaces are readily planed; since weathering granite passes from the massive rock to the mineral grain with few intermediate-sized particles, removal of waste is effected over low gradients; granitic piedmonts are usually underlain by homogeneous material and hence are readily smoothed; since the difference between the rock of hill and plain in granite areas is largely a matter of massiveness or jointing, a compartmented separation follows, with a well-formed piedmont angle; in pedimented granitic landscapes the uplands are generally relatively small and the waste accordingly limited; further, it tends to be moved and spread by many small streams rather than by a few large channels; lastly, granitic areas are generally stable and the scene of long and uninterrupted planation.

An exception to this general rule are the soft-rock pediments which in arid Australia commonly flank duricrusted residuals, as in the Barrier Ranges (Dury, 1966) and in the area south of Alice Springs. These are long, concave surfaces which pass without break into the hillslope and they are typically armoured with gibbers from the range capping. They, rather than granitic pediments, express the processes of slope retreat and piedmont regrading on soft, fine-textured rocks to which the term pediplanation has been given.

The low gradients of desert piedmonts reflect the ease of transport of loose sediment across little-vegetated surfaces, but there is a tendency for their steepness, which may exceed 10 per cent, to be masked by an overall smoothness. This property results from the way in which the rills shoal, multiply, and replace each other rather than incise and enlarge, owing to abundant coarse loads and to lack of consolidated bank materials, and also to the obstruction of channels by growing plants or wind-placed deposits during periods of drainage inactivity. The desert piedmont commonly shows a mesh of anastomosing rills which pass downslope into terminal depositional lobes.

Periodically, the slow balance of transport, deposition, and erosional regrading at the hill foot is disturbed by changes in climate or by a rejuvenation of local drainage. This can result in widespread stripping and dissection of pediments or, where fluctuations occur, in staged pediments or piedmont terraces as in the Macdonnell Ranges of central Australia (Mabbutt, 1966b).

Desert River Systems

Average annual run-off is less than 0·5 inch over the Australian arid zone as a whole, but the amount varies widely from place to place; because of high absorption and evaporative losses the character of the surface on which the rain falls is critical. Steep rocky slopes give rise to drainage systems of above-average density, in fact an inverse relationship with rainfall has been demonstrated for such areas (Melton, 1957). An upland will generate an aureole of ephemeral streams, but the adjoining lowlands with their mantle of pervious deposits will yield little run-off, and only the major outlets will persist beyond the piedmont to link into larger systems. Since down-channel losses through influent seepage and evaporation are not compensated by tributary recharge, the magnitude and frequency of flows will normally decrease with distance from the upland margin.

Dry stream channels in desert uplands are normally irregular in section and profile, and rocky sectors alternate with sand-filled reaches which become increasingly important downvalley. Rock holes and scour pools may form valuable surface-water storages. The larger channels out on the plains are trench-like (Fig. 1a), with large width-depth ratios indicating plentiful coarse-textured bedload (Schumm, 1963); in plan they are sinuous or braiding rather than meandering (Fig. 1d), and their gradients are much greater than those of comparable channels in humid regions. They normally contain a sandy fill which may be scoured, mobilised, and redeposited during floods. Large but temporary waterholes may occur at bends or rock bars, and underflow after short-lived floods can be an important source of shallow groundwater. The main channel will normally be flanked by flood channels within silty levee banks and occupy a sandy floodplain with sand dunes locally.

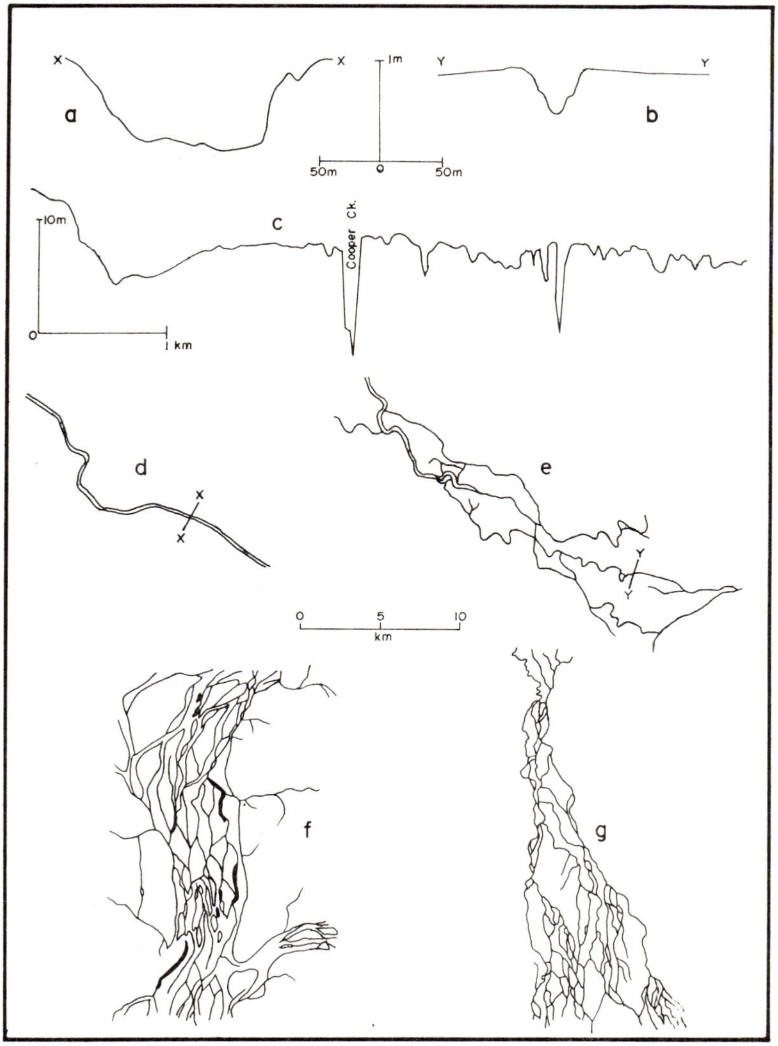

Fig. 1 Forms of desert rivers. a and b: cross-sections of Finke River channel at Idracowra and Charlotte Waters, N.T.; c: cross-profile of Cooper Creek floodplain at Windorah, Qld.; d and e: courses of Finke River at Idracowra and Charlotte Waters, showing location of sections 1a and 1b; f: anastomosing channels and billabongs (black) of Thompson River above Jundah, Qld.; g: distributary 'deltoids' of Warrego River below Cunnamulla

Nearer the drainage terminal, which may be an ephemeral floodout swamp or pan, the channel undergoes changes in form indicative of smaller discharges and of an increasing proportion of clay and silt in the alluvium carried. The channel diminishes but is deepened relative to its width (Fig. 1b), a cross-section adjusted to a suspended-load régime (Schumm, 1963), braiding gives way to close meandering, the gradient decreases, and finally the main channel may break into distributary terminals (Fig. 1e). Waterholes in these sectors are commonly long, deep and straight, and may be perennial since they are subject to smaller infiltrative loss through the finer alluvium. The relative importance of the bed-load and suspended load channels will vary with the alluvium supplied and hence with the geologic setting; where clay and silt predominate, deep anastomosing or meandering channels in clay plains may occur throughout.

Desert Lakes or Playas

Many playas dot the map of arid Australia, but as evidence of the derangement of drainage under desert conditions, not of abundance of surface water. Some act as foci of interior drainage systems and almost all occur near or in prolongation of drainage terminals, for they commonly mark the disintegration or abandonment of river courses and most are river-fed.

The largest, such as Lake Eyre, occupy tectonic depressions modified by wind erosion and deposition, by lacustrine and fluvial filling, and by wave action locally. Such lakes may be evaporative links in groundwater systems and are commonly saline (salinas); the permanent salinas which lie on a north-south structural line east of the Bulloo River between Thargomindah and Eulo in western Queensland are of this type. Other playa basins may be the result of uneven erosion or deposition by wind or water, although the division between erosional and barrier lakes is rarely clear. This is demonstrated by the lunettes, circular deflation basins with crescentic sand or clay barriers on their leeward rims, which are prominent near the lower Darling River. Playas which are purely river-fed may be non-saline (claypans).

Most playas show a zonation of sediment and salinity which marks stages in drying out on evaporation, with the finest deposits and highest salinity in the lowest parts.

Open playa floors with scant vegetation are liable to wind erosion and to the wind-sorting of the sediments. Sands are formed into fringing dunes and islands, with trapping by marginal shrub vegetation playing an important part, and the final materials are carried away. A limit to the lowering of a playa by deflation is generally set by the groundwater table, which in a large playa is never far below the surface, or by a protective clay seal.

Desert Dunes and Wind Action

In 'moderate deserts' such as those of Australia, wind generally plays the minor role in shaping the land surface, but erosion and deposition by wind are everywhere favoured by the sparse vegetal cover and by widespread unconsolidated surface deposits left exposed between episodes of streamflow. They are apparent in the widespread mounding of sand about the base of desert shrubs and more spectacularly in desert dust storms. Wind action naturally becomes more important during drought and wherever the ground cover has been disturbed by fire or by man and his grazing animals. In the dunefields beyond the ambit of the desert river systems, wind action has of course been paramount; the surface materials are wind-transported and the whole relief wind-moulded.

It would be incorrect to represent wind and water as independent agents operating in separate provinces or at different periods, for their activities are closely related. Most dunefields occupy former alluvial lowlands and most dune sands originated in alluvium; river banks and playa margins with abundant water-borne sand are favoured areas for current dune growth; dust storms are most frequent following episodic flooding during drought periods, when unvegetated fine alluvia are exposed to deflation, and the main duststorm areas are river plains with sufficient sand to cause the dislodgement and deflation of alluvial silt and clay. The economic consequences of wind action are greatest along the settled riverine margins of the dunefields, where alluvial soils are subject to blowing and where advancing dunes can threaten settlements and communications.

The areas of most active sand movement and dune growth are in the dry southeast, near Lake Eyre, whence the sand ridges become increasingly vegetated northwards into areas of greater rainfall. Over most of the arid zone the dune flanks are grassed below a crestal belt of live sand, with clumps of *Zygochloa* or deep-rooting shrubs, which is subject to shifts of form with short-term changes in wind régime. Along the north edge of the arid zone, as in the Pindan of the West Kimberleys, the dune ridges are rounded, heavily vegetated throughout, and have soil profiles, and their fossil character is beyond doubt. It seems likely that the major dune systems were formed in one or more past periods of greater aridity. Since neither the age or ages of the inland dune ridges nor their mode of growth are known in detail, it is not possible to reconstruct this past geography. It would seem from the trend of the dunes and their close relationship with the present dominant winds that the wind and pressure systems were not vastly different from those of today, but the arid zone was then larger and drier, for its rainfall throughout must have been somewhat lower than that of the driest areas of current dune growth, say 4 inches or less.

The apparent stability of the Australian dunefields should not be exaggerated. The 1956-66 drought in central Australia led to increased sand movement on vegetated flanks and swales as well as on dune crests, particularly in grazed areas. In Queensland, dunes in the Cooper floodplain near Windorah have been partially mobilised and have extended downwind, exposing at their southern ends cores of clay-rich sand formed in long periods of earlier stability. The dynamic balance here is a sensitive one.

MAJOR OUTLINES OF RELIEF AND DRAINAGE

Structure and Relief

The arid zone covers three-quarters of Australia and comprises much of two of its three major tectonic divisions, namely the Western Platform and the Central Basin (Fig. 2).

The Platform has a basement of Archean granite and gneiss which is exposed in updomed shield areas. Its most extensive occurrence is in the Yilgarn Shield in the south of Western Australia, but it also gives rise to extensive plains studded with granitic hills north of the Macdonnell Ranges and near Kulgera in the Northern Territory, and in the Gawler Block in the west of South Australia. About these nuclei are a number of cratonic blocks subjected to repeated epeirogenesis in their later geologic history; topographically the most important are the Kimberley and Pilbara (Hamersley) blocks, where flat-lying resistant quartzites form rugged plateaux with up to 2000 feet of relief. Between these positive tectonic areas are basins with varied tectonic history and relief. Folded sandstones in trough basins have been selectively eroded to give the prominent strike ridges of the Flinders and Macdonnell Ranges, and in places the crystalline basement has itself been thrust up along the margins of such fold belts to give rugged granitic ranges, as in the northern Macdonnells. Unfolded basin sediments form plains and tablelands, for instance the Palaeozoic limestones and shales of the Georgina Basin in the Barkly Tableland and the Tertiary marine limestones of the Eucla Basin in the Nullarbor Plain.

The extent, uniformity, and essential rigidity of this tectonic unit are expressed in the subdued monotony of its major relief. The overriding characteristic is a dominance of the horizontal, with broad interior plains cut across the large areas of homogeneous crystalline rocks during a long history of undisturbed subaerial planation, and high plateaux, low tablelands and plains formed by block and basin sediments preserved from folding by the underlying basement. Narrow fold welts give rise to island ranges within the plains, but there is an absence of the very strong relief characteristic of more mobile structural settings, and Mt Zeil (4900 feet) is the highest summit in the arid zone.

That part of the Central Basin falling into the arid zone consists mainly of the Great Artesian Basin, with part of the Murray Basin in the south. It is an area of marine Mesozoic claystones and soft sandstones with a partial cover of Tertiary sands. These were gently basined within the marginal upwarp of the Eastern Uplands and the overall structure is well expressed in the centripetal drainage linking a peripheral belt of predominant erosion with an inner area of Quaternary deposition which falls to 23 feet below sea level in Lake

Eyre. An absence of massive or resistant rocks or of strong deformation leaves a relief assemblage of low tablelands and cuestas, extensive undulating lowlands, wide alluvial plains, and a central portion with dunefields and playas.

By the middle of the Tertiary Era, both Platform and Basin had been extensively planed and their rocks deeply weathered and duricrusted, principally with laterite on the Platform and silcrete in the Basin. Older land surface remnants preserved on rocks resistant to erosion and weathering remained as uplands above the duricrusted surface, as in the Isa Highlands and the Macdonnell Ranges. A period of earth movement followed. The Platform underwent mainly vertical movements involving broad warping and marginal faulting and flexure, as in the Darling Range, giving the large-scale pattern of interior plateau, marginal escarpment, and narrow coastal plain. Locally, strong block-faulting occurred along old structural lineaments, as in the formation of the Yorke Peninsula and its flanking gulfs and the Mount Lofty Range. Within the Central Basin there was differential local movement, mainly in the form of broad shallow periclines elongated north-south.

These movements initiated the dissection of the duricrusted land surface. On the Platform in Western Australia the Old Plateau has largely been replaced by the New Plateau by the etching away of a layer of weathered rock up to 80 feet thick, with low lateritic breakaways separating the two land surface elements. In the Central Basin the larger duricrusted remnants form tablelands; with more advanced dissection the hard silcrete is reduced to mesa cappings and finally to gibber mantles on plains eroded at a lower level. Most of this area is less than 1000 feet above sea level, and relief, limited by the amplitude of local earth movements, rarely exceeds 325 feet.

Advanced weathering, the removal from the area of the products of erosion, and the continuity of slopes and extant drainage lines indicate that the duricrusted surface in the arid zone formed under a more humid climate than today, and with out-going drainage. Younger land surfaces formed below the duricrust stand in strong contrast; their rocks are little-weathered or are masked by crusts of calcrete or gypsite indicative of excess evaporation, their drainage systems are endoreic or disorganised, and their lower parts are buried beneath fluvial and aeolian deposits. They mark the onset of arid conditions across Australia.

Drainage Patterns of Arid Australia

In some of the world deserts, rivers entering from the humid borderlands bring important supplemental water supplies; this is unfortunately not so in Australia, where the divide between interior and direct external drainage follows closely the limit of the arid zone (Fig. 2).

The association of aridity and endoreism, long ago noted by de Martonne (1925), is a complex relationship. In part it expresses the tendency for topo-

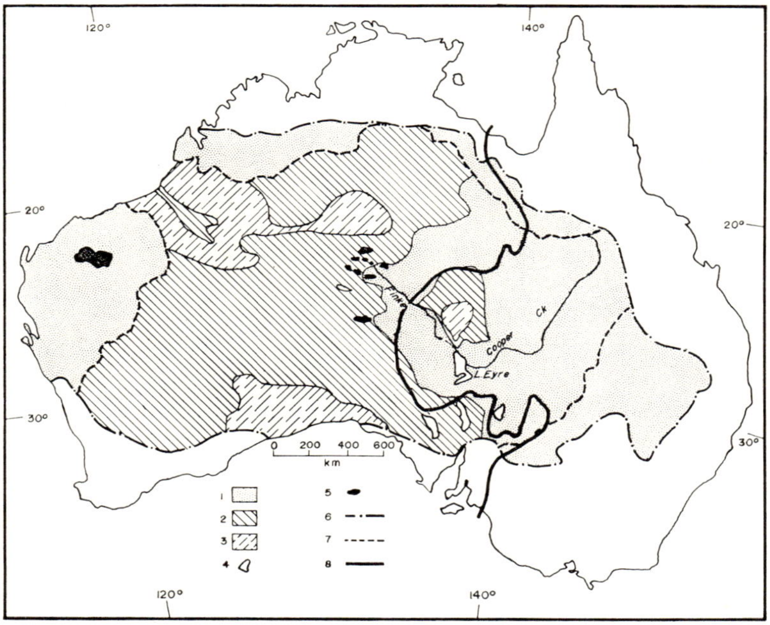

Fig. 2 Degree of drainage organisation in the Australian arid zone: 1. co-ordinated drainage; 2. unco-ordinated drainage; 3. riverless areas; 4. large playas; 5. areas above 3000 feet; 6. limit of arid zone; 7. divide between direct external and internal drainage of arid zone; 8. eastern limit of Western Platform

graphic barriers additionally to be climatic divides, whereby under marginally arid conditions the drainage limits of a basin may also form the rainshadow fringe of the desert. This is best seen in the northeast of the arid zone, where the high ground of the Great Divide is also the climatic boundary and only a small part of the Cooper Creek catchment extends beyond. In part, the link between aridity and interior drainage expresses the tendency for exoreic drainage of low gradient to disintegrate and become truncated under climatic stress, as has occurred on the wide plains of the Western Platform. Finally, aridity allows the persistence of tectonic derangement of drainage which might otherwise subsequently have found an exit seawards, as with the Bulloo River in the Central Basin.

Classification of desert drainage as 'exterior' or 'interior' is of limited usefulness, however, in that it fails to consider how little of a topographic catchment may be effectively occupied by its drainage nets. Arid Australia has undergone a long period of desiccation during which integrated drainage systems initiated in a moister Tertiary Era have become truncated and disorganised in varying degree. The degree of drainage disintegration has been classified by Hills (1953) and later writers in the following or similar terms:

co-ordinated external drainage
unco-ordinated interior drainage
disorganised interior drainage
riverless areas.

The distribution of these drainage classes in arid Australia is shown in Fig. 2, whence it appears that the degree of disorganisation is less a function of climatic contrasts within the arid zone than of differences in major relief and surface materials as controlled by geologic build and geomorphic history.

Co-ordinated drainage survives in all parts of the arid zone drained externally. This is mainly in the northwest, with a combination of strong relief, relative high rainfall, and the favourable structural fall of the basined and downwarped margin of the Platform. In this area the Fitzroy River attains almost a seasonal régime. In the north, too, there is external drainage where vigorous exoreic channels have pushed the divide southwards from the Pine Creek Upwarp (Hays, 1968), the structural margin of the Barkly Tableland and the limit of the arid zone locally. External drainage in the Murray Basin, in contrast with the structurally and topographically similar Great Artesian Basin to the north, is probably the expression of heavier rainfall southwards along the Great Dividing Range.

The only important area of co-ordinated interior drainage is the Great Artesian Basin, in which the structural disposition of peripheral upland and centripetal slopes on uniform soft sedimentary rocks has favoured the survival intact of large drainage units. It also exemplifies an additional factor, namely the pattern of greater and more effective rainfall northeastwards, towards the upland catchments. Only the central dunefields of the Simpson Desert are excluded. There are two independent systems, that of the Cooper, Diamantina, Finke and Macumba, tributary to the main tectonic depression occupied by Lake Eyre, and the smaller Bulloo system in the south. In the southeast the Paroo and Warrego Rivers cross the southern edge of the Great Artesian Basin to connect with the Murray Basin drainage.

Disorganised interior drainage describes the stage of disintegration into separate systems within a single topographic basin, each system with its own terminal and separated from others by areas lacking surface drainage. It occurs in two major settings.

In central Australia, with its extensive plains and island ranges, each upland may have a separate drainage net which fails to reach the lowland axis. The Amadeus Basin provides a good example, for the large salina which occupies its centre is not reached by the drainage rising in the higher ground to north and south. Such areas demonstrate how the disintegration of drainage is a progressive phenomenon in that the deposition of coarse-textured alluvia about the terminals of a shrinking drainage net tends still further to diminish run-off in the lowland tracts. Most of the sand deserts, with limited uplands and entering drainage, have been placed in this category.

On the Platform farther west, with its extensive sandplains and featureless divides, a Tertiary drainage which formerly reached the Bight has degenerated into a number of interior systems each tributary to a playa. The playas form series aligned along the original valleys, but their fluvial origin is made clear only when they connect during heavy flooding. The barriers which separate the 'river lakes' are alluvial only, and groundwater flow may continue along the system as indicated by the limited salinity.

The main riverless areas combine low relief and high permeability; they include parts of the extensive sand deserts and also the limestone Nullarbor Plain.

MAJOR PHYSIOGRAPHIC TYPES OF DESERT

In Fig. 3 is mapped the distribution in arid Australia of six main physiographic types of desert, each of which is characterised by a particular combination of the dynamic settings discussed in the first section. The main control of their distribution is major structure. These physiographic types provide a convenient framework in which to describe the forms of desert relief. Their extent is given in Table 1.

Mountain and Piedmont Desert

Since a desert régime allows the clearest expression of bedrock character on bare, steep slopes, it is to be expected that the forms of desert mountains will vary markedly with lithology and structure. They include the domes, tors and pinnacles of granite ranges, as in the Musgraves and the northern Macdonnells, prominent ridges and cuestas predominantly of quartzite, as in the Flinders Ranges, massive plateaux in the Hamersley Ranges, and rugged uplands on metamorphic belts, as in the Isa Highlands. A rock outcrop constitutes a desert in miniature under any climate and there is great similarity between non-glaciated mountains over a wide climatic range, but structural details are nowhere more intricately exploited than under aridity.

Lack of moisture brings some changes in the relative resistance of rock types. Granitic rocks, which are physically resistant but susceptible to chemical rotting, tend to stand out in desert relief; compact limestones are also more prominent in the absence of solution weathering, for instance they are important relief-builders in the eastern Macdonnells; impervious claystones which would break down into subdued profiles in a humid area here remain firm at the surface and undergo close stream dissection into 'badlands'.

This desert type is the source of most of our desert drainage and of its water supplies and was a favoured habitat for the desert Aboriginal (Mabbutt, in press). Unfortunately, no Australian desert mountain rises high enough to yield significantly greater rainfall nor to store precipitation as snow, and so provide an exception to the ephemeral character of Australian desert rivers.

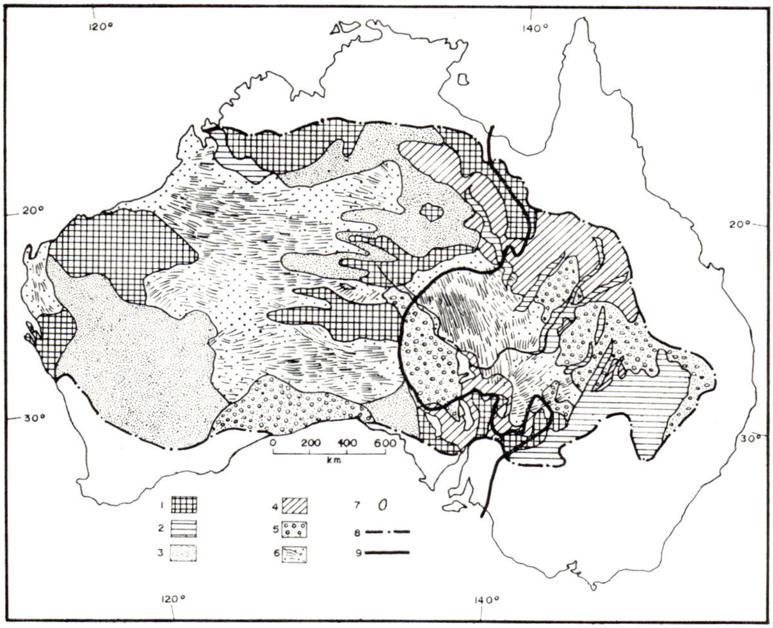

Fig. 3 Desert physiographic types: 1. mountain and piedmont desert; 2. riverine desert; 3. shield desert; 4. desert clay plains; 5. stony desert; 6. sand desert; 7. large playas; 8. limit of arid zone; 9. eastern limit of Western Platform

TABLE 1

EXTENT OF AUSTRALIAN DESERT PHYSIOGRAPHIC TYPES

	Sq. Miles	Percentage of Arid Zone
Mountain and piedmont desert	223,880	15·0
Riverine desert	50,180	4·5
Shield desert	289,500	23·0
Desert clay plains	165,980	13·0
Stony desert	154,400	12·5
Sand desert	405,300	32·0
	1,289,240	

Riverine Desert

Nearly all desert types contain floodplains and large channels, but the term riverine desert is here restricted to areas dominated by the alluvial plains of major rivers. It has its greatest development in the Channel Country of southwest Queensland, as along Cooper Creek, and in the northern tributaries of the Darling River. This reflects jointly the persistence of integrated drainage on a larger scale than elsewhere in the arid zone, and also a relatively favourable climate. Alluviation has resulted from subsidence in the centre of the Basin, and the distribution of the plains has been controlled by minor earth movements; most of them occupy synclines and are locally diverted or constricted by anticlinal structures.

The major river plains of the Channel Country are characterised by low gradients, of the order of 1:5000, and by fine-textured alluvium as yielded by the rocks of the Central Basin. The region is named for the multiple enmeshed channels of the larger rivers (Fig. 1f) which occur in straight tracts between 1·5 and 3 miles wide and which are of two types. By far the more numerous are deep, suspended-load channels with small meanders which anastomose in regular but rather rectangular patterns to form a net of lozenge-shaped reticulae. These are crossed by sinuous, shallower floodways which feed from and into them. At intervals, and particularly towards the margins of the channel tract, several anastomosing channels link in straight reaches up to 8 miles long which apparently mark slight flattenings of the plain; these are occupied by perennial billabongs which can be 50 feet deep and 325 feet or more wide. The channel tract is flanked by discontinuous levees of reddish sandy silt which contrasts with the darker clays of the main plain, and similar deposits also occur in scrolls between and adjacent to the anastomosing channels. The tract may be as much as 6 feet above the level of the backplain (Fig. 1c). A major plain may contain more than one such tract and air photos commonly show the existence of prior traces distinguished by their reddish alluvium and preserved channel patterns. Locally, former courses are marked by lines of lunettes, although nowhere on the scale of the lower Darling plains.

Beyond the channel tract is the backplain formed of alluvial dark cracking clays, a featureless plain, often several yards wide, diversified by silt islands—higher parts of the plain which are rarely flooded—and by sand dunes. The latter may occur in aligned series of crescentic or reniform islands, indicating their origin as prior channel deposits; transitions from irregular mounds to looped crescentic dunes and from white to pinkish-red sand denote stages in development. Elsewhere, particularly along the lee margins of channel tracts, the dunes occur as small fields of short longitudinal ridges. The close genetic association of river plain, playa, and dune is general.

Near the outer margins of the plains are swamp or sump basins of circumdeposition which are fed by distributaries from the main channel or by backchannels associated with billabongs in the lowest parts of the backplain. They are regularly flooded, are floored with heavy expansive clays, and are strongly gilgaid, often with rectangular gilgai nets about the feeder terminals.

Locally, the channel tract feeds into the backplain by fan-shaped nets of anastomosing distributaries which Whitehouse (1944) has termed 'deltoids' (Fig. 1g), noting that they are associated with a slackening of gradient below a critical value of 1:5000. They occur at structural widenings of the plain, as below the Cooper-Wilson confluence, and also towards the lower reaches of the river systems where the channels are increasingly opposed by tectonic slopes, sand dunes, and the barriers of their own alluvium, as in the Bulloo Swamps and along the Warrego below Cunnamulla.

The features of these riverine plains represent an adjustment to enormous variations in discharge. In most summers there are substantial flows, sufficient to fill the main channels and billabongs and to flood into swamp basins; floods of this scale in the Cooper at Windorah have recorded discharges of about 200,000 acre-feet. At intervals of about five years there is extensive flooding covering all but the highest parts of the floodplain, and such a flood in the Cooper at Birdsville in the 1954-5 season had a discharge of 10 million acre-feet. With the exception of the marginal Fitzroy and Darling Rivers, with catchments partly outside the arid zone, the rivers of the Channel Country approach a seasonal régime more closely than any other of our desert rivers.

Shield Desert

This desert type comprises non-mountainous granitic terrain developed on the Western Platform. As represented on the Yilgarn Block in Western Australia it is a broadly undulating plateau mainly between 1000 and 2000 feet above sea level and consisting of an older lateritic element and a younger etchplain, the Old and New Plateaux. Apart from the lateritic breakaways, the main relief consists of low ridges of schist and quartzite and groups of small granite hills exhumed from the weathered profile. The New Plateau is best developed in the western area of external drainage and the Old Plateau becomes relatively more extensive inland, forming broad divides.

Much of the Old Plateau surface is sandplain derived from lateritic sands and yields little surface drainage. Stream systems head at the breakaway margins of the Old Plateau and lead to the 'river lakes' by way of valleys with deep fillings of alluvium cemented in its upper layers by chalcedonic limestone. These tracts, known locally as 'opaline country', are a source of abundant shallow groundwater which becomes increasingly saline down valley, as does the alluvium. The surface limestones are generally dissected into platforms as much as half a mile in length and standing about 5 feet above the alluvial tracts which wind between them; they generally open out into a broad delta terrace at the entrance to the salina.

The salt lakes occupy former valleys cut below the Old Plateau, and clearly the disintegration of drainage followed the inception of its dissection. Jutson (1934) has noted that these lakes commonly lie close to the foot of the escarpment bordering the valley on the west and are flanked by lowlands in the east (e.g. Lake Way near Wiluna) and has suggested that they have migrated westwards, upwind, through the deposition of windborne lake-derived material on their leeward margins. The lakes tend to be rimmed by sand dunes, particularly on the east, and locally by crusted gypseous dunes (kopi), as at Lake Annean south of Meekatharra.

Aeolian sand becomes increasingly widespread on both plateau elements eastwards, where shield desert eventually gives place to sand desert, but as sandplain rather than as dunes except on playa margins.

A feature of alluvial and residual earth soils on the New Plateau is the occurrence from shallow depth of a hard siliceous pan which accentuates the effective aridity of the area (Litchfield and Mabbutt, 1962).

Shield desert in central Australia shares many of the features described, but with differences which stem from its development downslope from granitic mountain deserts which here serve as the main sources of drainage. Lateritic crusts, originally restricted to lowland tracts, have largely been stripped away and survive only in a few benches and tabular low hills. The two main surface elements are gently undulating plains cut on little-weathered granite, locally with 'opaline tracts' and dotted with larger hills than those in Western Australia, and stable alluvial plains with red earth soils which occur downslope from the erosional plains. Both elements are crossed by poorly defined, unchannelled drainage depressions, and large parts of them are subject to sheetflow.

Desert Clay Plains

The characteristic feature of this desert type are plains with dark grey cracking clay soils (black soil plains) formed on or derived from limestone or calcareous claystone. The type area is the Barkly Tableland in the Georgina Basin, part of the Platform, but similar though more strongly undulating country known as 'Rolling Downs' is developed farther east in the Central Basin. The Barkly Tableland consists of extensive open grasslands mainly between 500 and 1000 feet above sea level. The residual black soil tracts, generally the higher parts, are slightly gilgaid and extensively strewn with chert gravel; the alluvial soils, which occur in the main floodplains, as of the Georgina River, in broad tributary flood-outs, and in shallow 'swamp' depressions are relatively stone-free but more strongly gilgaid. The main floodplains in the east show on a small scale many of the features of the Channel Country, to which they are tributary.

Tertiary earth movements warped the western part of the Tableland into a shallow basin, the focus of an extensive interior drainage; this has since largely degenerated, although the centripetal pattern survives about the margins, but central plains of freshwater limestones mark the former lake terminals. Similar Tertiary limestones formed in the deeper valleys farther east, and these have since been dissected into low tablelands flanking the present floodplains, as along the upper Georgina River.

The black-soil plains are diversified by other surface types, including plains of sandy lateritic soils on sandstones (these also occurring as small islands within the black-soil areas) and stony plains on some dolomitic limestones. Red-earth alluvial plains occur within and above tributary flood-outs east of the Georgina River, indicating earlier periods of coarser-textured alluviation, and there are minor areas of sandplain.

Stony Desert

This desert type is characteristic of the Central Basin, with an extension on argillaceous rocks of the Amadeus Basin in the west. The characteristic landforms, as mentioned in the geomorphic history, are mesas and tablelands with silcrete cappings and with slopes formed in weathered (pallid) rock, undulating lowlands eroded on rock beneath the weathered profile, floodplains like those of the Channel Country, and sandplain and dunefields.

The stony tablelands with silcrete gibbers, which give this desert type its name, are best developed in the south and west, as extensive featureless surfaces. The silcrete mantle, or desert pavement, is not merely a lag deposit, for it is typically underlain by a deep, almost stone-free clay subsoil. Although there has been some concentration of stone at the surface through deflation and wash, the main accumulation must be the result of upward extrusion through montmorillonitic clays from the silcrete horizon beneath, in the manner of gilgai upheaval (Mabbutt, 1965). There are few gilgais save in shallow drainage hollows, probably because of the mulching effect of the gibber cover.

The development of the silcrete horizon is closely related to the existence of an arenaceous horizon above the essentially non-quartzose Mesozoic rocks, and for this reason the silcrete is thinner in the northeast of the Basin, where the role of relief builder is assumed by a hard reddish brecciated horizon, locally more than 30 feet thick, above the pallid rock. It gives rise to wooded tableland surfaces with only partial cover of silcrete boulders and rather dusty red earth soils, and to prominent breakaways with large blocks detached along well-developed vertical joints.

The residual plateaux survive mainly on the flanks of anticlines, locally as cuestas where dips are above-normal, and the escarpments face in to axial lowlands whilst the upland surfaces decline gently down-dip to the depositional plains which characterise the synclinal tracts.

The soft-weathered pallid claystones below the duricrust, which lend prominence to the lower escarpments, are subject to episodic saturation and provoke mass movements on a scale unusual in a desert. These comprise giant slips involving duricrust masses hundreds of yards in dimension, bouldery mudflows, and the slow creep which forms boulder terracettes on lower slopes.

The escarpments are fronted by foot-spurs with a thick mantle of silcrete fragments which become finer in size downslope, but where erosion has penetrated to relatively fresh rock on anticlinal tracts there are gently undulating open plains—'Rolling Downs'—with brown clay soils and strongly gilgaid surfaces with much silcrete stone.

The main floodplains through stony desert have the anastomosing tracts of the Channel Country, but generally with non-cracking calcareous silty clays down-valley and increasingly with red earths in upper sectors.

Large areas of aeolian sand are included in this desert type, as sandplain rather than dunefields save towards the margins of the Simpson Desert. Much of the sandplain is at the level of the stony tableland surfaces, as east of the Bulloo River below Quilpie, being derived from Tertiary sands associated with the silcrete. Sandplains flanking the river flats have formed on fluvially transported materials and may be traversed by lines of pans and lunettes marking abandoned drainage ways, as between the lower Warrego and Paroo Rivers. Longitudinal dune ridges occur within the sandplain, particularly in open situations along the lee margin of drainage tracts, reticulate dunes near pans and drainage floodouts, and less organised dune types of floodplain islands.

Sand Desert

Aeolian sand covers almost a million square miles of Australia, slightly less than half the arid zone. Although sand desert carries the popular connotation of extreme aridity, its distribution is not a simple expression of climatic gradients; it occurs mainly towards the centres of major lowlands in which it originated as plains of coarse-textured alluvia abandoned to wind action by drainage systems which were retracting under the stress of declining rainfall.

About half the Australian sand desert consists of sandplain in the form of flat or undulating spinifex grasslands on stable red clayey sands with a grit veneer. Sandplains are areas of shallow aeolian reworking and have clay-rich subsoils with non-aeolian horizons such as fine gravels at depths of a few feet. The main reason for restricted reworking appears to be the high content of binding clay, generally between 10 and 20 per cent as compared with 1 or 2 per cent in dune sands, for sandplains are commonly formed from sands derived from granitic rocks with a clay source and in some areas from sandy lateritic soils on the Tertiary weathered land surface, as on the Old Plateau. The limited reworking is reflected in moderate sorting, with Trask coefficients of 1·5 and greater compared with 1·4 and less for dune sands.

Factors other than clay content may additionally have controlled the formation of sandplain rather than of dunes. In the moister northern parts of the arid zone, as south of the Barkly Tableland, there are undulating sandplains with broad parallel rises less than a few feet high but resembling the regional dune systems in their trend and spacing. This suggests that incipient dune formation may here have been arrested by close vegetation under a relatively moist climate. Persistent throughgoing drainage activity has also prevented transformation into dunes locally, as along the northern margin of the Simpson Desert.

By far the most extensive dune type are the longitudinal ridges to which the Saharan term *seif* is commonly applied. These are parallel ridges with remarkably regular regional trend and spacing, generally between 30 and 100 feet high and with a crestal interval of between 300 and 400 yards. From their constant geometric relations, for instance the inverse relationship between height and spacing, they appear to be an equilibrium form. The ridges may continue for hundreds of feet with little change, save for occasional Y-shaped junctions.

On the continental scale (Fig. 3) the ridges trend in a giant arc through the Great Victoria, Simpson, and Great Sandy Deserts in an anticlockwise sense as indicated by evidence of dune extension such as abutment against obstacles, with junctions in almost all cases pointing in the direction of extension. Trend and extension are closely, but not exactly, related to the resultant dominant wind (Brookfield, in press). The lower, vegetated dune body is commonly asymmetrical in cross-profile, for instance the eastern flank is the steeper in most of the Simpson Desert: the crestal belt of live sand has changing forms, generally an alternation of blowouts and changing slip faces, some reflecting sand transport along the dune crest and others indicating movements oblique to it.

The swales are generally sandy, with uneven loose sand between the arms of junctions, but there are flat floors with claypans locally and areas of gravel outcrop; in most cases the sub-aeolian floor is little below the level of the swale (Mabbutt and Sullivan, 1968).

Longitudinal ridges appear to be the most evolved of the dune types, adjusted to a bi-directional wind régime without excessive sand supply, and they occur in open desert situations free from interference by relief or drainage. Their relative maturity is also indicated by the deep red iron-coatings of their sands and clay cores. Other dune types can be considered as subclimax forms hindered in their development (Mabbutt, 1968).

Linked crescentic dunes occur along the northwest margin of the Simpson Desert, bordering the floodplains of the Finke and Hugh Rivers which have acted as foci of dune growth. They are transitional in form between the linked dune chains currently growing at the river bank and the longitudinal ridges into which they pass eastwards by gradual increase in the length of the paired crescent arms and decrease in the frequency of transverse links; in colour, too, their sands are midway between the older and juvenile forms, and they constitute an instructive intermediate growth stage.

Reticulated dunes characteristically occur where dunefields are entered by major drainage, and in such areas the dune net incorporates small pans. Intersection of sets of dune trends may also be involved; about Lake Amadeus, for instance, the directional components of the pattern are represented separately in convergent longitudinal ridges in areas north and south. Short, keel-shaped dunes occur in intermont basins south of the Macdonnell Ranges, generally aligned in the regional dune trend, but locally unaligned.

All these less organised dune types occur where, through interference by drainage or relief, there has been less free play for dune-forming winds. To this factor may well be added that of wind régime, for these types have their fullest development towards the inner part of the arc described by the dune ridges, towards an area of more variable winds.

The dune sands are well-sorted fine sands with median diameters of about 0·02 inches on dune crests and 0·01 inches on slip faces, and with a tendency to bimodality in the swales. Large areas of dunes have hard cores due to cementation by illuviated clay, evidence of stability over a significant period.

The sand deserts generate no run-off, although they are entered by rivers; the Hale and Hay Rivers flood for considerable distances along the swales of the Simpson Desert for instance, their courses being slewed SSE on entering the dunes.

REFERENCES

Brookfield, M. (In press). On wind and the trend of desert dunes; a case study in central Australia.

Davis, W. M. (1938). Sheetfloods and streamfloods. *Bull. Geol. Soc. Amer.* **49**: 1337-416.

Dury, G. H. (1966). Pediment slope and particle size at Middle Pinnacle, near Broken Hill, New South Wales. *Aust. Geogr. Studies.* **4**: 1-17.

Hays, J. (1968). Land surfaces and laterites in the north of the Northern Territory. In J. N. Jennings and J. A. Mabbutt (eds.), *Landform studies from Australia and New Guinea.* A.N.U. Press. Canberra: 182-210.

Hills, E. S. (1953). Regional geomorphic patterns in relation to climatic types in dry areas of Australia. Desert Research Proc., Jerusalem, 1952. *Research Council of Israel Special Publication No. 2*, 355-63.

Jutson, J. T. (1934). *The physiography of Western Australia.* Geol. Surv. W. Aust. Bull. No. 61, Govt. Printer, Perth. 2nd ed.

Langbein, W. B. and Schumm, S. A. (1958). Yield of sediment in relation to mean annual precipitation. *Trans. Amer. Geophys. Union*, **39**: 1076-84.

Litchfield, W. H. and Mabbutt, J. A. (1962). Hardpan in soils of semi-arid Western Australia. *J. Soil Sci.* **13**: 148-59.

Mabbutt, J. A. (1965). Stone distribution in a stony tableland soil. *Aust. J. Soil Res.* **3**: 131-42.

——— (1966a). Mantle-controlled planation of pediments. *Amer. J. Sci.* **264**: 78-91.

——— (1966b). Landforms of the western Macdonnell Ranges. In G. H. Dury (ed.), *Essays in geomorphology*, Heinemann. London: 83-119.

——— (1968). Aeolian landforms in central Australia. *Aust. Geogr. Studies*, **6**: 139-50.

——— and Sullivan, M. E. (1968). The formation of longitudinal dunes: evidence from the Simpson Desert. *Aust. Geogr.* **10**: 483-7.

——— (In press). The Australian arid zone as a prehistoric environment. In J. Mulvaney and J. Golson (eds.), *Aboriginal man and environment in Australia.*

Martonne, E. de (1925). *Traité de géographie physique, Tome II: Le relief du sol.* Armand Colin. Paris.

Melton, M. A. (1957). *An analysis of the relations among elements of climate, surface properties, and geomorphology.* Office of Naval Research Tech. Report No. 11. New York.

Ollier, C. D. (1963). Insolation weathering: examples from central Australia. *Amer. J. Sci.* **261**: 376-81.

——— (1967). Spheroidal weathering, exfoliation and constant volume alteration. *Z. Geomorphologie.* **11**: 103-8.

Schumm, S. A. (1963). A tentative classification of alluvial river channels. U.S. Geol. Surv. Circular No. 477.

Tricart, J. and Cailleux, A. (1960). *Le modelé des Régions Sèches: Fascicule I. Le milieu morphoclimatique. Les mécanismes morphogénétiques des régions sèches.* C.D.U. Press. Paris.

Whitehouse, F. E. (1944). The natural drainage of some very flat monsoonal lands (the plains of Western Queensland). *Aust. Geogr.* **4**: 183-206.

3

METEOROLOGY AND CLIMATOLOGY

W. J. GIBBS*

AN ADEQUATE treatment of the subject of this chapter would require a compendium of many volumes. Taking into account the wide range of interests represented by the participants of the symposium, the chapter aims at a generalised explanation of why Australia's arid zone lacks rainfall and why the arid zone is one of high solar radiation, large range of ground and air temperature at the land-air interface, and high potential evaporation. To achieve this aim it is necessary to consider the general circulation of the earth's atmosphere and the consideration of this subject borrows heavily from a recent monograph by Lorenz (1967).

In such a brief space one can do no more than sketch a general outline of the meteorology and climatology of the arid zone. Indeed a complete climatology would consist of a mass of statistics in the form of tables, maps, and diagrams. A selection of data has been included more to illustrate the type of material available than to describe the climatology of the arid zone.

No attempt is made here to describe the meteorology and climatology of the present arid zone during recent geological time. However some consideration is given to factors which might produce a change in climate.

The bibliography concluding the paper provides references for those wishing to obtain more detail of meteorological or climatic aspects.

THE HADLEY CELL

The astronomer Edmund Halley presented a detailed and methodical account of the trade winds as observed in three separate oceans and sought a common cause for them (1686). He postulated that at low levels the air must move towards the most strongly heated equatorial regions and reasoned that the westward component of the trades was due to a tendency of air to move in sympathy with the westward movement of the diurnal maximum of air temperature near the earth's surface.

George Hadley (1735), in agreement with Halley, concluded that the distribution of solar heating would lead to a general rising motion in lower lati-

* Commonwealth Bureau of Meteorology, Melbourne, Australia.

ACKNOWLEDGMENTS. Thanks are due to Messrs H. R. Phillpot and K. T. Spillane of the Bureau of Meteorology for helpful discussion during the preparation of this paper.

tudes and sinking motion in higher latitudes, producing equatorward motion at low levels and poleward motion aloft, but he rejected Halley's idea that air moved in sympathy with the westward movement of the diurnal temperature maximum. He noted that the earth's surface moves more rapidly eastward with decreasing latitude and postulated that air initially having no east-west motion relative to the earth and moving equatorward would tend to conserve its absolute east-west speed and therefore move westward relative to the earth. Calculations showed him that his proposition would result in much greater westward speeds than those observed, and he assumed that the frictional drag of the earth's surface was responsible for the discrepancy.

Hadley was in error in supposing that the absolute east-west component of air motion tends to be conserved—it is the absolute angular momentum that does so and causes the apparent Coriolis deflection of bodies in motion on the earth. But the proposition that a circulation exists in which air rises over equatorial regions, moves to higher latitudes where it sinks and returns to its original latitude is widely accepted by meteorologists as a feature of the general circulation of the atmosphere, a feature commonly named the Hadley cell, to which reference will be made later.

The way in which Hadley's ideas have been refined and amended by observation and theoretical investigation is described by Lorenz (1967). There is now general support for three meridional cells in each hemisphere, the two Hadley cells in latitudes between about 30°N and 30°S; two polar cells with air sinking over the poles and rising about latitudes 60° with equatorward flow at low levels and poleward flow aloft; and indirect cells (sometimes called Ferrel cells after their proposer) between latitudes of about 30° and 60°.

ZONAL COMPONENT OF THE GENERAL CIRCULATION

As Hadley postulated, movement of air on the rotating earth induces changes in wind velocity measured relative to the earth's surface. Hadley had erroneously believed that the east-west component of the absolute wind velocity tended to be conserved; it is the absolute angular momentum which exhibits this tendency. Apparently Hadley had also not known of the interdependence of fields of wind and atmospheric pressure and temperature, a knowledge of which threw light on the theory of the general circulation. The development of ideas regarding meridional circulations such as the Hadley cell was limited by inability to measure vertical motion of the atmosphere and the relatively small magnitude of the south-north components of horizontal wind velocity in comparison with the magnitude of the west-east components. Even with the relatively accurate and comprehensive wind-measuring network we have today, direct observation of meridional cells is far from satisfactory.

Mean values of the zonal or west-east component are, however, usually at least of an order of magnitude larger than the south-north component, and the development of upper air observing networks over the last two or three decades has produced direct observational evidence of the zonal component, which in the mean exhibits the same main features in each hemisphere and at all longitudes. The picture for the months of July and January is shown at Fig. 1, which is based on observations in Australia and on information provided by Mintz

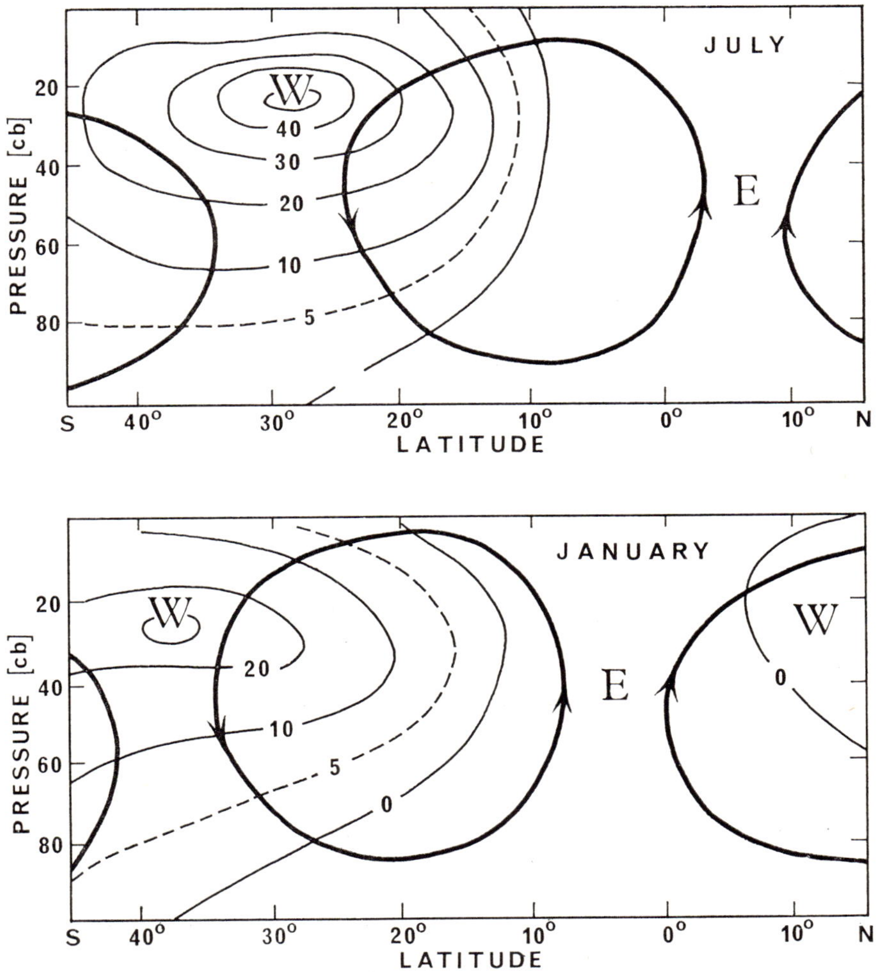

Fig. 1 Zonal and meridional circulations for Australian latitudes after Mintz and Palmen. Heavy lines show direction of meridional circulation. Zonal components are indicated in m. $sec.^{-1}$

(1954) and Palmen (1964). The figure shows that the main feature of the mean zonal circulation in the latitudes with which we are concerned is the westerly jet stream of over 40 m. sec^{-1} (80 kt) in winter and 20 m. sec^{-1} in summer. Also included in the figure are the tentative positions of the Hadley cells and the northern portions of the indirect cell in January and July.

It should be emphasised that the situation represented in Fig. 1 is idealised and is the mean for the months in question. Imposed on this mean circulation are perturbations or waves in the westerly current and, generally to a lesser extent, in the easterly. At times these perturbations can become considerable, such as when intense depressions or cyclones occur.

The importance of these perturbations in the transport of momentum, heat and moisture necessary to maintain the necessary balance of these quantities has become apparent in the last few decades as the result of investigations such as those of Rossby (1947), Priestley (1949), and Starr (1948). Indeed doubt has been expressed regarding the significance of the role of the meridional circulations in transporting these quantities. It has at least become clear that the meridional circulations alone are insufficient to provide the necessary transports.

Two interesting developments have thrown some light on the question of meridional cells and zonal currents. One is the work with laboratory models such as Fultz's rotating dishpan and the other is the use of large computers to process a mathematically specified model.

Fultz (1959) describes experiments with a rotating dishpan which contains water and is heated near the rim. Powder or dyes in the water and a camera which rotates with the dishpan enables the determination of water movement. The water simulates the atmosphere, the rim of the dishpan simulating the equator and the centre of the dishpan simulating the pole. Motions in the water reproduce the main features of the general circulation, inspiring confidence that similar physical processes are at work. It has been found that with low rates of rotation of the dishpan the circulation is symmetrical and free from perturbations, and closely resembles the Hadley model with one meridional cell. If rotation rate is increased the westerly current is perturbed. This suggests that the rate of the earth's rotation and the amount of heating in equatorial regions are of critical importance in determining the nature of the general circulation of the atmosphere.

In the United States Phillips and Smagorinsky have used large computers to model the atmosphere. A model of the atmosphere is constructed using mathematical descriptions of dynamical and thermal processes in the atmosphere, then a set of initial conditions for a grid of points covering a hemisphere is fed into the computer which by stepwise integration determines the behaviour of the model. Manabe and Hunt (1968) have reported a recent experiment. Their model reproduced the Hadley, polar, and indirect cells, although the polar cell was considerably larger and the indirect cell much smaller than usually represented. The appearance of these cells in the mathematical model gives some measure of confidence as to their reality, especially as their existence as such was not expressly provided for in the initial conditions. The great advantage of the laboratory and computer models is that observation of the results of varying initial conditions can lead to significant increases in our knowledge of the importance of various features of the general circulation.

SOME UPPER AIR OBSERVATIONS IN THE AUSTRALIAN REGION

Papers by Gibbs (1952, 1953), Phillpot (1959) and others present analyses of zonal wind and temperature distributions over the Australian region. These do not exhibit significant differences from other studies whose scope includes the Australian region, perhaps not surprisingly, because the source data are the same or similar. Statistical summaries of upper air data for the Australian region have also been published (Maher and McRae, 1964, 1966).

Table 1 presents some wind data for selected stations from Maher and McRae (1964). It will be seen that the west-east wind components show reasonable agreement with Fig. 1. The wind maximum in July appears to lie between the latitudes of Charleville (26°S) and Woomera (31°S) at an altitude of about 12 km with a value exceeding 50 m. sec^{-1}.

In January many stations between the latitudes of Charleville (26°S) and Cambridge (43°S) have mean west-east components at 12 km in excess of 20 m. sec^{-1}. It therefore appears that the less concentrated maximum shown for January in Fig. 1 is substantiated by the observations in Table 1.

That variation of mean west-east component with longitude is slight is illustrated by a comparison of data from stations with approximately the same latitude, e.g. Cocos Islands and Darwin, Port Hedland and Cloncurry.

A feature of Table 1 at variance with Fig. 1 is the positive mean west-east component from 1.5 to 6 km in January at Lae. This is the north-west monsoon which is evident in summer over New Guinea and Indonesia. Its effect is also fairly evident in the January mean west-east components for Darwin but not at Cocos Islands at the same latitude, illustrating that the monsoon is a regional feature of the circulation associated with the juxtaposition of the Australian continent and south-east Asia.

The mean south-north components appearing in Table 1 make it apparent that there is not the same close correspondence with Fig. 1 as there was with the mean west-east components. The mean south-north components for Lae, Cocos Islands, and Darwin show general agreement with the sense of circulation of the Hadley cell in Fig. 1, but other stations at latitudes higher than 20°S either show south-north components consistent with a more northerly location of the indirect cell or with a perturbation in the mean zonal current

Table 1

WEST-EAST (we) AND SOUTH-NORTH (sn) COMPONENTS OF UPPER WINDS (m. sec^{-1})

Alt. km	Lae 7°S Jan. we	sn	July we	sn	Cocos Is. 12°S Jan. we	sn	July we	sn	Darwin 12°S Jan. we	sn	July we	sn	Pt Hedland 20°S Jan. we	sn	July we	sn
1·5	5	−1	−3	1	−3	1	−8	1	1	0	−7	1	−3	−1	−2	−2
3	5	1	−6	1	−1	1	−1	−1	−3	2	−2	−1	−4	3	6	−1
6	2	0	−7	−2	−1	0	−1	−1	0	0	1	−1	−2	3	19	−2
9	−2	0	−7	−2	−2	0	−3	−2	0	−1	1	1	4	3	26	1
12	−6	−1	−7	−4	−1	−2	−4	−7	−3	−1	2	−2	8	4	27	2
15	−11	−1	−5	−4	−5	−3	−7	−3	−9	−1	1	−1	5	4	23	2
	Cloncurry 21°S				Alice Springs 24°S				Giles 25°S				Charleville 26°S			
1·5	−4	−1	−1	1	−6	1	−2	1	−7	−3	0	−1	−2	−2	3	1
3	−4	3	7	2	−3	3	8	1	−2	0	8	−2	−2	2	8	2
6	0	3	19	2	0	5	22	−1	2	4	22	−1	−1	4	23	2
9	8	3	27	2	8	4	39	1	9	3	43	1	10	3	46	2
12	13	4	29	2	19	2	44	2	18	1	45	3	21	1	51	3
15	10	4	23	3	17	3	34	2	16	2	33	2	17	2	39	5
	Woomera 31°S				Guildford 32°S				Adelaide 35°S				Cambridge 43°S			
1·5	1	−2	5	−1	1	−1	6	−1	3	−1	6	−1	6	−3	6	−1
3	3	2	8	0	6	−2	9	0	6	−1	9	−1	11	−2	7	1
6	5	4	17	3	11	−1	17	1	8	1	13	1	22	−2	10	−1
9	10	5	36	6	17	−2	31	2	12	2	24	1	21	−2	13	−2
12	17	5	48	6	26	−3	47	2	18	3	38	2	23	−1	19	0
15	18	3	29	3	24	−3	35	3	18	2	30	3	17	0	22	−1

with a preference for a particular location. The possibility of the latter occurrence cannot be discounted, but the northward displacement of the indirect cell does not appear likely.

The fact that the mean south-north components are generally small tempts one to suggest that the errors of observation render them unreliable. It is true that the standard deviation of these components ranges from few m. sec^{-1} at low levels to as much as 20 m. sec^{-1} at jet stream levels, but the data do not have the appearance of random deviations from zero. The safest conclusion appears to be that the mean south-north components of Table 1 do not support the existence of a Hadley cell in the Australian region.

The question of the Hadley cell can be approached from another angle, that of the temperature pattern over the Australian region; this will be explored later.

REGIONAL AND SYNOPTIC CIRCULATIONS

Of course the zonal circulation represented in Fig. 1 is subject to a considerable amount of perturbation. The monsoons of eastern and southern Asia and the weaker but similar circulation over New Guinea, Indonesia, and the northern Australian coastline in the southern summer are examples of perturbation of the general circulation. Perturbations of smaller scale, such as cyclones (including tropical cyclines) modify, often to a considerable extent, the general circulation pattern. Of these synoptic scale disturbances the most significant for the arid zone of Australia are tropical cyclones and upper air 'cut-off' cyclones. In Australia tropical cyclones occur relatively infrequently (see Brunt and Hogan, 1956), but they contribute significantly to the total long-term rainfall of Australia's arid zone, a rough estimate being 30 per cent. 'Cut-off' cyclones as described by Foley (1956) and McRae (1955) are responsible for a considerable part of the remainder of the rainfall, although convective cells, usually in the form of thunderstorms, are also significant rainfall producers.

SUBSIDENCE AND ITS EFFECTS

If there is no heat gained from or lost to its environment, the temperature of a rising parcel of air will fall at approximately 1°C for every 100 m ascent; a rise of the same amount will occur for a 100 m descent; the reason is the adiabatic expansion and compression of the parcel as it ascends or descends. If condensation occurs during ascent of latent heat by condensation, the resultant lapse rate is called *pseudo-adiabatic* and is a function of air temperature and amount of water vapour in the air. Therefore with no heat received from, or lost to, the environment decrease of temperature due to ascent occurs either at the adiabatic or pseudo-adiabatic lapse rate while increase due to descent is always at the adiabatic lapse rate. An example is the well-known *föhn* effect, where moist air which has risen with condensation over a mountain range has a higher temperature, level for level, on the leeward than on the windward side, the difference being greatest at the lowest level. Furthermore, and more importantly for the subject under discussion at this symposium, the absolute humidity of the air on the leeward side of the mountain remains unchanged during descent if it receives no moisture from its environment.

Applying this model to the meridional circulation shown in Fig. 1 we would assume that the temperature of the air in the ascending limb of the Hadley cell, if the air is near saturation (which is usually so with surface air in the tropics), should fall at the pseudo-adiabatic lapse rate and that in the descending limb of the cell should rise at the adiabatic lapse rate. Furthermore, the absolute humidity in the descending limb should remain constant at that of the air of the top of the ascending limb. If we look at the mean temperature above the level of condensation (about 850 mb) for stations in the region of the ascending limb we find that they are very close to the pseudo-adiabatic, although the median values of absolute humidity are considerably below saturation level. On the other hand if we look at the mean temperature for stations in the region of the descending limb we find the temperature lapse closer to the pseudo-adiabatic than the adiabatic lapse rate. Indeed the lapse rate of the mean temperatures is often even less than the pseudo-adiabatic in the layer between 800 and 900 mb. (Table 2 gives some data for Cloncurry.) At the same time absolute humidities are usually very low, especially at 800 mb and above, although not always as low as would be expected if the meridional circulation were as shown in Fig. 1.

TABLE 2

COMPARISON OF TEMPERATURE AND MIXING RATIO OF SATURATED ASCENDING PANELS OF AIR AT CLONCURRY* IN JANUARY AND JULY

Pressure cb	Temperature (°C)	Cloncurry Temperatures (°C) Jan.	July	Mixing Ratio (g. kg)	Mixing Ratios Cloncurry (g. kg) Jan.	July
100	23	33	23	17·0	11·3	4·6
90	19	25	14	15·2	10·5	4·2
80	15	16	10	13·3	9·0	2·3
70	10	9	6	10·8	5·9	0
60	4	2	0	8·3	3·2	0
50	−4	−2	−8	5·7	—	—
40	−15	−16	−18	3·0	—	—
30	−32	−30	−33	1·0	—	—
20	−55	−51	−54	0·1	—	—
10	−93	−78	−74	—	—	—

*Latitude 21°

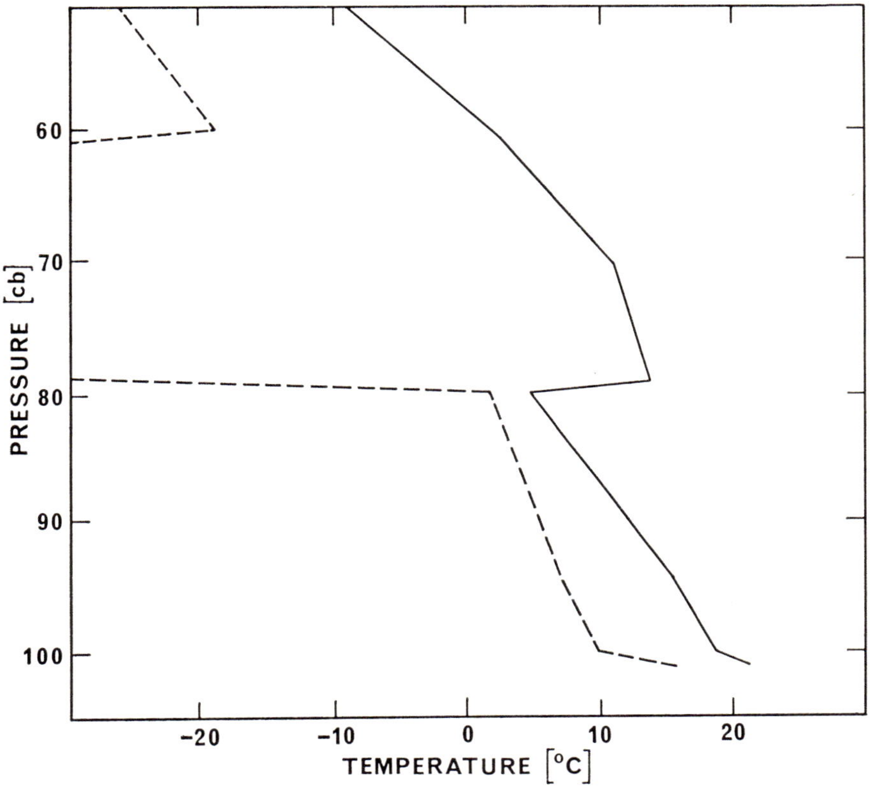

Fig. 2 Air temperature (full line) and dew-point temperature (broken lines) at Townsville 2315 hr 4 August 1968 GMT

Further light is shed on the small temperature lapse in the 800-900 mb layer by an examination of Fig. 2, which shows the temperature and humidity profiles for one day at Townsville. That this inversion of temperature and abrupt decrease in humidity is characteristic of soundings in Australia is shown by Table 3. This type of inversion is at its lowest altitude and most frequent about latitudes 35° to 40° in summer and 25° to 35° in winter. Note how well these latitudes agree with those of the region of subsidence shown in Fig. 1.

The inversion, called the 'trade wind' or 'subsidence' inversion, is formed, according to Riehl (1954), by subsiding, adiabatically-warmed air overlying a surface layer of air (usually about 1,500 m in depth) which has subsided but has had heat and moisture exchanges with a land and/or ocean surface. The air above the inversion might be considered to be that in the downward limbs

TABLE 3

MEAN PRESSURE AND MEAN TEMPERATURE INVERSION (°C) AND FREQUENCY OF OCCURRENCE

Station	Lat.	Mean Pressure (cb)		Mean Temperature Inversion (°C)		Frequency %	
		Jan.	July	Jan.	July	Jan.	July
Cocos I.	12	79	78	2	2	45	53
Darwin	12	71	83	3	3	7	31
Pt Hedland	20	61	77	1	3	24	48
Cloncurry	21	67	81	1	3	30	72
Alice Springs	24	15	81	2	3	15	61
Giles	25	54	76	2	2	47	74
Charleville	26	66	81	2	2	50	80
Woomera	31	79	82	2	2	53	63
Kalgoorlie	31	80	81	2	3	61	77
Adelaide	35	83	82	3	2	50	42

of the Hadley and indirect cells, while that below has been cooled and moistened by contact with the land or ocean surface, the lower layer being thoroughly mixed by turbulence. It follows that the inversion is not an impervious barrier but a surface through which air descends and is modified. Air above the inversion is extremely clear, no doubt because aerosols are absent, condensation in the upward limb being an effective cleanser. Furthermore the relatively very low humidities would prevent the growth of any hygroscopic nuclei which may be present. On the other hand air below the inversion is usually hazy, hygroscopic salt nuclei over the sea being one cause, dust and smoke (especially from grass fires) over the land being another. Over the sea and coastal areas a cloud layer is often present just below the subsidence inversion, and heat loss from the top of the cloud as the result of net long-wave radiation into space tends to intensify the inversion.

Observational evidence of the existence of the Hadley and indirect cells is not very convincing. The meridional motions (south-north and vertically) of the order of a few m. sec^{-1} or cm. sec^{-1} are overshadowed by the much more rapid zonal currents and the perturbation therein, but theoretical studies suggest that the Hadley and indirect cells exist and the subsidence inversion and the dry air above dovetail nicely with the suggested location of the downward limbs of the Hadley and indirect cells. Palmen (1964) has calculated that with the situation shown in Fig. 1 a parcel of air in the circulation of the Hadley cell would travel a little more than once round the earth; it is not surprising therefore that direct observational evidence of the Hadley cell is difficult to find. However, van de Boogaard (1964) in a study of the circulation of one day between 40°N and 60°N claims to have found evidence of the existence of the classical meridional circulations.

HEAT TRANSFERS AND SURFACE TEMPERATURE

The attenuation of the solar beam above the subsidence inversion is very small because of the purity of the subsiding air and general absence of cloudiness. Below the subsidence inversion the air is moister (although usually still very dry indeed inland) and is often heavily polluted with smoke or dust or has high concentrations of sea-salt nuclei over the sea or in coastal areas where breezes are onshore. The layer below the subsidence inversion is relatively shallow, however, so that in the absence of cloud in that layer, which usually applies inland, the attenuation of the solar beam is not great, even below the subsidence inversion, especially when the sun is at high elevation.

Hounam (1963) has calculated the intensity of solar radiation at the earth's surface and finds that the maximum for the Australian continent is centred inland about latitude 30° in January and along the northern coastline in July (see Fig. 3). It will be seen that these maxima are north of the latitudes of most frequent and lowest subsidence inversions, but it must be remembered that the maximum solar radiation at the top of the atmosphere is at about latitude 20°S in January and 20°N in July. The figure of 600 cal. cm^{-2} day^{-1} which the Australian arid zone receives in January is equivalent to 7×10^4 kw hour for the energy received in one day on a horizontal surface 100 m square. This is about a third of the energy of the nuclear bomb exploded at Hiroshima. In July the daily figure is about half that amount.

The general purity and dryness of the air above the arid zone also means that it transmits long-wave terrestrial radiation with very little attenuation. This means that radiative heat loss from the earth's surface is large; Hounam (1963) has estimated that this is at a maximum over the continent between latitudes 25°S and 30°S in January and at about latitude 20°S in July.

These high values of incoming solar radiation and outgoing terrestrial radiation produce strong contrasts between the temperature of the earth's surface between the hours of sunshine and darkness. During the hours of sunshine the temperature of the earth's surface will increase until the rate of heating by solar radiation is equalled by the rate of heat transfer from the surface. (This transfer is downwards into the ground and upwards to the air by conduction, and through the air by long-wave radiation.) The time of maximum ground temperature generally occurs in the early afternoon when the intensity of solar radiation is already decreasing. The fall of the temperature of the ground surface is followed by a reversal of heat flux by conduction, and heat begins to flow to the ground surface from below and from the air. These transfers are small, however, and are overshadowed by the heat loss by terrestrial radiation, as is the decreasing intensity of the solar radiation. This heat loss continues until shortly after sunrise, when the rate of loss due to terrestrial radiation is equalled by the contribution from solar radiation and from the air above and the earth beneath.

It will be seen that the temperature of the air near the earth's surface will be influenced by heat transfers from and to the surface. Over the arid zone these heat transfers are of such magnitude that the diurnal variation of air temperature in the lowest 100 m. is closely associated with that of the earth's surface, the magnitude of the variation decreasing with increasing height above the surface.

Air temperatures are also determined to some extent by the trajectory over the previous day or so, this effect being particularly evident near coastlines.

Records of air temperatures for a comprehensive network of stations in the arid zone are held on punched cards or magnetic tape in the Bureau of Meteorology. Selected statistics and discussions of surface air temperature distributions have been included in publications too numerous to mention here but which will be found listed in catalogues (for example Bureau of

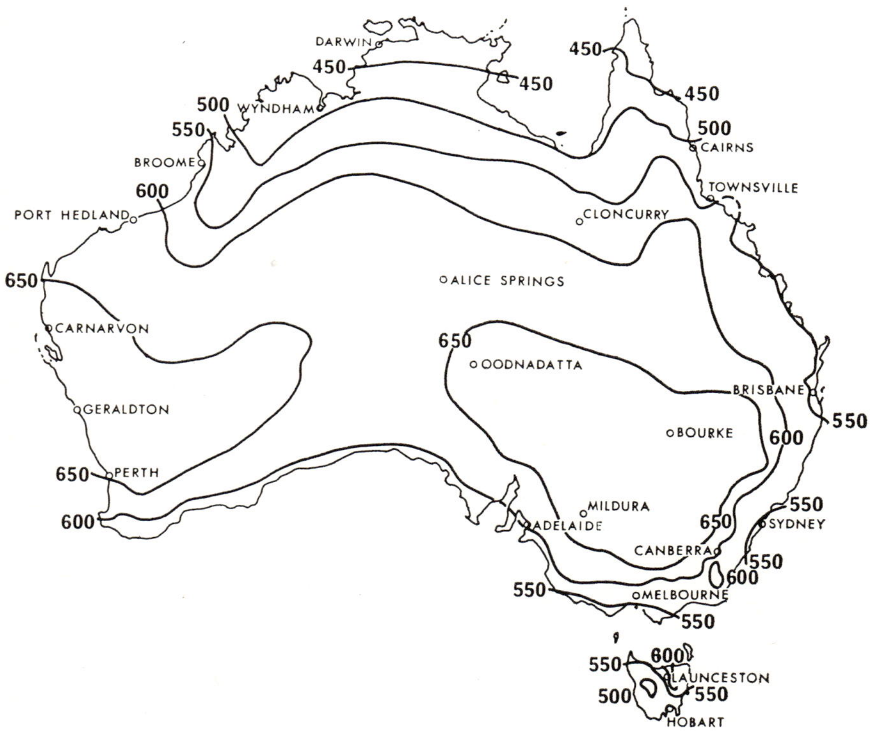

Fig. 3a Solar radiation (cal. cm^{-2} day^{-1}) over Australia in January

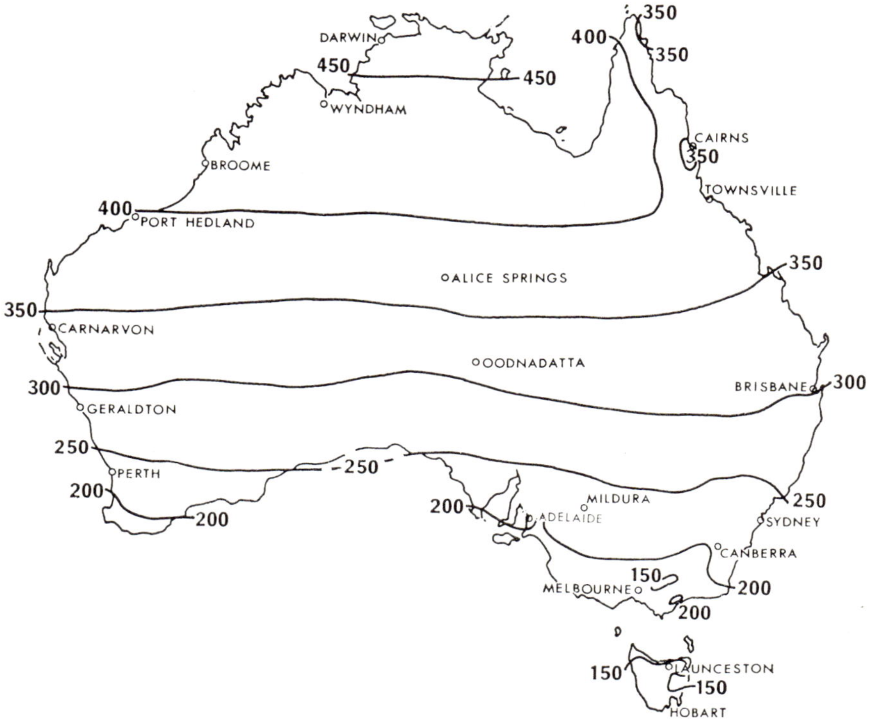

Fig. 3b Solar radiation (cal. cm^{-2} day^{-1}) over Australia in July

Meteorology, 1960, 1966). Table 4 gives some statistics for a few arid zone stations.

Table 4 has some interesting features. Although the stations listed have a 16° range of latitude and a 23° range of longitude, in January their mean maximum and mean minimum temperatures do not differ greatly from station to station, with the exception of Eucla, a coastal station on the Great Australian Bight. Even more remarkable, their highest maxima and lowest minima over all years of record show practically no variation from station to station, the exception being Alice Springs, which with its greater elevation has a lower minimum. Surprisingly, Eucla has the highest maximum, a circumstance warranting further investigation. The 90°F range between highest maximum and lowest minimum emphasises the thermal stress on plant and animal life in the arid zone.

TABLE 4

MAXIMUM AND MINIMUM SCREEN TEMPERATURES (°F) FOR SOME ARID ZONE STATIONS

		Jan.	Apr.	July	Oct.
Alice Springs	Mean max.	95·3	81·3	66·9	94·7
24S 134E	Mean min.	69·8	53·8	38·9	58·5
1901 ft	Highest max.	117·0	Lowest min.		19·0
Daly Waters	Mean max.	97·5	93·2	83·9	100·7
16S 133E	Mean min.	75·6	67·1	54·0	71·0
691 ft	Highest max.	116·1	Lowest min.		30·2
Eucla	Mean max.	78·3	75·0	64·6	73·1
32S 129E	Mean min.	62·5	56·2	44·2	51·9
15 ft	Highest max.	123·2	Lowest min.		29·0
Marble Bar	Mean max.	106·2	97·0	80·6	100·1
21S 120E	Mean min.	78·9	69·5	52·4	68·7
595 ft	Highest max.	120·5	Lowest min.		34·0
Marree	Mean max.	97·7	80·2	65·6	83·5
30S 138E	Mean min.	67·8	53·5	40·0	54·9
160 ft	Highest max.	115·9	Lowest min.		27·0
Wiluna	Mean max.	99·0	84·4	66·0	85·6
27S 120E	Mean min.	72·4	58·1	40·8	56·1
1700 ft	Highest max.	116·0	Lowest min.		26·0
Windorah	Mean max.	100·8	86·0	69·7	90·7
25S 143E	Mean min.	74·5	59·6	43·0	61·0
390 ft	Highest max.	116·0	Lowest min.		26·0

RAINFALL

The two prerequisites for rainfall—a moist atmosphere and ascending motion—are usually both lacking over the arid zone. The 3 pm percentage surface relative humidity over the arid zone is usually in the twenties except in the northern fringes in the summer and the southern area in the winter when air from adjacent oceans carries more moisture. As mentioned, it is the infrequent perturbation to the circulation which interrupts the normal subsidence and advects more moisture which is responsible for most of the rainfall of the arid zone. In addition moist air below the subsidence inversion in northern and eastern sections of the arid zone in summer months leads to convective instability and scattered thunderstorm rains.

The contribution of dew to the total water supply of the arid zone is not significant. Much of the water in dewfall during hours of darkness has evaporated from the earth's surface during the previous day.

Rainfall data are so numerous for Australia that it is not possible to list all the references, most of which can be found in Bureau of Meteorology catalogues of publications (1960, 1966). A comprehensive tabulation of frequencies of monthly and annual rainfall has also recently been published (Bureau of Meteorology, 1968).

Papers discussing the rain process and theoretical aspects of rainfall statistics have also been listed in a bibliography (Bureau of Meteorology, 1963).

As pointed out by Gibbs (1964) and many others, rainfall totals for hourly, daily, and monthly intervals are not normally distributed. Indeed it appears that the square roots of rainfall have a distribution close to normal, although this must be qualified by the fact that the distributions are often truncated, possessing a number of zero values (see for example Table 5).

Space does not permit an adequate discussion of this subject, but it is necessary to point out that the use of mean or 'average' values of hourly, daily, monthly, or annual values for the arid zone is unwise. A much more useful set of values are those relating to percentiles or the values which are exceeded by a stated proportion of hours, days, months, or years.

Figure 4 is a map of the median or 50 percentile value of annual rainfall for Australia on which the area defined as arid for the purpose of this symposium is shaded.

Figure 5 shows the frequency distribution of annual rainfalls for selected stations. The shape of many of these curves is similar, indicating that the standard deviation of the square roots is also of the same order.

TABLE 5

PROBABILITY THAT RAINFALL IN ALICE SPRINGS 24°S 134°E WILL EXCEED STATED AMOUNT

(%)

Rainfall in.	January Hour	January Day	January Month	July Hour	July Day	July Month	Year
0·0	1·5	18·0	90·0	1·3	8·0	60·0	100·0
0·1	0·4	8·0	76·0	0·1	2·0	40·0	100·0
0·3	0·2	5·0	66·0	0·05	1·0	31·0	100·0
0·5	0·08	3·0	56·0	0·01	0·6	24·0	100·0
1·0	0·02	1·0	42·0		0·1	19·0	100·0
2·0		0·2	25·0			6·0	99·5
3·0			12·0			2·0	98·9
4·0			7·0			0·7	95·0
8·0			0·2				65·0
12·0							28·0
16·0							10·0
24·0							1·5

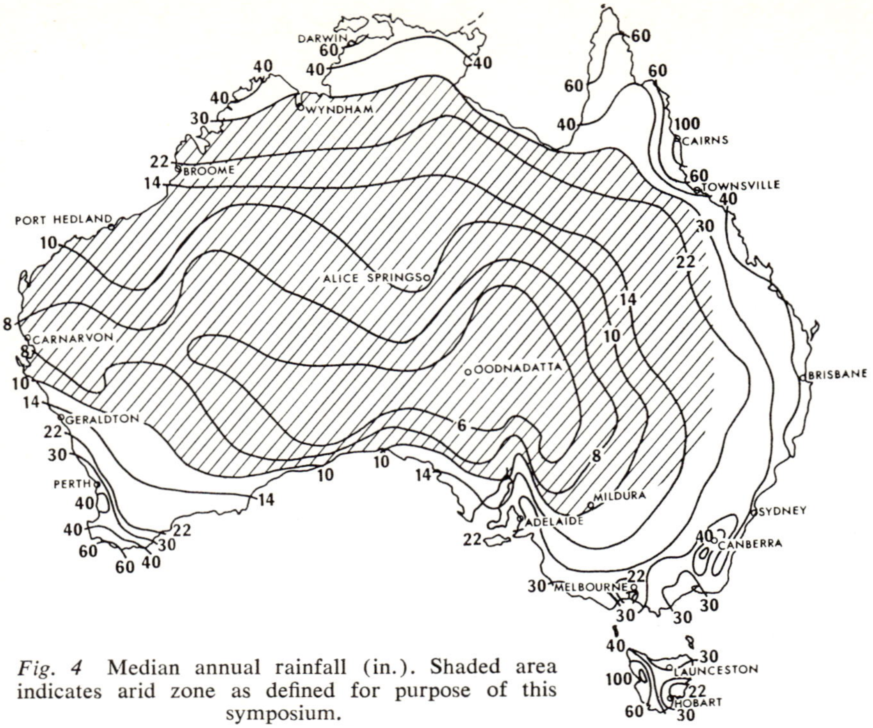

Fig. 4 Median annual rainfall (in.). Shaded area indicates arid zone as defined for purpose of this symposium.

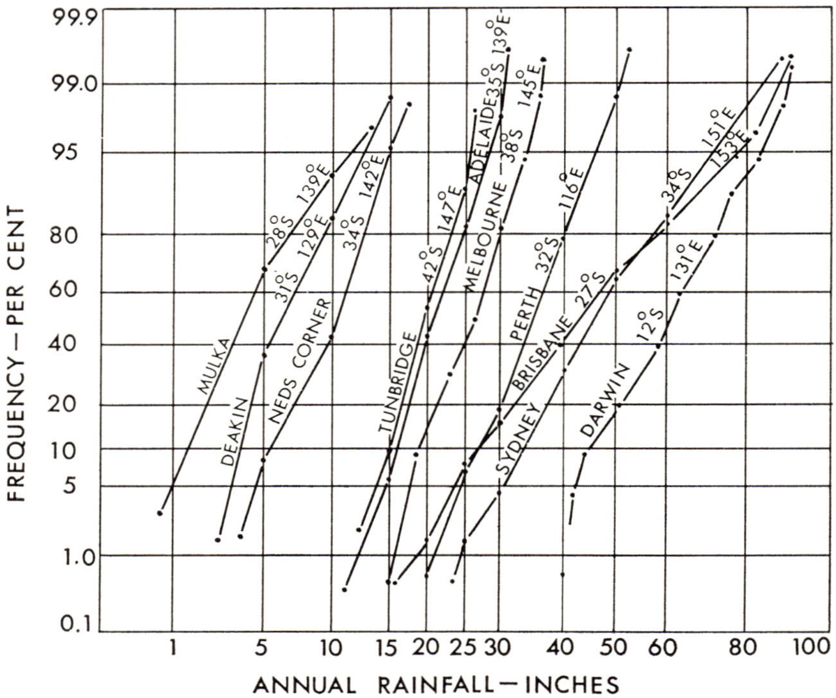

Fig. 5 Frequency distribution of annual rainfall (in.) at selected Australian stations. Scale of abscissa is square root of rainfall. That of ordinate —percentage probability.

TABLE 6

LOWEST AND HIGHEST AND 10, 50, AND 90 PERCENTILE VALUES OF RAINFALL

Selected stations		Jan.	Apr.	Jul.	Oct.	Year
				Inches		
Noonkanbah	L	0·22	0	0	0	5·77
18°30′S 124°50′E	10	1·60	0	0	0	9·26
(58 years)	50	4·15	0	0	0	16·70
	90	10·51	2·50	0·75	0·65	26·05
	H	15·69	12·59	2·85	2·00	31·79
Daly Waters	L	0·58	0	0	0	8·99
16°16′S 133°23′E	10	1·96	0	0	0	16·57
(93 years)	50	5·84	0·31	0	0·38	24·22
	90	10·10	2·89	0·15	2·22	35·60
	H	23·16	10·07	0·68	3·85	45·95
Coober Pedy	L	0	0	0	0	1·20
29°01′S 134°45′E	10	0	0	0	0	2·12
(34 years)	50	0·11	0·06	0·07	0·29	4·81
	90	0·91	0·77	0·54	1·47	10·26
	H	6·54	1·45	1·51	1·92	12·94
Eucla	L	0	0	0·11	0·04	4·42
31°43′S 128°53′E	10	0·01	0·08	0·29	0·11	6·69
(75 years)	50	0·22	0·78	0·82	0·55	9·79
	90	1·96	2·24	1·70	1·55	13·59
	H	3·75	8·07	2·45	2·91	17·05
Camooweal	L	0	0	0	0	5·91
19°55′S 138°07′E	10	0·76	0	0	0	8·04
(74 years)	50	2·29	0·12	0	0·21	14·14
	90	7·68	1·87	0·71	1·18	24·13
	H	16·16	3·51	3·51	4·93	32·27
Tibooburra	L	0	0	0	0	1·88
29°26′S 142°01′E	10	0	0	0	0	3·59
(74 years)	50	0·32	0·18	0·28	0·34	6·84
	90	2·37	1·99	1·50	2·25	14·82
	H	4·07	3·17	3·18	4·31	25·81

Tables 5 and 6 show some distinctive characteristics of arid zone rainfall. Table 5 shows that over 98 per cent of hours and 80 per cent of days are rainless at Alice Springs. Moreover 10 per cent of January and 40 per cent of July are without rain. At the same time the totals for the drier months are considerably smaller than the highest hourly totals. Similarly the highest totals for indivivdual months are greater than the lowest annual rainfalls.

Table 6 shows variability of monthly and annual totals for selected stations. In the north, rainless months are a feature of the winter and in the south are more common in the summer, in sympathy with the movement of the zone of maximum subsidence. The asymmetrical nature of the distributions is indicated for annual as well as monthly values, demonstrating the wisdom of using median rather than mean average values.

The sequence of occurrence of rainfall in Australia has been widely studied (for example O'Mahony, 1961). In brief it can be said that, while serial correlation is evident in hourly and daily rainfall, there is little evidence of a significant departure from random occurrence of monthly and annual totals.

EVAPORATION AND THE WATER BALANCE

The rate of evaporation from the earth's surface depends on the amount of solar energy it receives and the availability of water for evaporation. Water lying on the surface of the earth is evaporated most freely, the amount of evaporation being approximated by measurements made in an evaporation tank or by estimates derived from formulas such as those of Penman (see Hounam, 1961). If the water is transpired through the surface of plants or

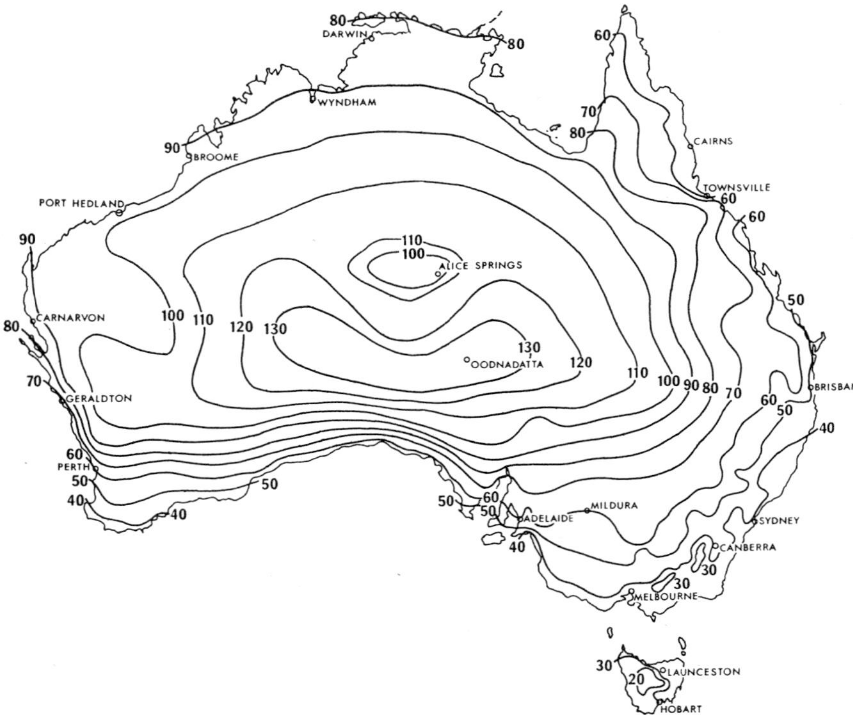

Fig. 6 Average annual potential evaporation (in.)

rises through the earth before evaporation at the surface, the rate of evaporation will in general be much lower, although in a green pasture or a growing crop completely covering the ground, the evaporation is close to that of a free water surface. Figure 6 shows the total potential evaporation over a year for Australia. A comparison of this figure with Fig. 4 shows that those areas in Australia where annual precipitation exceeds annual potential evaporation are very small. Further details of evaporation for arid zone stations are given in a recent publication (Bureau of Meteorology, 1968). Because in the arid zone the earth's surface is usually dry and the vegetation restricts evapotranspiration, the actual evaporation is very much below the potential evaporation. The actual evaporation is extremely difficult to measure, especially over an area with mixed vegetation types.

A comparison of Figs. 4 and 6 may suggest that storage of water in the arid zone is impossible. However, it is necessary to investigate the daily balance between precipitation and evaporation to assess the amount of water available to run in streams or to percolate below the earth's surface. Table 2 shows that rainfall over the arid zone is characterised by infrequent occurrence, the rain often being heavy. This means that when rain does fall the daily evaporation is likely to be considerably less than the precipitation total. Thus water is available for replenishment of underground storages as well as for the periodical filling of small and large inland lakes.

Recent severe droughts over Australia have stimulated a study of this subject. Foley (1957) has catalogued droughts for all years of record until 1955. Gibbs and Maher (1967) have found that a drought usually occurs when the rainfall falls within the first decile range. They point out that the only useful definition of a drought is the situation where supply of water falls below critical demand; and as the demand is generally a function of man's activities, drought can be considered to be man made.

CLIMATIC CHANGE

The features of the general circulation contributing to the dryness of Australia's arid zone suggest that a climatic change to wetter conditions in this region could result only from:

(1) a change in speed of rotation of the earth,
(2) a change in axis of rotation of the earth,
(3) a change in the amount of energy received from the sun.

The experiments of Fultz previously referred to show that in a dishpan the rate of rotation significantly affects the circulation in the fluid contained therein. It is reasonable to suppose that a slowing down of the earth's rotation would significantly change the general circulation of its atmosphere. Whether it would produce a significant change in the rainfall of the arid zone is difficult to determine but if the latitudes of the downward limbs of the Hadley and indirect cells were much higher than at present it seems likely that Australia's arid zone might receive a higher rainfall.

A change in the climate of Australia's arid zone resulting from a shift in the axis of the earth's rotation would only be significant if the shift was greater than about 20° latitude.

A change in the intensity of solar radiation, even if no change in the general circulation resulted, would conceivably produce a change in air temperature. This almost certainly would induce changes in the general circulation which could well alter the rainfall of the arid zone.

Artificial means of changing the climate of the arid zone are likely to be unsuccessful unless they induce a change in some features of the general circulation. Even if, as has been proposed from time to time, large lakes were created in Central Australia, if they remained in the subsidence region the increased evaporation from their surfaces is not likely to alter the climate significantly.

Cloud seeding is unlikely to effect a major climatic change. The more optimistic estimates of the results of cloud seeding claim a 15-20 per cent increase in natural occurring rainfall. This would not significantly alter aridity. During drought years, in the absence of naturally occurring rain, cloud seeding can have no effect.

REFERENCES

Boogaard, H. M. E. van de (1964). On the characteristics in the meridional plans of some physical properties of the atmosphere between the equator and 40°N based on a single day's data. *Proc. Symp. Tropical Met. N.Z. met. Service.*

Bureau of Meteorology (1956). Climatic averages Australia.

——— (1960). List of publications.

——— (1963). Selected bibliography to the end of 1962 Australia.

——— (1966). Supplementary list of publications.

——— (1968). Review of Australia's water resources—annual and monthly rainfall and evaporation.

Brunt, A. T. and Hogan, J. (1956). The occurrence of tropical cyclones in the Australian region. *Proc. Trop. Cyclone Symp. Bureau of Meteorology*: 15-8.

Foley, J. C. (1956). 500 mb contour patterns associated with occurrence of widespread rains in Australia. *Aust. met. Mag.* No. 13: 1-18.

——— (1957). Droughts in Australia. *Bull. Bur. Met. Aust.* No. 43.

Fultz, D. (1959). Studies of thermal convection in a rotating cylinder with implications for large scale atmospheric motions. *Met. Monogr.* No. 21.

Gibbs, W. J. (1952). Notes on the mean jet-stream over Australia. *J. Met.* **9**: 279-84.

——— (1953). A comparison of hemisphere circulations with particular reference to the Western Pacific. *Q. Jl. R. met. Soc.* **79**: 121-36.

——— (1964). Space-time variation of rainfall in Australia. In *Water resources, use and management*, 71-9. Melbourne University Press.

——— and Maher, J. V. (1967). Rainfall deciles as drought indicators. *Bull. Bur. Met. Aust.* No. 48.

Hadley, G. (1735). Concerning the course of the general trade winds. *Phil. Trans. R. Soc.* **29**: 58-62.

Halley, E. (1686). An historical account of the trade winds near the tropics with an attempt to assign the physical course of said winds. *Phil. Trans. R. Soc.* **26**: 153-68.

Hounam, C. E. (1961). Evaporation in Australia. *Bull. Bur. Met. Aust.* No. 44.

——— (1963). Estimates of solar radiation over Australia. *Aust. met. Mag.* No. 41: 1-14.

Lorenz, E. A. (1967). The nature and theory of the general circulation of the atmosphere. W.M.O., Geneva.

McRae, J. N. (1955). A case of cyclogenesis associated with an upper cut-off low. *Aust. met. Mag.* No. 11: 36-46.

Maher, J. V. and McRae, J. N. (1964). *Upper wind statistics—Australia.* Summary, Bureau of Meteorology, Melbourne.

——— (1966).*Upper air statistics—Australia.* Summary, Bureau of Meteorology, Melbourne.

Manabe, S. and Hunt, B. G. (1968). Experiments with a stratospheric general circulation model. U.S. Weather Bureau. *Mon. weath. Rev.* **96**: 477-539.

Mintz, Y. (1954). The observed zonal circulation in the atmosphere. *Bull. Am. Met. Soc.* **35**: 208-14.

O'Mahony, G. (1961). Time series of some Australian rainfall data. *Bur. Met. Study* No. 14.

Palmen, E. (1964). General circulation of the tropics. *Proc. Symp. Tropical Met. N.Z. met. Service.*

Phillpot, H. R. (1959). *Winds at 30,000 and 40,000 ft. in the Australian-New Zealand-Fiji area.* Project Report 59/2645 Bureau of Meteorology, Melbourne.

Priestley, C. H. B. (1949). Heat transport and zonal stress, between latitudes. *Q. Jl R. met. Soc.* **75**: 28-40.

Riehl, H. (1954). *Tropical Meteorology*, 43-71, 372-80. McGraw-Hill.

Rossby, C. G. (1947). On the distribution of angular velocity in gaseous envelopes under the influence of large scale horizontal mixing processes. *Bull. Am. met. Soc.* **28**: 53-68.

Starr, V. P. (1948). An essay on the general circulation of the earth's atmosphere. *J. Met.* **5**: 39-43.

4

WATER RESOURCES

N. H. FISHER*

INTRODUCTION

ARID AUSTRALIA, as defined for the purpose of this symposium, comprises an area of approximately 1,953,000 square miles: Of the twelve main drainage divisions now commonly used (Anon., 1967) for describing Australia's surface water resources, arid Australia includes (Fig. 1) all of the Lake Eyre, Bulloo-Bancannia, and Western Plateau Divisions, nearly all of the Indian Ocean Division, except the south-western corner, the southern half of the Timor Sea Division, the south-western part of the Gulf of Carpentaria Division, the western fringe of the North-east Coast Division, the north-western third of the Murray-Darling Division, and the northern extension of the small South Australian Gulf Division. By far the greater part of this, nearly 1½ million square miles, is an area of internal drainage, where streams are ephemeral or non-existent, most have no defined destination, and those that do, such as the Cooper and Diamantina systems terminating in Lake Eyre, reach it only on the rare occasions of the largest floods. The unreliability of run-off and the high rate of evaporation prevalent throughout the area enforce a widespread dependence on groundwater for pastoral and domestic purposes.

THE STATE OF KNOWLEDGE

Surface Water

The assessment of surface water resources depends mainly on three sets of measurements—rainfall, evaporation, and stream flow. Rainfall recording stations are naturally most sparsely distributed over the arid areas, largely because of lack of observers in these relatively unpopulated sections. They are sufficient to give a general picture of the pattern of precipitation but rarely adequate for a detailed assessment of rainfall in any given localised area. Pluviograph stations are shown on Fig. 1 but it has not been practicable to show the positions within arid Australia of all 350 to 400 rain gauges reporting to the Meteorological Bureau.

* Bureau of Mineral Resources, Geology and Geophysics. This chapter is published with the permission of the Director of the Bureau.

ACKNOWLEDGMENTS. The author thanks T. Quinlan and J. N. Casey for reading the manuscript and suggesting improvements, and many other colleagues in the Bureau of Mineral Resources for helpful discussion and comment.

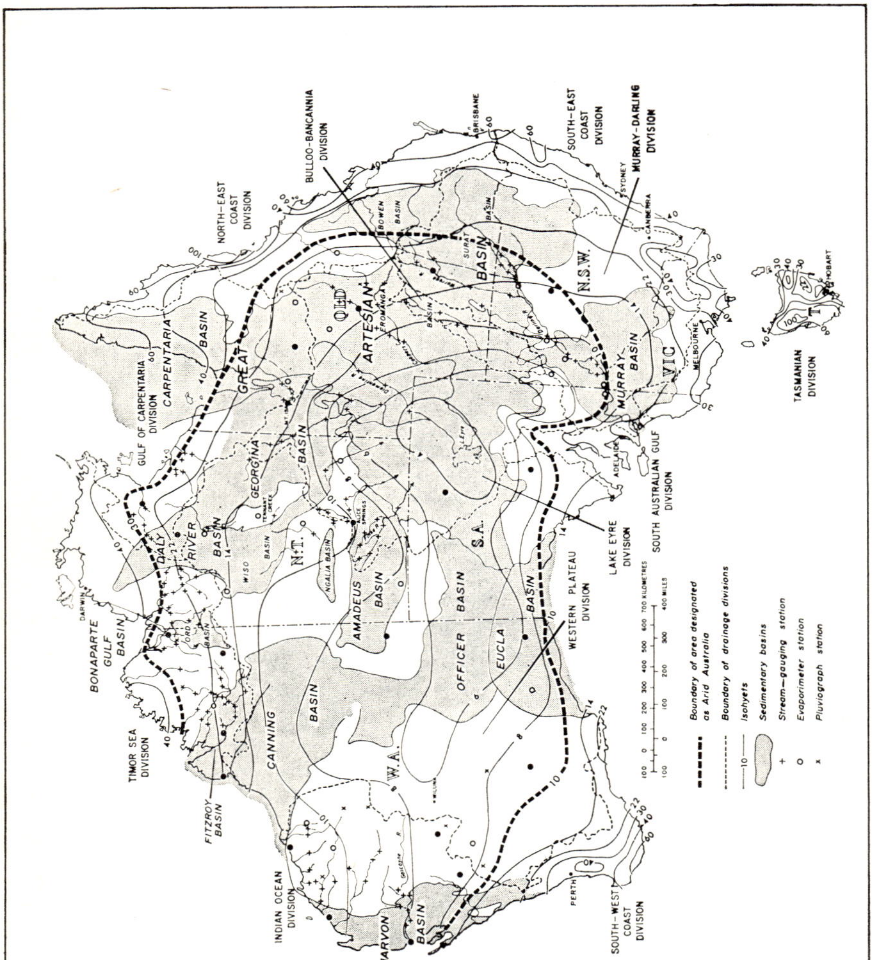

Fig. 1 Map showing limits of 'arid Australia', rainfall distribution, drainage divisions and sedimentary basins, and location of stream-gauging stations, pluviographs, and evaporimeters

The arid area is characterised throughout by extreme variability or unreliability of rainfall. Approximately two-thirds of it is contained within the 10-inch isohyet,[1] but annual rainfall ranges from nothing up to more than sixty inches, and in some parts years may go by with virtually no rain at all. The northern part of the arid zone comes within the area of influence of the north-west monsoon and can usually rely on summer rain though it is erratically distributed in time, and its quantity and reliability decrease southwards.

Australia's arid zone lies entirely in a region of high temperatures, so evaporation is everywhere high, commonly of the order of 100 inches per year or more and in many places exceeding 120 inches per year. Distribution of evaporimeters (Fig. 1) is naturally sparsest in the arid zone but because of the universal high rate of evaporation and constant climatic conditions is enough to enable a reasonable estimate to be made of evaporation at any point.

The number of stream-gauging stations in Australia is being approximately doubled during the 10-year period 1965 to 1974 as a result of recommendations by the Australian Water Resources Council. Within the arid area, however, stations are comparatively few in number and most of them have not been operating for more than a few years so that, in conjunction with rainfall data, they enable only a general estimate to be made of annual run-off. For many river basins run-off can only be estimated by comparison with adjacent rivers on which gauging stations exist.

Table 1 gives the number of stations in each of the drainage basins within the arid area, and an indication of the length of time that they have been operating (data to end of 1967: Anon., 1967-8).

As part of the Australian Water Resources Council's program of intensified investigation of water resources and also as part of Australian activities in connection with the International Hydrological Decade, a number of representative basins (generally less than 100 square miles in area) have been nominated for special study. Of those so far proposed nineteen are within the arid area, although five of these are marginal to it.

Groundwater

Assessment of groundwater resources is a more difficult and complex undertaking. In addition to meteorological and surface hydrological data, the geology must be thoroughly understood, data from all bores examined, and investigation bores put down to enable aquifer characteristics, reserves, and yields to be determined: quality of water varies over a wide range and is commonly the factor that determines whether the water is usable.

As a result of the geological mapping activities since 1950 of the Bureau

[1] Lines of equal median annual precipitation—the amount in inches per annum equalled or exceeded in 50 per cent of years.

TABLE 1

SUMMARY OF SURFACE WATER RESOURCES IN THE ARID AREA

Drainage Division and River Basin	Drainage Area (sq. m.)	Annual Rainfall Zone (in.)	Gauging Stations* (No.)	Estimated Average Annual Discharge (000 acre-feet)
Indian Ocean				
Murchison†	26,340	7–15	2 (0)	478
Wooramel	18,980	6–8	0	338
Gascoyne	30,490	7–8	12 (0)	480
Lyndon-Minilya	18,600	8–9	0	374
Ashburton	31,680	9–11	4 (0)	777
Onslow Coast	6,330	10–12	1 (0)	223
Fortescue R.	21,240	9–12	5 (0)	640
Pt Hedland Coast	13,300	10–12	4 (0)	472
de Grey R.	21,800	8–12	2 (0)	743
Timor Sea				
Fitzroy	33,400	14–30	14 (2)	5,000
Lennard	5,430	22–25	4 (1)	1,500
Isdell†	7,440	25–30	1 (0)	4,000
Pentecost	11,030	22–28	1 (0)	3,000
Ord	21,050	20–28	15 (2)	4,300
Keep	4,740	25–30	4 (0)	900
Victoria	30,000	15–30	27 (7)	5,000
Gulf of Carpentaria				
Roper†	20,000	22–30	15 (2)	4,000
Towns	1,900	22–30	0	500
Limmen Bight	6,400	19–30	0	1,200
Rosie	2,100	24–31	0	400
McArthur	7,100	20–31	6 (0)	1,400
Robinson	4,400	23–31	4 (0)	700
Calvert	4,100	22–31	0	600
Settlement	6,811	23–31	0	300
Nicholson	22,155	17–30	0	200
Leichhardt	12,199	15–30	2 (0)	150
Flinders	41,604	14–31	8 (7)	400
Norman†	19,566	20–32	0	780
Murray-Darling‡				
Darling	44,740	8–16	19 (14)	19
Warrego	28,100	12–21	13 (10)	700
Paroo	29,400	11–14	5 (2)	220
Bulloo-Bancannia				
Bulloo	30,200	6–16	8 (3)	330
Lake Eyre				
Georgina	89,000	3–16	13 (0)	700
Diamantina	60,000	3–16	4 (2)	815
Coopers Creek	110,000	3–20	11 (8)	1,800
Finke	43,500	3–10	14 (0)	300
Todd	25,800	6–10	19 (3)	?
Western Plateau	1,051,500	3–20	17 (0)	?

*Figures in parentheses are the numbers of gauging stations that have been in operation for 10 years or more.

†Minor portions of these basins lie outside the adopted boundary of arid Australia.

‡Parts of the Burdekin River system (Belyando), Fitzroy (Nogoa), and Culgoa-Condamine (Neebine Creek) are inside the 'arid' boundary but are omitted from the table.

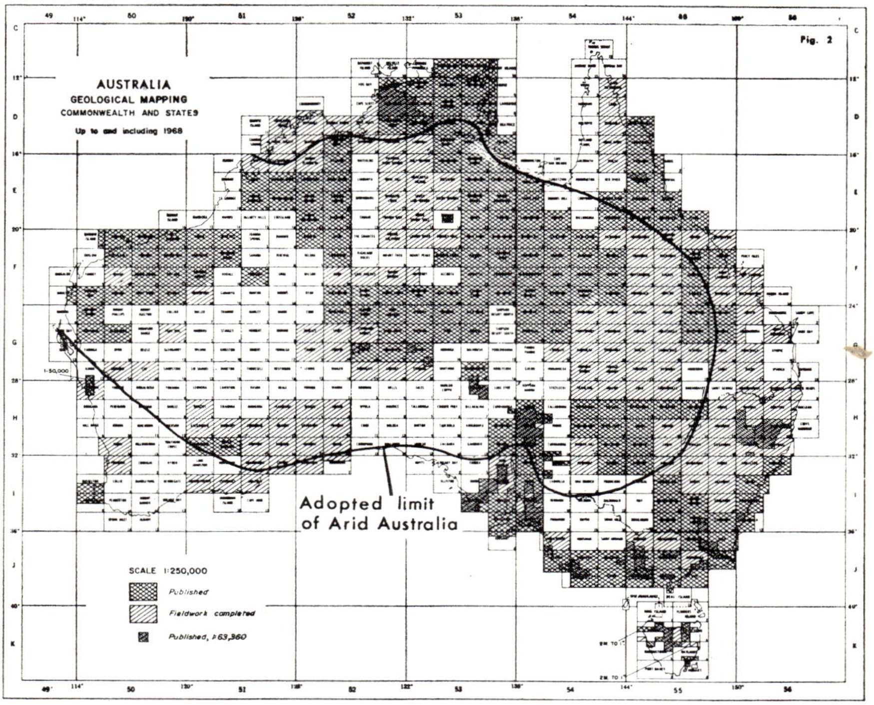

Fig. 2 Map showing the state of geological mapping at 1:250,000 scale at the end of 1968

of Mineral Resources and the State Geological Surveys, geological maps at 1:250,000 scale are now available of a large part of arid Australia (Fig. 2). The mapped area includes nearly all the Queensland, New South Wales, and Northern Territory arid areas except for about 30,000 square miles in Queensland, 16,000 square miles in New South Wales, and 80,000 square miles in Northern Territory. In Western Australia geological maps are available or in preparation for about 60 per cent of the arid area and in South Australia more than 25 per cent.

The quality of data on existing bores differs considerably. In general present legislation enables water authorities to receive good records of bores currently being drilled, though there are many exceptions to this, particularly with regard to shallow bores. Records of past bores, however, are in many cases frustratingly inadequate, though in the Great Artesian Basin, for example, a program of gamma-ray logging has made it possible to fill in many of the previous gaps in knowledge of the sections penetrated. The recent upsurge of oil exploration has resulted in the sinking of many well-documented deep bores which have added vastly to knowledge of stratigraphy and of the characteristics of the sedimentary formations.

Information on water quality is reasonably good in that an analysis is made of the water in nearly all bores when they are first put down. Comparatively few records have been kept, however, of changes in water quality except where special investigations have been made of the hydrology of an area, but arrangements are now in hand for regular resampling of many of the bores as part of the program of intensified investigation of water resources.

Sources of Surface Water in Arid Australia

Except for a narrow north-eastern and a rather wider northern fringe, most of arid Australia lies within the zone for which average annual run-off is calculated to be less than half an inch per year (Anon., 1967); within this area, stream flow is so irregular and shortlived that its main usefulness is in groundwater recharge. High evaporation and lack of suitable reservoir sites, combined with irregularity of flow, make surface conservation impracticable. Parts of some of the main drainage basins in the arid area, however, which abut on to better watered climatic zones, do carry important amounts of surface water. Table 1, compiled from data in the Review of Australia's Water Resources 1963 (Anon., 1965) and Stream-Gauging Information (Anon., 1967-8), lists the drainage basins, their rainfall zones, calculated annual discharge and the number of gauging stations.

Indian Ocean Drainage Division

The arid area in this Division includes most of the Murchison, the Wooramel, Gascoyne, Minilya, Ashburton, Fortescue, de Grey, and several smaller river basins. Of thirty stream-gauging stations recorded up to the end of 1967 (Anon., 1967-8) none were in operation before 1960, and only twelve before 1965, most of these being on the Gascoyne River. From these rather meagre data and the rainfall records, the possible annual average discharge of the principal rivers listed above has been estimated to range between 340,000 and 780,000 acre-feet but the flow in any one year in almost any of these rivers may range from nothing up to several millions of acre-feet. None of these waters are conserved and usage is confined to local pumping for irrigation of plantations along the banks of the lower reaches of some of the streams, e.g. production of bananas and beans on the Gascoyne near Carnarvon.

Timor Sea Drainage Division

Within the area designated as arid zone this Division includes important resources of surface water—the Fitzroy, Ord and Victoria rivers and their tributaries; also the Lennard, Isdell, Durack, and Pentecost rivers.

Most of the catchment of these rivers lies within the 15-inch rainfall belt and much of it within the area calculated to have an annual run-off ranging from two to ten inches. Rain is seasonal monsoon and run-off is in the form of high summer floods. The Victoria and Fitzroy rivers are estimated to discharge each annually about 5,000,000 acre-feet and the Ord 4,300,000 acre-feet. Only on the Ord River are conservation works in existence or in planning stages and there is a small irrigation development at Camballin on the Fitzroy. A considerable number of gauging stations have been installed along these rivers and their tributaries (Table 1 and Fig. 1), though few of them have been in operation for more than ten years.

Gulf of Carpentaria Drainage Division

The basins of all the rivers from the Roper in the north-west to the Norman in the south-east lie within the area described as arid, although because of the heavy summer rains, many of the rivers discharge large amounts of water. Except on the Roper, McArthur and Robinson rivers there are few gauging stations. Annual discharge of the Roper River is estimated at 7,000,000 acre-feet per year, and the Towns, Limmen Bight, Rosie, McArthur, Robinson, Calvert, Settlement, Nicholson, Leichhardt, Flinders and Norman rivers have possible average discharges ranging from 150,000 to 1,400,000 acre-feet per year. The main conservation works are the Moondarra dam on the Leichhardt River and the Rifle Creek dam which supply Mt Isa, and the Corella dam, which supplies the town of Mary Kathleen, on Corella River, a tributary of the Cloncurry River which is part of the Flinders River system.

Lake Eyre Drainage Division

This Division has an area of 440,000 square miles and is one of the largest areas of internal drainage in the world. The main river systems within it are Coopers Creek, Diamantina, and Finke. The first two are characterised by very low stream gradients and extensive systems of braided channels which spread flood waters out over large areas in south-western Queensland, hence the name Channel Country. The Lake Eyre Drainage Division lies entirely within the area with less than half an inch of annual run-off and the average for the whole area is only 0·15 inches (Anon., 1967). Scarcity of suitable storage sites, irregularity of flow, and high evaporation rate make surface conservation unlikely. The upper reaches of some of these streams contribute to the recharge of the Great Artesian Basin, and the lower reaches to shallow groundwater sources.

Bulloo-Bancannia Drainage Division

This small internal drainage basin has similar characteristics to those of the Lake Eyre Division.

Western Plateau

This is a huge area—over 1,000,000 square miles—in which there is no regular drainage pattern, only ephemeral local streams, and it has no significant run-off. Average annual rainfall is less than ten inches over most of the area, and is higher only along the northern fringe which comes under monsoonal influence. The whole area, like the two preceding ones, must rely mainly on underground water for regular and well-distributed supply.

North-East Coast Drainage Division

The 'arid' area includes parts of the basins of the Belyando and Nogoa rivers, tributaries respectively of the Burdekin and Fitzroy rivers, the two largest rivers in the Division, but this section of the Division is on the western fringe of the summer rainfall belt and has an annual run-off only of the order of one inch. The upper part of the Nogoa River, included within the 'arid' boundary, comprises the greater part of the catchment area for the Fairbairn dam on the Nogoa, near Emerald, now under construction.

Murray-Darling Drainage Division

The arid area includes the Darling River and its northern tributaries, an area in which annual average rainfall increases from about eight inches in the south-west to twenty-two inches at the north-eastern end.

Average annual run-off of this section is less than half an inch per year and its characteristics are generally similar to those of the even more arid Bulloo-Bancannia and Lake Eyre Divisions to the west. Possible average annual discharge is estimated at 700,000 acre-feet for the Warrego River and 220,000 for the Paroo. Measurements at Bourke and Menindee show that discharge in the River Darling decreases appreciably downstream due to evaporation and seepage.

Groundwater Resources

In assessments that have been made (Anon., 1965), Australia's groundwater resources are considered under three main headings; water in unconsolidated sediments—alluvial and aeolian deposits mostly of Cainozoic age; porous rocks in sedimentary basins; and the igneous and metamorphic rock areas in which water occurs mainly in fractures and joints and in weathered material.

Most systematic investigation by the water authorities has been done in alluviated valleys and coastal sand aquifers near cities, where yield is high and water can be made available for irrigation and town supply. Comparable investigations within the arid zone have been made at Alice Springs, Tennant Creek and Wiluna, and a number of towns in Western Australia also derive or augment their supplies from groundwater supplies. A smaller scale investigation has been made of the Willowra and Tea Tree alluvial basins northwest of Alice Springs.

Aquifers in sedimentary basins such as the Great Artesian, Murray, Canning, Amadeus, and Georgina basins, contain large reserves of water over wide areas. Quality of water is variable and in many areas suitable only for stock. In few of the sedimentary basins in Australia is the state of knowledge adequate to form more than a general picture of their water resources or of the amounts that can safely be harvested, but the extent of the aquifers enables information to be extrapolated over large areas.

Quantity of water in storage in fractured rocks is generally small, quality ranges from very good to highly saline, and the geology, rainfall, evaporation, and surface hydrology of each individual area have to be closely studied to enable estimates of available water to be made. Quantities are usually sufficient only for stock, domestic supplies and home garden irrigation, not for irrigation on any larger scale.

Sedimentary Basins

The main resources or potential resources of groundwater in the arid zone are in the aquifers of the sedimentary basins which lie partly or wholly within the zone; these include most of Australia's major basins, the Great Artesian (except the south-eastern part), the Georgina, Daly, Wiso, Amadeus, Ngalia, Bonaparte-Ord, Canning (including the Fitzroy), Officer, Eucla and Carnarvon and part of the Murray Basin. In none of these has the hydrogeology been studied sufficiently intensively to enable reserves of water in storage or safe yields to be estimated accurately; the best known is the Great Artesian Basin, because of its importance in providing water supplies to the grazing industry over an enormous area (670,000 square miles) and the number of bores (over 20,000) which have been put down to exploit it.

Great Artesian Basin. Geological mapping in recent years and gamma-ray logging of water bores have added greatly to our knowledge of water distribution and occurrence within the basin, but detailed studies of the hydrogeology of the basin based on this additional knowledge have yet to be carried out. Current assessments are based mainly on the 1954 report of the Queensland Committee of Investigation (Anon., 1954).

According to graphs in this report, present flow of water from the basin in Queensland is about 200,000,000 gallons per day and eventual safe yield (by year 2010) is estimated at 110,000,000 gallons. Eventual discharge (flow, pump, and springs) from the whole basin is expected to be of the order of 170,000,000 gallons per day or about 230,000 acre-feet per year. Individual bores yield flows up to 1,000,000 gallons per day. Salinity is usually less than

3,000 ppm.,[2] but because of the high content of alkali carbonate, the water is generally regarded as unsuitable for irrigation.

The main intake areas are along the north-eastern edge of the basin at the head of the Flinders River and its tributaries (according to Whitehouse: Anon., 1954), farther south at the head of the Belyando, Nogoa, Maranoa, and Warrego rivers, around the south-eastern margin of the basin and possibly in the Finke River basin and in the Boulia area in the north-west. The general direction of flow of water through the basin is south-westerly (Griffin, 1962) with local variations according to the topography of the intake areas, a westerly to northerly flow in the Carpentaria Basin, and a general southerly flow in the Surat Basin.

In Queensland, where the stratigraphy has been the subject of special study in recent years, the most widely tapped aquifers (Vine, pers. comm.) belong to three major age groups:

Lower Jurassic, the Hutton Sandstone (300-800 feet thick) throughout most of the Eromanga and Surat basins and, to a lesser extent, the Precipice Sandstone (0-400 feet) in the Surat Basin.

Upper Jurassic to Lower Cretaceous, the Hooray Sandstone (0-900 feet) of the Eromanga Basin (including marginal facies) together with some of its correlates: the Gubberamunda Sandstone (200-800 feet) and Blythesdale Formation (400-600 feet) of the Surat Basin, and Wrotham Park Sandstone/Gilbert River Formation (0-300 feet) of the Carpentaria Basin.

Lower to Upper Cretaceous, the Winton and Mackunda Formations (0-5,000 feet) of the central Eromanga Basin.

In addition, all porous sandstone formations of Triassic and Jurassic age contain major aquifers, and porous sandstone beds in argillaceous units of Triassic, Jurassic and Cretaceous age are aquifers of lesser importance.

Total reserves of water within the basin must amount to more than 20,000 million acre-feet.

Georgina, Daly, and Wiso Basins. These basins, which have a total area of about 170,000 square miles (Stacey, 1968) consist largely of sediments, dominantly calcareous, of Upper Proterozoic to Ordovician age, overlain in places by comparatively thin Cretaceous beds. The Georgina Basin includes the important pastoral lands of the Barkly Tableland. In this area according to an excellent study by Randal (1967) the groundwater system can be divided into two sections, a south-easterly one corresponding to the surface basin of the Georgina River, and a north-westerly one comprising the area of internal drainage between Alroy Downs and Anthony Lagoon.

Most of the bores produce more than 1,500 gallons per hour, but potential yield is not accurately known. The water is generally suitable for watering

[2] Salinity is expressed as parts per million total dissolved solids.

stock, but quality varies considerably, and salinity is higher in the north-western internal drainage section (up to 11,000 ppm. but generally in the range 2,000-5,000) than in the south-eastern section (up to 2,500 ppm., normally 500-1,500). Much of the groundwater is characterised by high sulphate or fluoride content and is unsuitable for human consumption, though very little is unusable by stock. Use for agriculture is limited by the high sodium content relative to magnesium and calcium. Recharge data are inadequate to determine safe yield either locally or regionally, but Randal considers the area could support much greater withdrawal than the 360 or so bores in the area produce at present.

Assuming an average thickness of 400 feet, which is conservatively low, for the rock-body containing aquifers and an effective porosity of 5 per cent, the water content over the 40,000 ± square miles of the Barkly Tableland would be in the order of 500 million acre-feet.

Conditions over the other parts of the Georgina Basin and the Wiso and Daly basins are generally similar to those on the Barkly Tableland. The principal aquifers are the Middle Cambrian limestones in the Daly and Wiso basins (Randal and Brown, 1967; Randal, 1969), although water also occurs, somewhat erratically, in the fracture and joint systems of the Antrim Plateau Volcanics to the north-west, the Cambrian-Ordovician dolomites and sandstones of the Georgina Basin (Smith, 1969), and the Devonian Dulcie Sandstone in the south-western part of the Georgina Basin. Bores generally yield 1,000 gallons per hour or more, and salinity is usually less than 1,000 ppm., though more saline water occurs in places, for instance in parts of the northern Wiso Basin beneath the Lower Cretaceous rocks or in the plains country to the south-east the water has had to travel considerable distances and is correspondingly saline —up to 5,000 ppm.

Ngalia Basin. Few bores have intersected potential aquifers in this small asymmetric basin, which covers about 6,000 square miles and includes sandstone and dolomite formations, mostly separated by unconformities, and ranging from Proterozoic to Carboniferous in age (Wells *et al*, 1968). Most exploration for water has been in alluvium overlying basement rocks, and bores have been very erratic in both quantity and quality, with a high proportion of failures though some success has been achieved in the Tertitary(?) sediments in the central part of the basin. Little can be said at present about the water resources of the sedimentary beds.

Amadeus Basin. The Amadeus Basin (Wells *et al.*, 1969) covers an area of approximately 60,000 square miles and includes thick sequences of Proterozoic, Cambrian, Ordovician, possibly Silurian, and Devonian sediments, as well as overlying Permian and Jurassic sediments and Tertiary and Quaternary deposits. These younger formations yield good supplies of water in places, and important Palaeozoic aquifers are the Cambrian-Ordovician Goyder Formation and Pacoota Sandstone, the Devonian Mereenie Sandstone, and some of the formations of the Devonian-Carboniferous Finke Group.

The Mereenie Sandstone, which now provides most of Alice Springs water supply, is a particularly good aquifer consisting of generally well-sorted sand with an intergranular porosity around 20 per cent and generally good though variable permeability. Quality where the supply has been obtained is good, less than 1,000 ppm. and in places less than 500 ppm. It occurs over an area of about 10,000 square miles and ranges in thickness up to 3,000 feet. Quinlan (1968) has estimated that the Mereenie Sandstone contains 1,000-2,000 million acre-feet of water. Rates of movement or recharge of this aquifer are not known and considerable investigation would be necessary to determine how much water could safely be drawn off each year. It is thought to be feasible to construct bores producing up to 50,000 gallons of water per hour.

Bonaparte-Ord Basin (Traves, 1955; Veevers and Roberts, 1969). In these two basins, which occupy an area of about 10,000 square miles, water occurs in Proterozoic sediments, in which, though hard to drill, good bore sites can probably be found if knowledge of the local geology is adequate: in the Antrim Plateau Volcanics in joints and fractures, where supplies of good water are common, although failures occur; and in limestone and sandstone beds in the Palaeozoic sediments, whose aquifer characteristics are little known because they are generally near the coast or in poor pastoral country. Quality of the water is poorer in the limestone than in the sandstone or the volcanics and may be suitable only for stock.

Canning-Fitzroy. The Canning Basin, and the subsidiary Fitzroy Basin which lies along its north-eastern side, cover an area of about 150,000 square miles. Much of it is occupied by the sand dunes of the Canning or Great Sandy Desert, and in this part most water supplies have been obtained from shallow wells (Veevers and Wells, 1961; Casey and Wells, 1964).

About twenty deep bores have been put down in the basin in the search for oil and a few deep and many shallow water wells. Supplies generally of good quality water have been found in the Permian sandstones of the Grant and Liveringa Formations and the Poole Sandstone, respectively more than 8,000, 1,850-3,000, and 220-1,600 feet thick (Guppy *et al.*, 1958) and in Devonian limestones. Large flows of artesian water have been obtained, particularly near the coast, from coarse sandstone formations of the Mesozoic sequences which overlie the Permian. Individual sandstones in these formations are usually not more than a few hundred feet in thickness.

Carnarvon Basin. The Birdrong Formation (up to 380 feet thick) at the base of the Cretaceous section forms the most widespread single aquifer of the basin, extending over almost half of the total basin area of about 45,000 square miles (Condon, 1965-8). Over much of its extent it is artesian, and flows of more than 80,000 gallons per hour have been reported. Quality is good towards the intake areas in the eastern part of the basin, but salt content increases westward and the water contains hydrogen sulphide. The Palaeozoic sequence in the Carnarvon Basin contains many sandstone aquifers which yield fair quality sub-artesian water. The formations in which these aquifers occur are commonly less than 600 feet thick, though some exceed 1,000 feet, and dip fairly steeply, thereby limiting the area within which any aquifer may be developed. These formations include the Munabia Sandstone in the Devonian, Moogooree Limestone in the Carboniferous, and the Coyango, Thambrong, Weedarra, Moogooloo, Mallens, Norton and Mungadan Sandstones in the Permian.

Officer-Eucla Basins. Information on groundwater in these two quite large basins (Eucla 78,000, Officer 118,000 square miles) is very limited. Water is obtained in shallow bores in Tertiary limestone in the Eucla Basin but is generally saline except near recharge areas. The underlying Cretaceous sandstones have been penetrated only in a few bores put down in connection with petroleum exploration and in some deep water bores along the Transcontinental railway line. Fairly good supplies of sub-artesian water were obtained in the deep bores in the Cretaceous basal sands, with salinities ranging from 3,000 to 5,400 ppm.

The Officer Basin is known even less; Proterozoic and probable lower Palaeozoic sediments are covered in places by comparatively thin Permian and Cretaceous beds. Flows in the Cretaceous up to 6,000 gallons per hour have been found during oil exploration. Quality in some is good, but ranges up to 3,000 ppm. (Jackson, 1966). In the Birksgate well in South Australia, about forty-five miles east of the Western Australian border, in the north-central part of the Officer Basin, water in the Cretaceous (110-520 feet depth) was fresh, in the Permian 'red beds' (555-1,417 feet) content of total dissolved solids ranged from 572 to 2,500 ppm., and in the Upper Proterozoic (1,643-3,125 feet) up to three times that of seawater.

Murray Basin. Only the northern part of the Murray Basin is included within the arid boundary. The basin contains up to 1,500 feet of Tertiary and Mesozoic sediments. Most boring has been done in the alternating sand-clay Tertiary sequences, which, although commonly yielding fairly copious supplies, in places artesian, contain water which is generally saline (7,000 to more than 15,000 ppm.) except near sources of recharge. The underlying Mesozoic formations contain better quality water but have been penetrated in comparatively few boreholes (Griffin, 1962).

Unconsolidated Sediments

Most investigations of this source of groundwater have been made in the higher rainfall areas. Formations coming under this description are widely scattered over arid Australia and provide domestic and stock water in many places. Some of them have potential for irrigation on a local scale but need special investigation to determine quality, depth, yield, recharge, storage, and hydrogeological characteristics.

Quinlan (1965) refers to two main types:

1. Alluvial Basins—small basins filled with unconsolidated alluvium and sediment. They are associated with single drainage systems and store moderate quantities of groundwater. Recharge is by infiltration along the bed of the drainage channels, and the ratio of the groundwater stored to recharge is less than 10. The groundwater which is used at Lorna Glen and Wiluna, W.A., is withdrawn from basins of this type.
2. Bolsoms, or large basins filled with unconsolidated lacustrine and marine sediment and with piedmont deposits, formed by the coalescing of individual alluvial fans. Commonly they are basins of internal surface and groundwater drainage, which terminate in salt lakes at the foot of the piedmont deposits. Recharge is usually available from several drainage systems and the ratio of groundwater storage to recharge is 100. The bolsom to the north of the Western MacDonnell Ranges, N.T., is 100 miles long and 30 miles wide.

With regard to quality of water he comments:

1. The recharge mounds adjacent to the surface drainage channels which cross small alluvial basins contain groundwater suitable for domestic consumption. Away from the recharge mounds the quality will deteriorate rapidly until it is unsuitable for irrigation.
2. The piedmont deposits contain saline water in the vicinity of the salt lakes. The quality improves rapidly as the distance from the lakes increases, so that much of the groundwater is suitable for most purposes. Unless there is a physical connection between aquifers in the alluvium and those in the unconsolidated sediment, the groundwater stored in the latter will only be suitable for stock.

Until 1964 Alice Springs drew its water supply from an alluvial basin beneath the town which was sporadically recharged by the irregular flows of the Todd River, which has a catchment area above the town of about 200 square miles. This is one water source in arid Australia that has been well documented. Various estimates have been made of the safe yield, the latest of which (Quinlan, 1967; Quinlan and Woolley, 1969) was 156,000,000 gallons per year. Probably on the average twenty times as much as this flows down the Todd River past the Alice Springs Town Basin per annum and the total discharge of the Todd Basin is about 3,340,000,000 gallons per annum, or 12,500 acre-feet. Quality of the water drawn off for town supply from the Town Basin averaged 300 to 400 ppm.; much of the water in storage contains 200 to 1,000 ppm.

As a result of a survey of the lands of the Alice Springs area by CSIRO Division of Land Research (Perry *et al.*, 1962) attention was drawn to the possibilities for irrigation of several small basins, lying mainly north and north-west of Alice Springs and filled to a depth of 600 feet or more with Mesozoic and Tertiary sediments, largely unconsolidated. Further information collected on several of these basins (Perry *et al.*, 1963) indicated that six selected basins —Narwietooma, Hamilton, Tea-Tree, Harpers, Ingallana, and Stirling—with a total area of 11,260 square miles, contain 4,000,000 acre-feet of water and that mean annual recharge is of the order of 70,000 acre-feet. Water quality in most existing bores ranges between 800 and 2,000 ppm.

Fractured Rocks

Metamorphic, igneous, and indurated sedimentary rocks, mostly of Precambrian age, occupy more than 250,000 square miles of the arid zone in the Shield area of Western Australia and considerable areas in the Kimberleys, the western and southern parts of the Northern Territory, the Cloncurry-Mt Isa district, north-eastern Queensland outside the Artesian Basin, and in South Australia (R.A.W.R., 1965: Map 4, Underground Water)[3] Water if present in these rocks is generally found in bores at less than 150 feet depth, its occurrence is difficult to predict, yield is limited in quantity, usually less than 1,000 gallons per hour, quality varies over a wide range, much of it being suitable, at best, only for stock, and supply is subject in places to fluctuation with rainfall. In the arid zone it is particularly difficult to find good water in these rocks; the better quality water occurs where individual aquifers are in a position to receive local recharge.

CONCLUSION

The main usable water resources of arid Australia are the rivers along the north and north-western coast, with sporadic flood flows fed by summer rains. Much further investigation will be needed of conservation sites and availability of irrigable land. The Ord River scheme constitutes a large scale testing ground for suitable crops and the economics of irrigated production in the north.

Rivers in the eastern and central part of the arid zone also flow occasionally and at times carry large volumes of water, but unsuitable topography and high evaporation rates make surface conservation generally difficult.

Very large reserves of groundwater exist in the aquifers of the sedimentary basins, lesser reserves in unconsolidated sediments, and local supplies only in the metamorphic and igneous terrains. Water quality and the factors which govern it are generally the most important consideration with respect to

[3] A new series of four maps comprising the principal groundwater resources and the groundwater resources of (1) shallow unconsolidated sediments; (2) porous consolidated rocks of the principal sedimentary basins; (3) fractured rocks is being compiled and is programmed for printing in the latter half of 1970.

groundwater in the arid zone. Comparatively little is known yet about recharge, transmissibility, safe yield, amounts of water in storage, even, in many parts, of the occurrence of sources of groundwater. An enormous amount of investigation and study will be necessary if the groundwater resources of arid Australia are to be utilised to the full and properly managed, and this is essential if maximum development of the country is to be achieved.

REFERENCES

Anon. (1954). Artesian water supplies in Queensland. *Dep. Coord. Gen. Pub. Works Qd.*

Anon. (1965). *Review of Australia's water resources, 1963.* Dep. Nat. Dev. Aust.

Anon. (1967). Surface water resources, *Atlas of Australian resources.* Dep. Nat. Dev. Aust.

Anon. (1967-8). Stream gauging information, Australia, 1965; and supplements 1966, 1967. Dep. Nat. Dev. Aust.

Casey, J. N., and Wells, A. T. (1964). The geology of the north-east Canning Basin, Western Australia. *Rep. Bur. Miner. Resour. Geol. Geophys. Aust.* 49.

Condon, M. A. (1965-8). The Geology of the Carnarvon Basin, Western Australia. *Bull. Bur. Miner. Resour. Geol. Geophys. Aust.* 77, Pts 1-3.

Griffin, R. J. (1962). The underground water resources of New South Wales. *Geol. Surv. N.S.W. Rep.* 16.

Guppy, D. J., Lindner, A. W., Rattigan, J. H. and Casey, J. N. (1958). The Geology of the Fitzroy Basin, Western Australia. *Bull. Bur. Miner. Resour. Geol. Geophys. Aust.* 36.

Jackson, P. R. (1966). Geology and review of exploration, Officer Basin, W.A. Unpubl. report for Hunt Oil Co.

Perry, R. A., Mabbutt, J. A., Litchfield, W. H., Quinlan, T., Lazarides, M., Jones, N. O., Slatyer, R. O., Stewart, G. A., Bateman, W. and Ryan, G. R. (1962). General report on lands of the Alice Springs area 1956-57. *CSIRO Aust., Land Res. Ser.* No. 6.

——— Quinlan, T., Jones, N. O. and Basinski, J. J. (1963). Preliminary assessment of groundwater suitable for irrigation in the Alice Springs area, and its agricultural significance. *CSIRO Land Res. Surv. Tech. Pap.* 21.

Quinlan, T. (1965). The utilization of the local water resources of the Arid Zone. Bur. Miner. Resour. Aust. Rec. 1965/128 (unpubl.).

——— (1967). The Alice Springs Town Basin, a case history. Bur. Miner. Resour. Aust. Rec. 1967/17 (unpubl.).

——— (1968). Groundwater in the Northern Territory. Bur. Miner. Resour. Aust. Paper (unpubl.).

——— and Woolley, D. R. (1969). Geology and hydrology, Alice Springs Town and Inner Farm Basins, N.T. *Bull. Bur. Miner. Resour. Geol. Geophys. Aust.* 89.

Randal, M. A. (1967). Groundwater in the Barkly Tableland. *Bull. Bur. Miner. Resour. Geol. Geophys. Aust.* 91.

——— (1969). Groundwater in the Northern Wiso Basin and environs, N.T. Bur. Miner. Resour. Aust. Rec. 1969/16 (unpubl.).

——— and Brown, M. C. (1967). The geology of the northern part of the Wiso Basin, Northern Territory. Bur. Miner. Resour. Aust. Rec. 1967/110 (unpubl.).

Smith, K. G. (1969). Geology of the Georgina Basin. *Bull. Bur. Miner. Resour. Geol. Geophys. Aust.* 111.

Stacey, K. (1968). *Consumption and production of petroleum from the Australian mainland 1968-1990.* Aust. Nat. Univ. Dep. Geog. Occ. Pap. 7.

Traves, D. M. (1955). The geology of the Ord-Victoria region, Northern Territory. *Bull. Bur. Miner. Resour. Geol. Geophys. Aust.* 27.

Veevers, J. J. and Roberts, J. (1969). Upper Palaeozoic rocks, Bonaparte Gulf Basin of Western Australia. *Bull. Bur. Miner. Resour. Geol. Geophys. Aust.* 97.

——— and Wells, A. T. (1961). The geology of the Canning Basin, Western Australia. *Bull. Bur. Miner. Resour. Geol. Geophys. Aust.* 60.

Wells, A. T., Evans, T. G. and Nicholas, T. (1968). The geology of the central part of the Ngalia Basin, Northern Territory. Bur. Miner. Resour. Aust. Rec. 1968/38 (unpubl.).

——— Forman, D. J., Ranford, L. C. and Cook, P. J. (1969). Geology of the Amadeus Basin, Central Australia. *Bull. Bur. Miner. Resour. Geol. Geophys. Aust.* 100.

5

VEGETATION RESOURCES

J. H. LEIGH AND J. C. NOBLE*

BEFORE detailed studies can be made of the vegetation resources of arid Australia it is essential to know what communities were present when grazing by exotic animals first commenced, their distribution, and their present condition. Everyone familiar with these areas will realise the magnitude of such a task. It is made more difficult by the failure of Australian scientists to agree on a universally acceptable basis for mapping—understandable in that their requirements as phytogeographers, palaeontologists, taxonomists, ecologists, foresters, or agriculturalists have been divergent.

The aims of this paper are to describe in broad terms the vegetation regions, structural forms, and grazing lands; and to indicate the possibilities for the resource use.

On the basis of climate, Burbidge (1960) has shown that the Australian flora can be classified into three main regions: the Tropical, Temperate, and Eremaean provinces or zones. The boundaries between the zones are not sharp owing to different tolerances of characteristic taxa, and transitional zones have therefore been recognised (Fig. 1). The Eremaean Province, which covers the whole of arid Australia, has been subdivided in Western Australia (Gardner, 1959) into three subregions, the Northern, Middle, and Southern zones, on the basis of rainfall seasonality. The Northern zone receives summer rain, the Southern zone winter rain, and the Middle zone has no marked seasonal incidence. In general the climate of the whole Eremaean zone is characterised by extreme annual and seasonal variability in rainfall. Floristically the zones are not strongly distinct, because small mountain ranges provide refugia for both northern and southern elements.

The dominant species in each of these zones have been listed and it has been noted that edaphic rather than climatic factors control the distribution of associations. Crocker and Wood (1947) have discussed the difficulty in

* CSIRO Division of Plant Industry, Riverina Laboratory, Deniliquin, N.S.W.

ACKNOWLEDGMENTS. Permission to use the previously unpublished maps prepared by Dr R. M. Moore, Mr R. A. Perry, and Dr J. A. Carnahan is gratefully acknowledged. Appreciation is due also to Messrs R. W. Munyard and M. Woodward for the preparation of maps and slides and to Messrs W. E. Mulham and B. Dixon who provided valuable assistance with the manuscript. We are also indebted to the following people for providing previously unpublished material: Dr R. M. Moore, Messrs R. A. Perry, G. M. Chippendale, S. Stephens, E. C. Greig, G. C. Stocker, A. A. Clemson, J. L. Willis, W. H. Eastman, and Professor R. L. Specht.

independently assessing the effect of soils and climate on species distribution, since climate has a dominant role in pedogenesis. Furthermore climate has a strong influence on the direction of migratory highways following the release of climatic pressure on relic flora. Relief, drainage, and soil differences (especially major and minor mineral deficiencies) resulting from past geological and climatic history, grazing history, and variable physiological tolerances of different species have been shown to explain departures from expected 'climatic climax' vegetation.

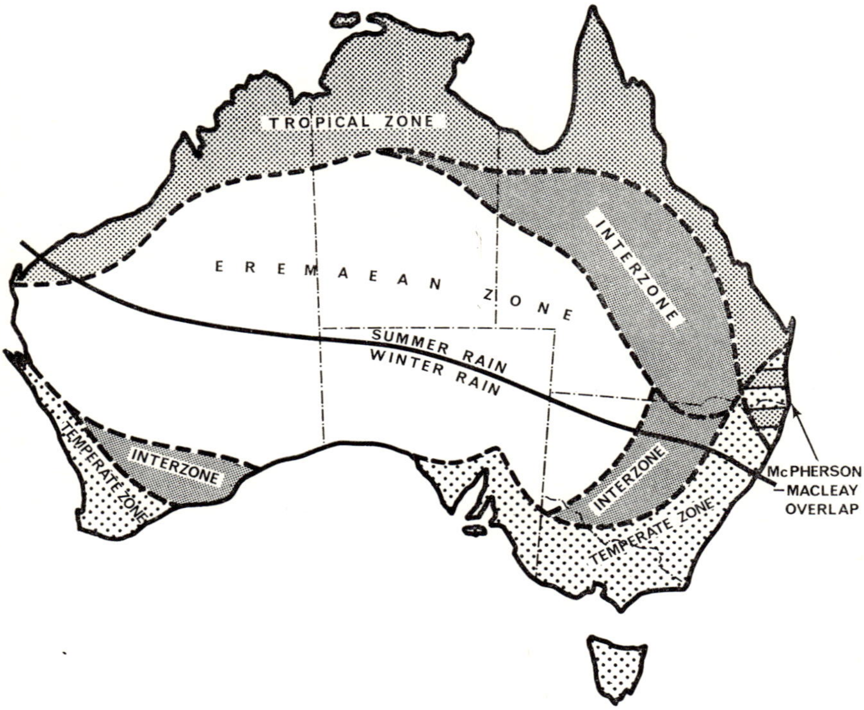

Fig. 1 The Eremaea and other principal floristic zones of the Australian Region. (*After N. T. Burbidge*)

Burbidge (1960) considers that Eremaean flora may have originated from a coastal sand dune and littoral type involving elements that migrated to the Australian region as early as the Cretaceous. The Eremaea, though comparatively impoverished on a floristic basis, contains many interesting palaeotropic representatives including the kurrajong (*Brachychiton populneum*) and the psammophytes, especially *Eremophila*. Other characteristics are the high development of phyllodineous species of *Acacia* and the dearth of large succulents such as the Cactaceae and Asclepiadaceae, although many of the

Chenopodiaceae and Portulacaceae are succulent or semi-succulent. Although the Myoporaceae is more richly represented in this province than in any other part of the world, the Poaceae (= Gramineae) is by far the largest family. Other prominent families include the Zygophyllaceae, Chenopodiaceae, Asteraceae, Portulacaceae, Mimosaceae, Goodeniaceae, Myrtaceae, Fabaceae, Caesalpiniaceae, Amaranthaceae, Borraginaceae, Solanaceae, Malvaceae, Proteaceae, and Brassicaceae. While the affinities of Eremaean representatives of some families lie with the taxa of the temperate areas of the Old World, some genera regarded as typically Eremaean in Australia, such as *Ptilotus*, *Tetragonia*, *Trianthema*, *Tribulus*, and *Salsola*, are found either in coastal habitats or restricted to low altitudes in Malaysia.

In the past thirty-five years many valuable maps have been produced of vegetation communities (Fig. 2) based on the Clements's (1928) climax concepts (with local modifications). Some are accompanied by soil and

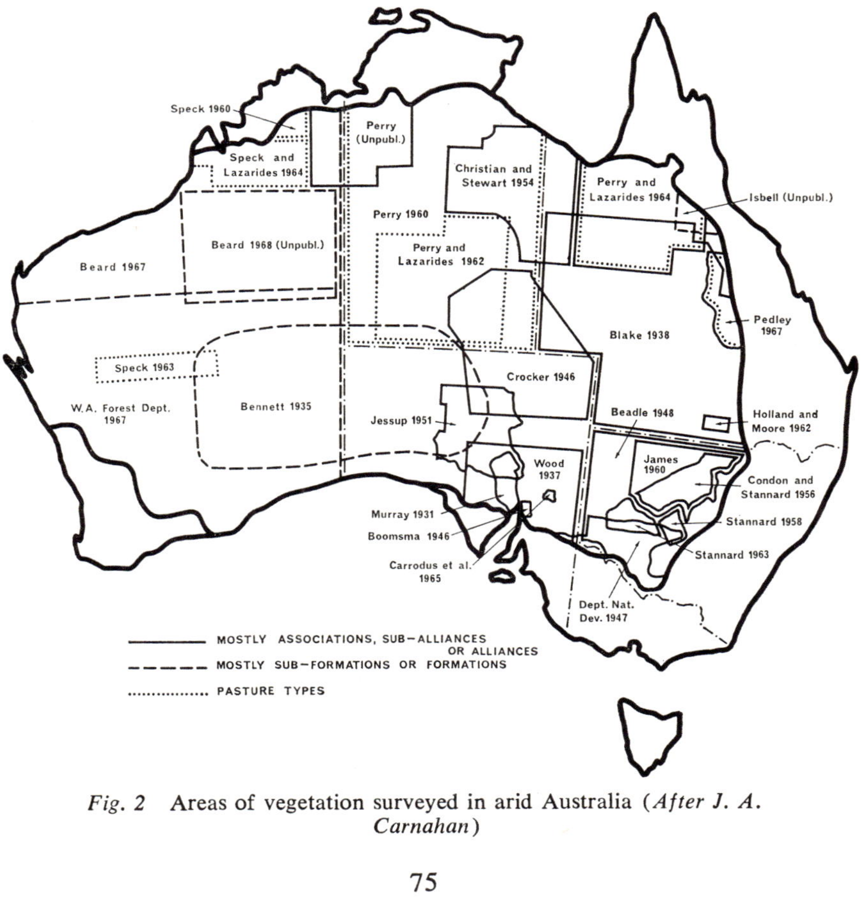

Fig. 2 Areas of vegetation surveyed in arid Australia (*After J. A. Carnahan*)

geological maps. To some extent the maps suffer from a lack of uniformity in concepts, for example the fundamental floristic unit 'association' has been interpreted in many ways, as has 'dominant species'. Confusion has resulted from indiscriminate naming of structural units in climatic terms often based on arbitrary acceptance or modifications of Clements's theory. In more recent years this confusion has largely been avoided by the use of units defined by Beadle and Costin (1952).

STRUCTURAL FORMATIONS AND GRAZING LANDS OF ARID AUSTRALIA

Both Williams (1955) and Cochrane (1967) have recognised seventeen structural subformations in arid Australia. Moore and Perry (unpublished data) have recognised ten subformations (Fig. 3). Their map also depicts the

TABLE 1

VEGETATION SUBFORMATIONS OF ARID AUSTRALIA: AREAS AND PERCENTAGES IN EACH STATE

WOODLANDS

	Tropical		Sub-humid		Shrub		Temperate		Mallee	
N.S.W.			7,840	2·5	35,420	11·5	2,940	1·0	9,080	2·9
S.A.							1,820	0·5	13,720	3·6
W.A.	14,420	15·0	4,480	0·5					83,440	8·5
N.T.	39,200	7·5	19,880	3·8						
Qld	62,160	9·3	68,320	10·2						
Vic.									2,680	3·0
Aust	115,780	3·9	100,520	3·4	35,420	1·2	4,760	0·2	108,920	3·7

SHRUBLANDS

	Steppe		Eucalypt		Acacia	
N.S.W.	31,112	10·1			72,757	23·5
S.A.	99,403	26·2			106,130	27·9
W.A.	30,423	3·1	43,042	4·4	358,836	36·8
N.T.	6,590	1·3	22,292	4·3	66,594	12·8
Qld	6,589	1·0	1,540	0·2	115,474	17·3
Aust.	174,117	5·9	66,874	2·3	719,791	24·3

GRASSLANDS

	Xerophytic Tussock		Xerophytic Hummock	
N.S.W.	182	0·1		
S.A.	280	0·1	87,199	22·9
W.A.	6,028	0·6	310,432	31·8
N.T.	37,714	7·2	232,277	44·6
Qld	147,471	22·1	42,015	6·3
Aust.	191,675	6·5	671,923	22·6

Arid areas in each State (%)

N.S.W.	51·6	N.T.	81·5
S.A.	81·2	Qld	66·4
W.A.	87·2	Vic.	3·0

Australia: 74·0
2,189,780 sq. miles

VEGETATION	GRAZING LANDS
WOODLANDS	
TROPICAL WOODLAND	TROPICAL TALL GRASS
SUBHUMID WOODLAND	XEROPHYTIC MID GRASS – NORTHERN
SHRUB WOODLAND	XEROPHYTIC MID GRASS – SOUTHERN
EUCALYPT SHRUBLAND (MALLEE)	XEROPHYTIC MID GRASS – SOUTHERN
TEMPERATE WOODLAND	TEMPERATE SHORT GRASS
SHRUBLANDS	
SHRUB STEPPE (SALTBUSH)	SALTBUSH – XEROPHYTIC SHORT GRASS – FORB
EUCALYPT SHRUBLAND	EUCALYPT SHRUB & HUMMOCK GRASS – FORB
ACACIA SHRUBLAND	ACACIA SHRUB & SHORT GRASS – FORB
GRASSLANDS	
ARID TUSSOCK GRASSLAND (MITCHELL)	XEROPHYTIC TUSSOCK GRASS
ARID HUMMOCK GRASSLAND (SPINIFEX)	XEROPHYTIC HUMMOCK GRASS

Fig. 3 Vegetation and grazing lands of arid Australia. (*After R. M. Moore and R. A. Perry*)

main grazing lands and, therefore, these two aspects are discussed together. The approximate area of these subformations or grazing lands and the approximate percentage they occupy in each state and in Australia are shown in Table 1. Climatic details, dominant species and land use for the woodlands of arid Australia are summarised in Table 2, for the shrublands in Table 3 and for the grasslands in Table 4.

Woodlands

Woodlands occur widely throughout Australia (12·5 per cent of total Australian land area) and are variable in form, ranging from tall to low trees and from dense to very open spacing. Five subformations have been recognised of which Tropical (3·9 per cent) and Sub-humid Woodlands (3·4 per cent) are most common.

Shrublands

Three subformations have been recognised, these occupying 32·5 per cent of the total Australian land area. Of these *Acacia* Shrubland is the most widespread (24 per cent) of all arid subformations, being found in all mainland states. Considerable variation occurs in the communities grouped under *Acacia* Shrubland, the communities ranging from low open woodland and open shrub woodland in the higher rainfall areas (17 in.) to low open tussock grassland in the drier areas (5 in.).

Shrub-steppe is the fourth most widespread subformation (5·9 per cent) and lies mainly south of the Tropic of Capricorn.

As in many other subformations, extensive botanical changes have occurred in these shrublands following grazing and indiscriminate lopping for drought feed. This has resulted in the disappearance of some of the palatable shrub species and a marked increase in other less palatable species. In some cases shrublands have been reduced to grasslands and severe erosion has occurred.

Grasslands

Two grassland subformations have been recognised, together constituting 29·1 per cent of Australia's land area. Of these, Arid Hummock Grassland is the second most widespread arid subformation (22·6 per cent). Arid Hummock Grassland frequently occurs associated with the *Acacia* Shrubland, these two subformations together constituting almost half of Australia's land area or 64 per cent of the area considered by this symposium.

Xerophytic Tussock Grass occupies 6·5 per cent of Australia's land area.

TABLE 2

WOODLANDS OF ARID AUSTRALIA: DOMINANT SPECIES

Subformation, Grazing Land, and Location	Rainfall	Trees and Shrubs	Grasses and Herbs	Land Use
TROPICAL WOODLAND. TROPICAL TALL GRASS. Tropical Australia.	Summer rain 17-30 in., pronounced dry season for 3 to 9 months between April and December.	Sclerophyllous eucalypts mainly *E. dichromophloia*, *E. microneura*, *E. argillacea* and *E. brevifolia*. In northern areas a tall shrub layer consisting of genera such as *Terminalia*, *Ventilago*, *Erythrophleum*, *Gardenia*, *Alphitonia*, *Hakea*, *Grevillea*, and *Melaleuca* is present. As rainfall decreases to the south shrubs become sparser or absent.	A continuous medium to tall grass cover 3 to 10 feet high exists below the trees, the main perennial grass genera being *Heteropogon*, *Chrysopogon*, *Aristida*, *Sorghum*, *Themeda*, *Sehima*, *Triodia* and *Plectrachne*. Annual grasses include *Sporobolus australasicus* and *Aristida* spp.	Despite the relatively unpalatable nature of the grasses these areas are widely used for cattle grazing at rates of about 1 beast to 50-120 acres.
SUB-HUMID WOODLAND. NORTHERN XEROPHYTIC MID-GRASS. Northern W.A., N.T., northern and eastern Qld, northern N.S.W.	Summer rain 17-25 in.	Variation occurs in terms of species composition, density and height of tree species, and presence or absence of associated shrubs. One of the main communities is *Acacia harpophylla* which is characterised by its low dense growth 15-25 feet high and a subordinate discontinuous strata of lower shrubs. Other widespread species include *Casuarina cristata*, and *Acacia cambagei* and *Geijera parviflora*.	Poorly developed herbaceous ground cover where tree cover is dense, becoming better developed in more open areas. The main perennial grasses are *Paspalidium* spp. and *Chloris acicularis*. In many areas extensive ringbarking, spraying and or clearing of *A. harpophylla* has occurred and improved grasses and legumes sown.	Nutritional quality and palatability of the herbaceous flora declines markedly after maturity. In *A. harpophylla* communities stocking rates are low (from 1 sheep to 5-10 acres; cattle at 1 beast to 35-75 acres).

TABLE 2—*continued*

Subformation, Grazing Land, and Location	Rainfall	Trees and Shrubs	Grasses and Herbs	Land Use
SHRUB WOODLAND. SOUTHERN XEROPHYTIC MID-GRASS. N.S.W.	Winter rain 8-16 in.	Mainly sclerophyllous eucalypts principally *E. populnea*, *E. similis*, *E. melanophloia* and *E. microtheca*. Other common tree species include species of the genera *Callitris* and *Casuarina*. Shrubs are prominent in some areas and include species of the genera *Eremophila*, *Petalostigma*, *Myoporum*, *Erythroxylon*, *Cassia*, *Dodonaea*, *Apophyllum*, *Acacia* and *Erythrophleum*.	A well developed continuous short to mid-grass cover occurs beneath and between the trees and shrubs. The most important grass genera include *Aristida*, *Stipa*, *Chloris*, *Neurachne*, *Eragrostis* and *Enneapogon*. Cool season annuals of the genera *Bassia*, *Calotis*, *Medicago*, *Erodium*, *Hordeum* and *Vulpia* are common.	Extensively grazed by sheep at rates ranging from 1 sheep to 5-11 acres.
TEMPERATE WOODLAND. TEMPERATE SHORT GRASS N.S.W., Vic. and northern S.A.	Winter rain 10-20 in.	Mainly sclerophyllous eucalypts the principal species being *E. camaldulensis*, *E. largiflorens*, *E. woollsiana* and *E. microcarpa*. Species of the genera *Callitris* and *Casuarina* occur in some areas. Shrubs are uncommon although weeds such as *Bassia quinquecuspis* and *Salsola kali* may sometimes be present.	A well developed continuous short grass and forb cover grows beneath and between the trees. The most important perennial grass genera include *Danthonia*, *Stipa*, *Chloris*, and *Poa*. Of the annual grass genera *Chloris*, *Sporobolus*, *Hordeum*, *Panicum*, *Avena*, and *Vulpia* are widespread and common. Cool-season annual forbs of the genera *Medicago*, *Trifolium*, *Helipterum*, *Calotis*, *Hypochaeris*, *Hedypnois*, and *Erodium*, flourish in most years.	These communities with the exception of *E. camaldulensis* cover or have covered most of the marginal wheat belt and higher rainfall sheep country. Few remain in their original state. Sheep are grazed at rates varying between 1 and 5 acres per sheep.

MALLEE EUCALYPT SHRUBLAND. SOUTHERN XEROPHYTIC MID-GRASS-TRIODIA. W.A., S.A., N.S.W., Vic.	Winter rain 8-17 in.	The main species of mallee are *E. oleosa*, *E. dumosa*, *E. socialis*, *E. sideroxylon*, *E. fruticetorum* and *E. viridis*. A sparse discontinuous secondary layer of xeromorphic shrubs may also occur, these shrubs belonging chiefly to the genera *Acacia*, *Atriplex*, *Kochia*, and *Bassia*. Allied to mallee but not necessarily assuming the multiple trunk habit are the salmon gum woodlands in W.A., the main species of which are *E. salmonophloia*, *E. salubris*, *E. longicornis*, and *E. gracilis*.	The herbaceous layer is dominated by *Triodia irritans*. Following good rains a profuse growth of ephemeral grasses and forbs occur but during summer the ground is relatively bare.	In areas of higher rainfall much of the area has been cleared for wheat cropping. In areas where the community remains, sheep are grazed at low stocking rates of approximately 1 sheep to 25-35 acres.

TABLE 3

SHRUBLANDS OF ARID AUSTRALIA

Subformation, Grazing Land, and Location	Rainfall	Trees and Shrubs	Grasses and Herbs	Land Use
SHRUB STEPPE. SALTBUSH/XEROPHYTIC SHORT GRASS: FORB. W.A., S.A., N.T., N.S.W.	Mainly winter rainfall 6-20 in.	The shrubs are normally halophytes with hairy, reflectant, semi-succulent leaves, growing to a height of 3-10 feet. The main genera are *Atriplex*, *Kochia*, *Rhagodia*, and *Bassia*. Much of the shrub has been reduced or eliminated by heavy grazing and its place taken by unpalatable spiny shrubs (chiefly species of *Nitraria* and/or *Bassia*), a low sparse grass cover or in extreme cases by complete scalding.	Following heavy rains a profuse growth of ephemeral grasses and herbs may occur but in summer the area between the shrubs is usually devoid of ground cover. The main perennial grass genera include *Danthonia*, *Stipa*, *Eragrostis* and *Aristida*.	Light grazing mainly by sheep, except in northern S.A. where grazed by cattle. Stocking rates vary from 1 sheep to 3-40 acres
EUCALYPT SHRUBLAND. EUCALYPT SHRUB AND TUSSOCK GRASS: FORB. Mainly in W.A. and N.T. and to a lesser extent in Qld.	Summer rain 10-30 in.	This is a modification of the Tropical and Sub-humid Woodland subformations described earlier but occurs in drier regions. The 'trees' or shrubs are low (10-25 feet high), the main species being *E. microtheca*, *E. odontocarpa*, *E. gamophylla*, *E. pachyphylla*, *E. brevifolia*, *E. pruinosa*, *E. terminalis*, and *E. argillacea*. Other 'tree' species include *Atalaya hemiglauca*, *Ventilago viminalis*, *Owenia acidula*, *Flindersia maculosa*, and a number of *Acacia* species.	The ground flora consists of a more open grassland dominated by the genera *Triodia*, *Aristida*, *Sehima*, *Chrysopogon*, *Themeda*, *Brachyachne* and *Bothriochloa* with a high component of annuals. This subformation occurs as a complex with Arid Tussock Grassland.	Extensive areas on their own have little or no grazing value but when they occur in association with other subformations cattle are grazed at low rates of about 1 beast per 125 acres.

Acacia Shrubland. Acacia Shrubland Short Grass:Forb. Most widespread of all subformations. Found in all mainland states.	Summer and or winter rainfall 5-17 in.	Considerable variation occurs in the communities grouped together under this subformation. The dominant 'tree' species include *A. aneura*, *A. brachystachya*, *A. pendula*, *A. sowdenii* and *A. kempeana*. Of these *A. aneura* is the most widespread and occupies the largest area. The 'trees' range from 6 to 30 feet in height and commonly occur as almost monospecific stands. Other associated tree species include *A. excelsa*, *A. loderi*, *Casuarina cristata*, *Heterodendrum oleifolium* and *Flindersia maculosa*. A discontinuous shrub layer may occur beneath the trees, the main species being chenopods, or a low shrub community may intersperse the extensive stands of *A. aneura*. In the latter case the main species belong to the genera *Eremophila*, *Dodonea*, *Hakea*, *Acacia*, *Lycium*, *Myoporum*, *Exocarpus*, and *Cassia*. In some areas the low tree and shrub cover has been reduced or eliminated and erosion is severe.	In the higher rainfall areas a continuous herbaceous layer is present. In the south this layer is dominated by members of the genera *Danthonia*, *Enneapogon*, *Stipa* and/or *Eragrostis* and in the north by the genus *Astrebla*. Under low rainfall the herbaceous layer becomes discontinuous and dominated by *Triodia* spp. or even absent.	Extensive grazing at low stocking rates by sheep in the south and into Qld, and for beef cattle in the north. Stocking rates vary from about 1 sheep to 5-30 acres and cattle from 1 beast to 125 acres.

Table 4

GRASSLANDS OF ARID AUSTRALIA

Subformation, Grazing Land, and Location	Rainfall	Trees and Shrubs	Grasses and Herbs	Land Use
Arid Tussock Grassland. Mitchell Grass Xerophytic Tussock Grass. Principally in Qld and N.T.	Summer rain 10-18 in.	On the more fertile soils shrubs are absent or rare but on the more stony soils scattered shrubs are often present. These include *Acacia farnesiana*, *A. cana*, *A. georginae*, *A. cambagei*, *Terminalia volucris* and *T. arostrata*.	This subformation is dominated by an open cover of low xeromorphic perennial grass tussocks generally 2 feet apart and 1½-3 feet high. Almost all of these grasslands are characterised by one or more of the *Astrebla* species, the main species being *A. lappacea*, *A. pectinata*, *A. squarrosa*, and *A. elymoides*. *Astrebla* species alone dominate most of the area but in some parts other species are important constituents or may even dominate. These include *Aristida latifolia*, *Dichanthium fecundum*, *Panicum whitei*, *Eragrostis setifolia*, and *E. xerophila*. The area between the tussocks is usually bare except after rainfall when ephemeral grasses and herbs occur.	These areas provide valuable grazing although their value is limited by the long dry season. Sheep in the southernmost areas are grazed at rates of 1 sheep to 5-12 acres. Cattle are grazed at one beast to 40 acres

ARID HUMMOCK GRASSLAND. SPINIFEX XEROPHYTIC HUMMOCK GRASS. Widespread and extensive arid community being found in W.A., S.A., Qld, N.T. and western N.S.W.	Infrequent.	5-20 in.	Trees and shrubs may be entirely absent but more commonly scattered trees and shrubs mostly 3-6 feet high occur. In some areas *Casuarina decaisneana* grow up to 60 feet high and along the northern margin Arid Hummock Grassland grades into Eucalypt Shrubland.	This subformation is characterised by low isolated hummocks of coarse perennial evergreen sclerophyllous grasses chiefly species of *Triodia* and *Plectrachne*. The 'normal' hummock is 3-5 feet in diameter, 1½-2 feet high. In some species (e.g. *T. longiceps*) the hummocks grow to as large as 20 feet in diameter and 6 feet high and in others (e.g. *T. basedowii*) the centre of the hummock dies and mature plants take the form of annuli 6-15 feet in diameter, about 1½ feet wide and 1½ feet high. Other grass genera include *Aristida*, *Cymbopogon*, *Zygochloa*, *Chrysopogon*, and *Eriachne*. Commonest herbaceous genera include *Calandrinia*, *Ptilotus*, and *Calotis*. Short-lived ephemeral grasses and forbs may occur on the bare ground between the hummocks after infrequent heavy rain or in the first few years after fires.	Little or no grazing value. Vast areas are unoccupied.

PRESENT AND POTENTIAL RESOURCE USE

Pastoralism

Sandplain, dunefield, rocky hills, and salt lakes comprise about half of Australia's arid zone. These areas are generally considered as unsuitable for grazing and a large proportion remains as unalienated Crown land. The remaining portion of the arid zone is used for grazing by either sheep or cattle under extensive management conditions (see chapters 7 and 8).

Most research on the effects of grazing on, and the value of, Australian arid zone plants has been concerned with browse (top-feed) species, most of which belong to the Mimosaceae, Myoporaceae, Chenopodiaceae, Sapindaceae, and Sterculiaceae. The most valued leguminous browse plants are mulga (*Acacia aneura*), yarran (*A. homalophylla*), myall or boree (*A. pendula* and *A. sowdenii*) and ironwood (*A. excelsa* and *A. estrophiolata*). Other valuable species include rosewood (*Heterodendrum oleifolium*), belah (*Casuarina cristata*), leopardwood (*Flindersia maculosa*), kurrajong (*Brachychiton populneum*), wilga (*Geijera parviflora*), and whitewood (*Atalaya hemiglauca*). The important chenopods include bladder saltbush (*Atriplex vesicaria*), old man saltbush (*A. nummularia*), bluebush (*Kochia sedifolia, K. pyramidata, K. planifolia, K. astrotricha*), and cotton bush (*K. aphylla*). Dietary preferences of stock and the relative palatabilities of browse species have been examined and some species have been shown to be toxic.

Edible trees are generally high in protein, low in phosphorus, and low in digestibility. Mulga is an important browse species and is considered an important drought reserve. Fallen leaves and fruits of mulga are eaten by stock and, during drought, it is common for trees to be cut or pushed over. Indiscriminate felling has been shown to affect regeneration of mulga. Fire and severe defoliation by lopping or overgrazing also cause mulga and similar species to die, although fire promotes the regeneration of some arid trees and shrubs (Moore, 1962). Exclusion of stock does not necessarily ensure the regeneration of mulga. Regeneration can occur on denuded areas providing mature trees are nearby, but in some areas excessive mulga regeneration can be a problem.

Saltbushes and bluebushes are generally high in nitrogen, sodium, and potassium. Saltbushes accumulate ions irrespective of soil mineral composition, and the high levels of sodium can significantly affect the water intake of sheep grazing saltbush pastures (Wilson, 1967). Natural regeneration of saltbushes and bluebushes depends on many factors, including the presence of mother plants, viable seed, and suitable rainfall and temperature conditions (Hall *et al.*, 1964).

Wilson (1969) reviewed the role of browse in the nutrition of grazing animals and pointed out that, despite many chemical analyses and pen feeding trials, there is little evidence of animal production responses to browse species. Their value appears to lie in their use as reserve forage during drought periods. Leigh *et al.* (1968) found no response in terms of body weight and wool growth by Merino sheep to the presence of small amounts of cotton bush (*Kochia aphylla*) in a semi-arid grassland.

Robards *et al.* (1967) studied the dietary selection of Merino sheep grazing a semi-arid *Danthonia caespitosa* grassland and doubted whether any other management system would be better than set-stocking on this community. The plant and animal response to the application of various fertilisers on this pasture is being investigated currently.

Forestry

Although few trees of arid Australia are valuable for timber production because of their extreme hardness, small size, and slow growth, many xerophytic species, when viewed collectively, constitute a valuable resource. Overseas arid zone foresters are interested in Australian tree species and considerable quantities of seed are sent to underdeveloped and densely populated arid zones for reafforestation. Most of these species are used for local building, fencing, shade and shelter requirements. Local timber has also been used for industrial charcoal, pit props, and fuel for steam power and domestic purposes (Cromer, 1954) in the major population centres in arid Australia that have been mining centres.

The history of forest and timber resources throughout arid Australia has changed from indiscriminate deforestation to controlled harvesting under the administration of State Forestry authorities. Since most arid trees and shrubs are sensitive to fire, burning is undesirable and conservation is the primary function of forestry in these areas. Much silvicultural research is required before satisfactory techniques can be devised for the future management of important tree communities (Bateman, 1962).

The more important and widespread species are:

RIVER RED GUM (*Eucalyptus camaldulensis*) which occurs as riverain forest in Victoria and New South Wales and as ribbon stands along perennial and ephemeral rivers in all mainland states. It generally occurs as monospecific stands and attains a height of up to 160 feet. The timber is used for general and heavy construction and for railway sleepers, and is particularly resistant to decay and insect attack (Anderson, 1968).

CYPRESS PINE (*Callitris* spp.) occurs in South Australia, Queensland, Victoria, and New South Wales on mainly sandy soils either alone or in various associations with eucalypt species. The feasibility of chipping operations has been considered for non-millable pine since recovery percentages are low

(46-50 per cent) because of faults and log taper, and knots tend to dry and fall out.

The variable regeneration of pine species, including the effects of amount of seed, litter fall, and fire, have been studied. The importance of grazing exclusion, particularly by sheep and rabbits has been stressed. On the other hand there is the problem in some areas of dense thickets of *C. columellaris* precluding optimum pasture growth.

Cypress pine is an important building timber, considerable quantities being transported to the capital cities.

MALLEE (*Eucalyptus* spp.) is found in Western Australia, South Australia, Victoria, and New South Wales. Extensive areas have been cleared for cropping but in most states such clearing requires government consent and special conditions apply. For example in New South Wales up to 300 acres may be cleared but they must be surrounded by a windbreak of undisturbed mallee not less than 300 feet wide and sandridges and areas of light sandy soil are excluded from clearing.

Mallee timber is used for firewood, fencing, and trellis timbers, but probably its main value is in soil stabilisation.

MULGA (*Acacia aneura*). The distribution and fodder value of this species has already been discussed. It is used extensively for firewood and fencing but, like mallee, is now protected. In Western Australia, for example, cutting is confined to dense areas only, seven live trees over four feet high to every square chain must be left, a minimum diameter limit of three inches is imposed for all cutting, and the cutting of green mulga for domestic firewood is prohibited.

Mulga is a valuable soil stabiliser even when dead. In an effort to minimise wind erosion and dust at Alice Springs, the Forestry and Timber Bureau has investigated the dead mulga resources within a radius of fifty miles with the aim of controlling firewood collection.

Drugs

Many biochemical compounds occur in the hundreds of arid species, and many Aboriginal drug plants have been shown to be physiologically active. Webb (1968) was cautious about the clinical value of many of the plants used by Aborigines but further investigations may discover economically important drugs.

Among the drug-producing plants which grow in arid areas two groups deserve special mention:

Duboisia hopwoodii (pituri) is a small shrub which grows on sandy country mostly where rainfall is less than 10 inches but extending almost to the 15-inch zone. The leaves of the plant are reported to contain two alkaloids, nicotine and neo-nicotine (Hatt, 1957), and could be used for the production of insecticides. The Aborigines chewed a preparation made from the leaves and twigs for its narcotic effects.

Appreciable quantities of medicinal oils can be distilled from various mallee eucalypt species including *E. fruticetorum*, *E. viridis*, *E. sideroxylon*, *E. dumosa*, *E. oleosa*, and *E. socialis*.

E. fruticetorum has a high percentage of cineole (eucalyptol) in the oil and provides one-third of the total eucalyptus oil produced, the bulk of which is exported (Penfold and Willis, 1961). Most of these oils are useful for medicinal purposes and exceed the minimum cineole content demanded by pharmaceutical houses. There is an urgent need for preservation of some species, particularly *E. fruticetorum*, and the establishment of plantations has been suggested as the only way of saving the decaying oil industry in Australia.

Other Essential Oils

In contrast to oil-producing *Eucalyptus* from the higher rainfall areas of Australia, no useful industrial or perfumery oils are obtained from arid zone eucalypt species, but many other genera are rich in aromatic oils. Australian sandalwood (*Santalum spicatum*, *S. lanceolatum*, *S. acuminatum*, and *S. murrayanum*) contains a valuable perfumery oil obtained from the roots and butts. Production has declined since World War II, mainly because the more accessible stands have been exploited for the export to Asiatic countries of logs for incense (Penfold and Willis, 1954). Attempts to regenerate these root parasitic species have not succeeded (particularly because of rabbits) and they may become extinct. The Forests Department in Western Australia has imposed a limit on the amount of sandalwood which can be collected (630 metric tons exported in 1967-8 from this state). Other oil-bearing species of potential economic value include budda or bastard sandalwood (*Eremophila mitchellii*) which is widespread, particularly in arid Queensland and New South Wales. The wood produces up to 3 per cent of reddish-brown oil with marked blending and fixative properties. Oils from cypress pine (containing chiefly guaiol and citronellic acid), cajiput (*Melaleuca leucadendron*), and *M. viridiflora* are being examined as a result of world-wide interest in wood oils by perfumers. Citronella oil was once extracted as a perfume base from species of *Cymbopogon* but is now manufactured synthetically.

Other Resource Potential

The mallee communities of arid regions are utilised periodically for the production of honey and beeswax. Apiarists have requested that certain areas be preserved since crops of up to 480 pounds per hive are obtained in good seasons, when these communities may be stocked at 100 hives per square mile. Alternative uses for arid timber include the production of rayon from dissolving pulp and the manufacture of homogeneous fibreboard.

The bark of species of *Acacia* has been used for tannin extraction in the past but little is produced now. There is a world shortage of vegetable tanning materials and this has stimulated interest in the establishment of shelter belts and plantations for tannin production, particularly with brown mallet (*Eucalyptus astringens*), dundas blackbutt (*E. dundas*), and dundas mahogany (*E. brockwayi*) (Brockway and Hillis, 1953). The Forests Department of Western Australia is establishing plantations of *E. astringens*, the first of which is to be harvested in 1978. Eucalyptus kino extracted principally from *E. camaldulensis* and consisting mainly of tannins (70-80 per cent) has been used as an astringent but commercial production has now virtually ceased.

A full evaluation of floristic development in the Eremaea is limited by current knowledge of palaeobotany, palaeopedology, and palaeoclimatology, and the remaining floristic relics are a valuable resource for future phytogeographical studies. Such relics include *Livistona mariae* and *Macrozamia macdonnellii* in central Australia and waddy (*Acacia peuce*) in south-western Queensland and the Northern Territory. A complete list of National Parks and equivalent reserves in Australia has been compiled by Mosley (1968). Under the International Biological Program all alliances within Australia are being listed to ascertain whether their existence will be ensured by existing parks and reserves. Such action is necessary because drought, tourism, and grazing threaten relic flora areas with extinction.

REFERENCES

Anderson, R. H. (1968). *The trees of New South Wales.* Govt. Printer, Sydney.

Bateman, W. (1962). Forestry of the Alice Springs area. *CSIRO Aust. Land Res. Ser.* No. 6: 237-9.

Beadle, N. C. W. (1948). *The vegetation and pastures of western N.S.W.* Govt. Printer, Sydney.

——— and Costin, A. B. (1952). Ecological classification and nomenclature. *Proc. Linn. Soc. N.S.W.* **77**: 61-82.

Beard, J. S. (1967). Some vegetation types of tropical Australia in relation to those of Africa and America. *J. Ecol.* **55**: 271-90.

Bennett, H. P. (1935). The work of the Mackay aerial survey expedition to western and southern Australia. *Aust. Geogr.* **11**(8): 3-7.

Blake, S. T. (1938). The plant communities of western Queensland and their relationships with special reference to the grazing industry. *Proc. R. Soc. Qd* **49**: 156-204.

Boomsma, C. D. (1946). The vegetation of the southern Flinders Ranges, South Australia. *Trans. R. Soc. S. Aust.* **70**: 259-76.

Brockway, G. E. and Hillis, W. E. (1953). Tanbark eucalypts of the semi-arid regions of south Western Australia. *Emp. For. Res.* **34**: 31-41.

Burbidge, N. T. (1960). The phytogeography of the Australian region. *Aust. J. Bot.* **8**: 75-211.

Carrodus, B. B., Specht, R. L., and Jackman, Margaret E. (1965). The vegetation of Koonamore station, South Australia. *Trans. R. Soc. S. Aust.* **89**: 41-52.

Christian, C. S. and Stewart, G. A. (1954). The land-use groups of the Barkly region. *CSIRO Aust. Land Res. Ser.* No. 3: 150-82.

Clements, F. E. (1928). *Plant succession and indicators.* Wilson, New York.

Cochrane, G. R. (1967). *A vegetation map of Australia with explanatory notes.* Cheshires, Melbourne.

Condon, R. W. and Stannard, M. E. (1956). Erosion in western New South Wales. *J. Soil Conserv. Serv. N.S.W.* **12**: 192-202.

Crocker, R. L. (1946). The Simpson Desert expedition, 1939. Scientific Reports: No. 8—The soils and vegetation of the Simpson Desert and its borders. *Trans. R. Soc. S. Aust.* **70**: 235-58.

——— and Wood, J. G. (1947). Some historical influences on the development of the South Australian vegetation communities and their bearing on concepts and classification in ecology. *Trans. R. Soc. S. Aust.* **71**: 91-136.

Cromer, D. A. N. (1954). Forest resources. In *Atlas of Australian Resources.* Department of National Development, Canberra.

Department of National Development (1947). Natural Vegetation: Murray Valley. In *Report of the Murray Valley Resources Survey Committee on resources and development of the Murray Valley.* Vol. 2.

Gardner, C. A. (1959). The vegetation of Western Australia. *Monographiae biol.* **8**: 274-82.

Hall, Elizabeth A. A., Specht, R. L. and Eardley, Constance M. (1964). Regeneration of the vegetation on Koonamore Vegetation Reserve, 1926-1962. *Aust. J. Bot.* **12**: 205-64.

Hatt, H. H. (1957). A survey of Australian phytochemistry. *Proc. Symp. Phytochem., Kuala Lumpur, 1957.* UNESCO.

Holland, A. A. and Moore, C. W. E. (1962). The vegetation and soils of the Bollon district in south-western Queensland. *CSIRO Aust. Div. Pl. Ind. Tech. Pap.* No. 17.

James, J. W. (1960). Erosion survey of the Paroo-Upper Darling region. *J. Soil Conserv. Serv. N.S.W.* **16**: 185-202.

Jessup, R. W. (1951). Soils, geology and vegetation of north-western South Australia. *Trans. R. Soc. S. Aust.* **74**: 189-273.

Leigh, J. H., Wilson, A. D. and Mulham, W. E. (1968). A study of Merino sheep grazing a cotton bush (*Kochia aphylla*)-grassland (*Stipa variabilis-Danthonia caespitosa*) community on the Riverine plain. *Aust. J. agric. Res.* **19**: 947-61.

Moore, R. M. (1962). The effect of sheep grazing on Australian vegetation. In A. Barnard (ed.), *The simple fleece: Studies in the Australian wool industry*: 170-83. Melbourne University Press.

Mosley, J. G. (1968). *National Parks and Equivalent Reserves in Australia. Guide to Legislation, Administration, and Areas.* Australian Conservation Foundation Special Publication No. 2.

Murray, B. J. (1931). A study of the vegetation of Lake Torrens plateau, South Australia. *Trans. R. Soc. S. Aust.* **55**: 91-112.

Pedley, L. (1967). Vegetation of the Nogoa-Belyando area. *CSIRO Aust. Land Res. Ser.* No. 18: 138-69.

Penfold, A. R. and Willis, J. L. (1954). The essential oil industry of Australia. *Econ. Bot.* **8**: 316-36.

——— and ——— (1961). *The eucalypts.* Leonard Hill Ltd, London.

Perry, R. A. (1960). Pasture lands of the Northern Territory, Australia. *CSIRO Aust. Land Res. Ser.* No. 5.

——— and Lazarides, M. (1962). Vegetation of the Alice Springs area. *CSIRO Aust. Land Res. Ser.* No. 6: 208-36.

——— and ——— (1964). Vegetation of the Leichhardt-Gilbert area. *CISRO Aust. Land Res. Ser.* No. 11: 152-91.

Robards, G. E., Leigh, J. H. and Mulham, W. E. (1967). Selection of diet by sheep grazing semi-arid pastures of the Riverine plain. IV. A grassland (*Danthonia caespitosa*) community. *Aust. J. exp. Agric. anim. Husb.* **7**: 426-33.

Stannard, M. E. (1958). Erosion survey of the south west Cobar peneplain. *J. Soil Conserv. Serv. N.S.W.* **14**: 30-45.

——— (1963). Erosion survey of the central east-Darling region. *J. Soil Conserv. Serv. N.S.W.* **19**: 17-28.

Speck, N. H. (1960). Vegetation of the North Kimberley area, W.A. *CSIRO Aust. Land Res. Ser.* No. 4: 41-63.

——— (1963). Vegetation of the Wiluna-Meekatharra area. *CSIRO Aust. Land Res. Ser.* No. 7: 143-61.

——— and Lazarides, M. (1964). Vegetation and pastures of the West Kimberley area. *CSIRO Aust. Land Res. Ser.* No. 9: 140-74.

Webb, L. J. (1968). Australian plants and chemical research. In *The Last of Lands.* Jacaranda Press, Melbourne.

Western Australia Forests Department. (1967). *Vegetation map of Western Australia.*

Williams, R. J. (1955). Vegetation regions (map and notes). In *Atlas of Australian Resources.* Dept. of National Development, Canberra.

Wilson, A. D. (1967). Observations on the adaptation of sheep to saline drinking water. *Aust. J. exp. Agric. anim. Husb.* **7**: 321-4.

——— (1969). A review of browse in the nutrition of grazing animals. *J. Range Mgmt* **22**: (in press).

Wood, J. G. (1937). *Vegetation of South Australia.* Govt Printer, Adelaide.

6

NATIVE ANIMAL RESOURCES

A. R. MAIN*

THIS CHAPTER considers populations of native animals that convert plant material into animal protein which may then be harvested for human use. Species such as grasshoppers or termites are excluded even though they may be important in the regional ecology.

The information on which to base a detailed analysis of the resource of native animals is not available. Consequently the present treatment is restricted to discussing general problems. Strictly an adequate judgment depends on having sufficient detailed knowledge of the biology of arid land animals—both native and introduced—so that comparisons can be made.

The rational end of such consideration is to compare native with introduced or domestic animals on the basis of sustained yield harvest.

All populations whose diet is plant material can be viewed from two economic points of view:

(a) They provide protein which can be harvested for human use.

(b) They so deprive economically useful species of food or so reduce their numbers or condition that a substantial reduction of the harvest ensues.

We may consider animals of type (a) to be potentially economically useful. The potential will be realised when the harvest actually yields a monetary profit. Animals of type (b) may always be economically important, but reduction in numbers is only feasible when the economic benefit derived from increased harvest is greater than the cost of reducing the pest numbers.

In arid Australia feral herbivores, cattle, sheep, and the ruminant-like macropod marsupials are all potential competitors for the limited plant material available. However, most interest centres on the relationship between sheep and macropods. Sheep fit category (a) but, depending on the point of view, macropods may be considered as belonging in either category. When considered in (a) they are a resource and when in (b) a pest.

In the restricted sense being used here the resource consists solely of the

* Professor of Zoology, University of Western Australia, Nedlands, Perth.

ACKNOWLEDGMENTS. Unpublished results reported in this paper were supported by research grants from the University of Western Australia, Australian Research Grants Committee, and CSIRO.

J. C. Sampson ran a computer program from which Table 2 was constructed.

TABLE 1

REQUIREMENTS AND WEIGHT OF KANGAROO AND WALLABY SPECIES AT TWO YEARS OF AGE

Species	Habitat	Requirements: Food	Requirements: Shelter	Observed density (acres per animal)	Minimum dietary nitrogen requirement per day (g)	Water intake as % of body weight	Growth to two years of age (kg)
Red	savannah: widespread where vegetation suitable	grasses dicotyledons	shrubs: small trees	70 to > 600*	2·2	5·2	15
Grey	woodland to savannah	shrubs, grasses	,,	—	—	—	
Euro	rocky hills, localised	spinifex grasses	rockpiles	26†	1·99	6·5	12–15
Rock wallaby	vicinity of rockpiles, very local	dicotyledons	rockpiles	—	0·6	5·9	3
Tammar	scrub and scrubby woodland, local	grasses, shrubs	low shrubs	4–5	—	3·7–5·0	2·8

* Newsome, 1965a; Newsome *et al.*, 1967.

† Ealey, 1967a.

larger kangaroos. Table 1 lists their biological requirements and also gives data for two of the smaller wallabies. To a large extent the geographical ranges and preferred habitats of these species do not overlap.

The red and grey kangaroo have no special requirement for cover but require better type diets. The euro prefers trash diets and uses less water provided the preferred cover (rockpiles) is present.

The tammar has indifferent cover requirements, handles low quality diet and saline water, but does not naturally occur in such arid places as we are considering. The rock wallaby is too restricted in occurrence and specialised in habitat to be worth exploiting.

Macropods supposedly offer advantages for exploitation as a resource because:

(a) they have low dietary requirements (Brown, 1968);
(b) some (e.g. euro) can sustain a positive nitrogen balance on poor quality diet, e.g. spinifex (*Triodia*);
(c) they have high lean meat to fat ratio (Tribe and Peel, 1963);
(d) at the time of European settlement overgrazing was not apparent and hence it is assumed that native game would not (because of their low dietary requirements) produce the gross pasture degeneration believed to be the result of sheep husbandry;
(e) macropods are pests in some pastoral areas. Since they have to be controlled they may as well be harvested.

FEASIBILITY OF EXPLOITATION

There are two sorts of factors which determine whether harvesting a resource is feasible. These are those relating to

(i) the size of resource; its numerical abundance and ability to withstand sustained harvesting;
(ii) the ease of exploitation: herding, shipping, and slaughtering so that the harvest yields the maximum return.

Size and Extent of the Resource

There are no numerical data on the population sizes of any of the larger macropods. Newsome (1965a) has shown maximum densities of nine per square mile for red kangaroos, and Ealey (1967a) has shown densities of one euro per twenty-six acres. Two authors, Ealey (1967a) and Frith (1968), have shown that in many places kangaroos are more abundant than sheep. In all cases kangaroos are less abundant in unsettled areas than they are in pastoral holdings.

I suggested (1968) that in the pre-European time over much of Australia overgrazing by native herbivores was regulated by seasonal drought—which depressed appetite and so prevented heavy grazing. This is consistent with the reports by explorers, who rarely reported aggregations of large native game.

The corollary of this is that, given good water supplies, population numbers should rise unless population increases were compensated for by movement away from areas of high population density to those of lower density. This generally does not happen. Consequently as soon as reliable water supplies are provided on pastoral holdings the number of macropods increases so that they become many times more abundant than formerly. Newsome (1965b) has shown that red kangaroos are more abundant on the better quality herbage and on grazed grasses.

Exploitation

Macropods have been harvested for a long period. However, within the last decade or so the nature of the harvest has changed from harvest of skins for the leather trade to harvest for meat. The meat trade is more heavily capitalised than the leather trade and it requires the slaughter of a greater number of animals so that the freezers and processing works can be run economically. The conditions outlined in the previous section provide a suitable environment for harvesting for meat. Generally the meat is used only for the pet food trade, since the conditions of slaughter and processing do not meet the standards of hygiene appropriate to processing meat for human consumption.

The harvesting of macropods for the meat trade is analogous to a fishing industry in so far as both begin without any knowledge of the size, age structure, or other attributes of the population.

In regulated fisheries the fishermen are required to furnish returns of catch, effort (days at sea, steaming time, or other appropriate measure), gear used, etc. At the same time the catch is sampled for size, sex, and maturity at the processing works, so that changes in the fishery can be documented, management procedures devised, and exploitation continued on a sustained yield basis. There is no comprehensive collection and analysis of data for the kangaroo trade.

In an unregulated or inadequately regulated fishery any of the following might occur:

(i) it becomes more difficult or takes longer to obtain the same number or weight of fish, i.e. the catch per unit effort declines;

(ii) there is a decline in the size of the fish caught;

(iii) the fishery moves to new grounds as soon as the catch per unit effort or size of fish caught makes it uneconomic to continue fishing in the old grounds.

It is characteristic of the kangaroo trade that shooters keep moving on from sites which are shot out to new and unexploited localities. Frith (1968) has quoted figures for the decline in average size of the carcasses arriving at processing works. There are no data on catch per unit effort, but presumably this also falls. In total these are indications that, if the shooting of kangaroos for meat is regarded as an analogue of a fishery, it is being overexploited and is not meeting the requirements for exploitation on a sustained yield basis.

The data are not available for the calculation of what could be harvested on a sustained yield basis in any locality. However, by taking the average figures for reproductive performance and seasonal conditions cited by Newsome (1965c) and using the approach of Watt (1968) to simulate on a computer what would happen to a population, it is possible to show that, on the average, harvest is not possible when only 27 per cent of joeys born survive to become joeys at heel (see Table 2).

TABLE 2

EXPECTED AVERAGE SIZE OF HARVEST OVER FORTY YEARS OF 2-YEAR-OLD RED KANGAROOS

Conditions	Survival of Joeys	Annual harvest
Drought	0·27	0
Drought slightly less than above	0·30	small
No drought induced mortality	0·90	70

Assumptions: population 400; sex ratio 1 : 1; mortality rate of 10 per cent for joeys at heel, juveniles, and adults.

In theory it would be possible to harvest animals before they died of natural causes. However, as Newsome (1965c), Ealey (1963), and Kirkpatrick and McEvoy (1966) have shown, this seasonal drought-induced mortality is principally restricted to young animals, which are the least attractive for economic harvest.

Using the same general or average figures as above but assuming no drought-induced mortality, then it is possible to calculate the maximum harvest (see Table 2).

None of these admittedly speculative calculations gives much basis for a sustained yield harvest in unhusbanded stocks of macropods.

If macropods could be drafted as flocks, processed under acceptable hygienic conditions, and the meat sold at higher prices for human consumption, then with present knowledge it seems that a local population could perhaps be exploited for longer before reduced economic returns induced change to a new locality. Nevertheless, an improved economic return will not contribute to establishment of a sustained yield harvest, which depends on the relationship between mortality induced by natural causes and that resulting from exploitation.

Comparisons

Species can be compared in two ways: time to harvest, and size at harvesting. Table 1 lists size of kangaroos (in arid Western Australia) at two years of age, i.e. just before sexual maturity. At this age kangaroos are considerably lighter in weight than sheep. Ealey (1967b: Fig. 9) compared the increase in body weight of the euro and sheep with respect to the time taken to reach adult size. The most striking difference is the much slower growth rate of the euro. Sexual maturity is achieved at 80 per cent of adult body weight in both species, but the euro takes approximately eight months longer.

McIntosh (1966: Table 4) has shown that the dietary intake of red kangaroos, raised since birth in captivity, was 50-60 per cent of that of sheep of approximately the same weight. In the field in Western Australia kangaroos are less than half the weight of those used by McIntosh (see Storr, 1968: Table 10). The weights of kangaroos recorded by Griffiths and Barker (1966) are somewhat heavier than those recorded by Storr (1968) but they are considerably lighter than sheep. Griffiths and Barker (1966) show that the weight of stomach contents of kangaroos is much less than sheep, thus supporting McIntosh's findings.

All the foregoing are consistent and agree with the interpretation that macropods have a lowered metabolism (Brown and Main, 1967; Brown, 1968). The net result can be summarised by saying that while kangaroos eat less they take longer to grow to a harvestable size which is generally smaller than a sheep of equivalent age. Small size in itself would be no problem if the population were dense and there was a high turn off of economically valuable meat, for example for human consumption as an expensive delicacy. The population densities of red kangaroos reported by Newsome (1965a) and Newsome *et al.* (1967) suggest that overall densities are not high. The selection of 2-year-old animals for harvest in the simulated population (Table 2) shows the size of the choicest harvest. Harvest of older animals will yield individuals of larger size but numbers will be reduced by natural mortality and it is debatable whether there will be an overall gain in the size of the gross harvest.

GENERAL NATURE OF THE PROBLEM

The existence of an industry based on shooting and processing kangaroo meat demonstrates that there is a resource which can be harvested but not on a sustained yield basis. It should not be lost sight of that in the final analysis the harvestable kangaroo stock is dependent—like sheep—on the ability of the country to support the diverse pasture and forage plants on which the stock depends.

In a very real sense the ability of the country to produce forage is the true resource. Production of forage is severely limited by water and the soil nutrients needed for growth. In arid regions it is easy to see the limiting effects of water, for it is only abundant after rain. The shortage of nutrients such as phosphorus and nitrogen is less obvious but likely to be more serious. Australian arid land soils are markedly deficient in phosphorus (Wild, 1958). Beadle (1966) has shown that some of the sclerophylly characteristic of Australian plants may be due to low phosphorus levels. Soil nitrogen may be fixed by nodulating legumes, soil bacteria, or blue-green algae; however, Australian arid soils are extremely deficient in nitrogen (Charley and Cowling, 1968). The production of animal protein on a sustained yield basis ultimately depends on the nitrogen budget. Main (1968) attempted to relate kangaroo densities to the nitrogen budget, but data are inadequate. Perry (1967) has reviewed the need for wide-ranging ecological investigations on all aspects of arid land usage.

With the foregoing in mind it is appropriate to consider the possible, and alternative, objectives for the utilisation of arid lands for animal production. They are:

(1) Keep all fauna and flora while at the same time running domestic stock. This is an outlook which sees the indigenous plants and animals as maintaining the nitrogen budget. As soon as some elements fail in the face of grazing of domestic stock there is a case for assuming impairment of the nitrogen cycle.

(2) Help the pastoralists maximise their turn off. This objective acknowledges that pastoralists have capital invested and that it is reasonable and proper to assist them.

Two sorts of solution are possible:

(a) with short-term aims, i.e. those with immediate effect such as pest control, increased watering points, introduction of new pasture plants. Many of these solutions do not take into account the long-term wellbeing of the country.

(b) long-term solutions: management of stock and their numbers in relation to the forage and grazing-induced changes in the forage plants to ensure that production continues indefinitely in the future. These aims require active management such as deferred grazing and pest control so that pastures can regenerate.

Pest control is needed for both short-term and long-term solutions so that either more fodder is immediately available for domestic stock or vegetation can regenerate under deferred grazing. Shooting is a relatively ineffective way of regulating a pest population. There is a general tendency for animals to become shy and less conspicuous, but large numbers may still be present. Poisoning, provided it can be done when a large proportion of the pest population is accessible to poison, produces a very effective control.

Harvesting for meat is a way to produce a massive reduction in the population size. It is most effective under conditions where a high proportion of the population is accessible to shooters, i.e. no part of the country is inaccessible to motor vehicles and hence there are no refuges. The size of the kill, and hence the regulatory effect on the population, is largely dependent on local conditions such as aggregation around water or vegetational cover. On the other hand, this sort of population regulation costs the pastoralist nothing. The supposed reduction of the pest is assumed to be sufficient to allow more fodder to grow but there is no certainty that populations at a density too low for economic harvesting are incapable of overgrazing pastures.

When animals are regulated by shooting or poisoning the carcasses are left to rot and this can be viewed as a waste of a protein resource. On the other hand, harvest for pet food can also be regarded as wasteful since protein shortage is widespread in human populations.

The foregoing are all true but we cannot decide that one method of regulation is more useful than another divorced from the question of the aims and policies of the pastoral industry.

The possibilities are that sheep alone, sheep plus kangaroos, or kangaroos alone will provide the more rational use of available fodder. The argument in favour of native herbivores is that while they have low dietary requirements a high proportion of their tissue is protein. But at present this protein is of low commercial value and if kangaroos alone were farmed they would have to be stocked in such numbers that overgrazing would be intense. Under such conditions the vegetation would deteriorate and no advantage would arise from the low nutritional requirements of individual kangaroos.

The complexities of analysing the effects of running sheep and kangaroos together in the same region can be indicated by considering the findings of Griffiths and Barker (1966) and Storr (1968). Storr showed that in a coastal situation in the Pilbara, when fodder was abundant, sheep preferred dicotyledons, red kangaroos preferred grasses, and euros the hard grasses, e.g. spinifex. During drought, when diversity of forage was reduced, all species ate much the same diet. Under conditions when choice could be exercised, sheep had a higher dietary nitrogen intake and obtained more water from their diet than kangaroos. In such a situation the total removal of kangaroos might lead to an increase in sheep numbers. These will be maintained first at the expense of the more palatable forage, shrubs and dicotyledons, and after these become sparse or extinct the less-preferred plants will be eaten, but for a given dry matter intake sheep will obtain less protein and water (because the diet is poorer in these). The general result will be sheep in poorer condition and less able to stand the periodic drought.

Heavy grazing by one species can so change the flora that it becomes depauperate. It is to overcome this sort of vegetational degeneration that deferred grazing is recommended. Deferred grazing recognises that the problem of running stock on a long-term production basis is one of managing the vegetation; deferred grazing does not work unless the vegetation gets regular and prolonged respite from grazing by both stock and macropods. This regulation of kangaroos recognises the principal responsibility as being to the vegetation: considerations of harvesting only militate against quick and total regeneration of the vegetation.

DISCUSSION

The point of view taken here clearly sets the commercial yield as the measure of success. In this regard many generations of selection have led to strains or breeds of domestic stock which under appropriate conditions maximise the likelihood of receiving an economic return from their husbandry. The unexploited genetic diversity still present in domestic stock makes it clear that further selection is possible so that more adapted stocks will be available in the future.

On the other hand the stocks of native animals are still wild. Undoubtedly there is a great genetic diversity present and even if some is lost in any process of domestication a considerable residue of variability will remain. However, the cultural climate for domestication is novel and it seems likely that Australian herbivores have characteristics, apart from their locomotion, which are different from both the present domestic stock or the populations from which they were originally derived. These differences are as follows:

(1) Each species population may be considered to have a strategy for survival. The evolutionary consequences of a successful strategy are that the population is said to become adapted to the particular selective agent. Should the strategy fail to overcome the selective agent then the population will go to extinction.

In our commercial society, populations of animals either produce wool or meat and hides. The strategy is to maximise the turn off of these economically valuable items. Inability to maximise yield satisfactorily leads to bankruptcy and hence extinction of the herd or flock. It should be noted, however, that a great number of genetically diverse sorts of herds and flocks were built up by subsistence-oriented societies rather than monetary-oriented societies. Hence a good deal of selection took place under conditions where the strategy was to feed or clothe a few families only.

(2) The Australian arid land native animals are quite different. In their native situation those that have survived are the ones with a strategy for handling seasonal and irregular drought. Animals can do this by:

- (i) being of small size, with high reproductive potential, so allowing population to contract to minor favourable habitats during drought and rapidly expand again under favourable conditions;
- (ii) being of moderate or large size and restricted to locally favourable habitats, where respiratory water loss is not needed to regulate body temperature, e.g. rock wallaby and euro in rockpiles. Rock wallabies are not known to leave rockpiles but euro populations can expand and become independent of rocky shelter. However, when they do this they undoubtedly require access to drinking water in hot dry parts of the year;
- (iii) by adjusting physiologically so as to:
  - (a) lengthen the time that drought conditions can be endured;
  - (b) be capable of immediate and maximum reproduction once drought conditions break, so allowing the population (i.e. size of stock) to be replenished quickly.

The macropods which have become pest species have been able to do so because their survival strategy has been closely and effectively based on the solution suggested in (iii) (a) and (b) above. The net result is that the macropod genotype is geared to survival, not production, in arid and unreliable conditions. Given conditions where water is abundant and fodder ample, i.e. drought is not an element, production will be high and continuous and would lead to overgrazing unless the pastures could be regenerated by typical farming methods; by definition this cannot be done in arid areas, so that, if harvest is desirable, it can only occur on an itinerant hunting basis when and where animals are in sufficient density.

CONCLUSIONS

In arid Australia harvest of kangaroos for meat on a sustained yield basis appears to be unlikely. Harvest on an itinerant hunting basis is possible.

Hunting is likely to cause excessive mortality and population reduction when a high proportion of the stock is accessible, as at watering points during drought or in well-fenced areas with many tracks. Under such conditions the hunting-induced mortality is incompatible with conservation of the kangaroo stocks.

When population density is so low that harvesting is uneconomic it may still be desirable to reduce the macropod population so that the vegetation can be regenerated. Under such circumstances the wellbeing of the vegetation and not the harvest of protein should be the prime consideration.

REFERENCES

Beadle, N. C. W. (1966). Soil phosphate and its role in molding segments of the Australian flora and vegetation, with special reference to xeromorphy and sclerophylly. *Ecology* **47**: 992-1007.

Brown, G. D. (1968). The nitrogen and energy requirements of the Euro (*Macropus robustus*) and other species of macropod marsupials. *Proc. ecol. Soc. Aust.* **3**: 106-12.

——— and Main, A. R. (1967). Studies on marsupial nutrition V. The nitrogen requirements of the Euro *Macropus robustus*. *Aust. J. Zool.* **15**: 7-27.

Charley, J. L. and Cowling, S. W. (1968). Changes in soil nutrient status resulting from overgrazing and their consequences in plant communities of semi-arid areas. *Proc. ecol. Soc. Aust.* **3**: 28-38.

Ealey, E. H. M. (1963). The ecological significance of delayed implantation in a population of the hill kangaroo (*Macropus robustus*). In A. C. Enders (ed.), *Delayed Implantation*, pp. 33-48. University of Chicago Press, Chicago.

——— (1967a). Ecology of the euro *Macropus robustus* (Gould) in north-western Australia. 1. The environment and changes in euro and sheep populations. CSIRO *Wildl. Res.* **12**: 9-25.

——— (1967b). Ecology of the euro *Macropus robustus* (Gould) in north-western Australia. IV. Age and Growth. CSIRO *Wildl. Res.* **12**: 67-86.

Frith, H. J. (1968). Controlled harvesting of kangaroos. Aust. Conservation Foundation News Letter No. 3: 3-5.

Griffiths, M. and Barker, R. (1966). The plants eaten by sheep and by kangaroos grazing together in a paddock in south-western Queensland. CSIRO *Wildl. Res.* **11**: 145-67.

Kirkpatrick, T. H. and McEvoy, J. S. (1966). Studies of Macropodidae in Queensland. 5. Effects of drought on reproduction in the Grey Kangaroo (*Macropus giganteus*). *Qd Jnl agric. & anim. Scis.* **23**: 339-442.

McIntosh, D. L. (1966). The digestibility of two roughages and the rates of passage of their residues by the Red Kangaroo, *Megaleia rufa* (Desmarest), and the Merino sheep. CSIRO *Wildl. Res.*, **11**: 125-35.

Main, A. R. (1968). Physiology in the management of kangaroos and wallabies. *Proc. ecol. Soc. Aust.* **3**: 96-105.

Newsome, A. E. (1965a). The abundance of Red Kangaroos *Megaleia rufa* (Desmarest) in Central Australia *Aust. J. Zool.* **13**: 269-87.

——— (1965b). The distribution of Red Kangaroos *Megaleia rufa* (Desmarest) about sources of persistent food and water in Central Australia. *Aust. J. Zool.* **13**: 289-99.

——— (1965c). Reproduction in natural populations of the Red Kangaroo *Megaleia rufa* (Desmarest) in Central Australia. *Aust. J. Zool.* **13**: 735-59.

———, Stephens, D. R. and Shipway, A. K. (1967). Effect of a long drought on the abundance of Red Kangaroos in central Australia. CSIRO *Wildl. Res.* **12**: 1-8.

Perry, R. A. (1967). The need for rangelands research in Australia. *Proc. ecol. Soc. Aust.* **2**: 1-14.

Roff, C. and Kirkpatrick, T. (1966). The kangaroo industry in Queensland. First supplement 1961-1965. *Qd Jnl agric. & anim. Scis.* **23**: 467-73.

Storr, G. M. (1968). Diet of Kangaroos (*Megaleia rufa* and *Macropus robustus*) and merino sheep near Port Hedland, Western Australia. *J. Roy. Soc. W. Aust.* **51**: 25-32.

Tribe, D. E. and Peel, L. (1963). Body composition of the Kangaroo (*Macropus* sp.). *Aust. J. Zool.* **11**: 273-89.

Watt, K. E. F. (1968). *Ecology in resource management: a quantitative approach.* McGraw-Hill. New York.

Wild, A. (1958). The phosphate content of Australian soils. *Aust. J. agric. Res.* **9**: 193-204.

7

LAND USE AND PRESENT CONDITION

J. C. NEWMAN* AND R. W. CONDON†

INTRODUCTION

FOR THE PURPOSES of this paper the term 'rangelands', as described by Perry (1967), has been adopted as synonymous with 'arid Australia'. Rangelands are defined as 'those areas of the continent where rainfall on a given piece of land is inadequate for economic crop production or pasture improvement'. This area is larger than that normally depicted as arid but is considered a more meaningful concept when considering pastoral development. The area, comprising 74 per cent of Australia, is depicted in Fig. 1 (Perry, 1967).

As a consequence, the rangelands are dependent for production on the existing native pasture. Continuity of production depends upon the conservation of the climax community or the manipulation of the community to form a more desirable disclimax, be it more palatable, productive, or stable. Development is a matter of obtaining full utilisation of the pastures by ensuring they are available to the grazing animal and that they can be protected periodically from grazing. In practice it entails providing sufficient watering points, adequate fencing, and perhaps treating timber or weed species which restrict production, and pasturing only as many animals as can be supported without jeopardising the continuing productive capacity of the range. That this has not always been the situation in arid Australia can be seen from the present condition of the Australian rangelands.

Climate

A large range of climatic types is included in the area but they can be simply divided into four general groups depending on aridity and period of rainfall:

1. Winter rainfall arid.
2. Winter rainfall semi-arid.
3. Summer rainfall arid.
4. Summer rainfall semi-arid.

The winter rainfall areas normally receive a reasonable proportion of their rainfall in summer, though it is insufficient to be effective. The greater efficiency of the winter rainfall renders these areas more productive. The summer rainfall area is characterised by a more marked winter drought period.

Rainfall varies from below five inches per annum in the Lake Eyre region of South Australia to over thirty inches per annum in the Kimberley area. In

* Soil Conservation Service, N.S.W.
† Western Lands Commission, Sydney.

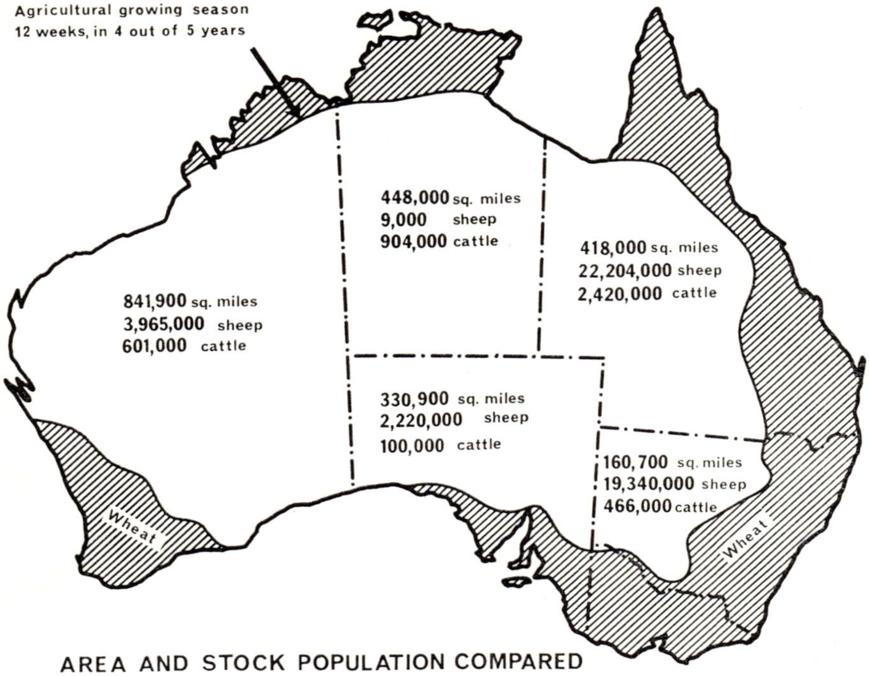

AREA AND STOCK POPULATION COMPARED WITH AUSTRALIAN TOTALS

	ARID AND SEMI-ARID ZONE	% OF AUSTRALIA
AREA	2,200,000 Sq. MILES	74
SHEEP	47,700,000	30
BEEF CATTLE	4,491,000	35

Fig. 1 Map of Australia showing arid and semi-arid zone and sheep and beef cattle populations (after Perry 1968)

the latter, the country is subject to long winter drought but the growing season is long enough to ensure native pasture growth.

In Australia we have no extreme desert (Meigs, 1953). Deserts are 'moderate' and, except for the sandy deserts, such as the Simpson, and some of the gibber plain areas, are all used to some extent for pastoral purposes. Those areas which are not used pastorally are in pristine condition and are not considered in this paper.

The whole rangeland area is characterised by warm to hot summers and mild winters. Evaporation from standard evaporimeters varies from six to nine feet per annum. The area is near flat to gently undulating, with occasional low ranges. Elevation varies from 40 feet below sea level at Lake Eyre to over

4,000 feet in the Macdonnell Ranges in the Northern Territory. The general surface varies from several hundred feet to 1,000 feet. This is in contrast to most of the arid areas of the world, which generally are more elevated.

The ranges, such as the Barrier Range and the Macdonnell Ranges, have only a minor orographic effect on climate but have local microclimatic effect, particularly in respect to run-on.

Condition

The concept of condition has been defined in different ways to suit different requirements. Range condition can be assessed on the basis of the degree of degradation from the original vegetation or, where the original vegetation may not be the most desirable, it can be assessed on the relative vigour and abundance of desirable and undesirable plants and on the extent of soil erosion. These latter facts are bulked into the general term 'health'. Range condition is sometimes considered as indicating the present productivity compared to the maximum productivity.

We have endeavoured to adopt the second of these concepts—that which has been termed the health of the range. The concern in the context of the paper is pastoral deterioration or improvement and the first definition assumes that climax vegetation is the most productive. This applies over most of the arid and much of the semi-arid lands but need not apply to all vegetation units, particularly those on the moister fringe: for example increased productivity can be obtained without loss of stability from dry forest stands or woodlands by thinning timber, in particular mallee lands and woodlands of *Eucalyptus populnea-Callitris collumellaris* (syn. *hugelii*).

DISTRIBUTION OF PRESENT FORMS OF LAND USE

By definition land use on the rangelands is confined to pastoral activity, apart from mining activity and other non-agricultural use which will not be considered here.

The Australian merino industry is based largely on native pasture and shrub browse and the beef cattle industry is heavily dependent on the rangeland pastures. Perry (1967) has estimated that 47·7 million sheep and 4·5 million beef cattle are carried on the 2·2 million square miles of the Australian rangelands. This represents 30 per cent and 35 per cent respectively of the total Australian stock numbers. The distribution of livestock in the rangelands is shown in Fig. 1.

Tenure

The major part of the rangelands in Australia is held under some form of Crown lease tenure. The fact that the land is leased highlights the fact that the value and ultimate use has not been fully assessed. In Queensland, for example, the pastoral lease provides a reservoir of land from which to supply the needs of closer settlement or to build up existing areas to living area standard.

The leases are held by individual pastoralists or by companies. There is only very limited freehold land and no tribally-owned land.

Types of lease vary considerably in detail from state to state but the general principles are similar. Leases may be in perpetuity having fixed rentals or may have rentals periodically reappraised, or they may be of fixed term with either fixed rentals or rentals subject to reappraisal at seven- to fifteen-year intervals.

Most leases include conditions to encourage development and require a certain value of improvements within a certain time and require that improvements be maintained. Improvements generally refer specifically to fencing and may contain residential requirements. Some leases require a certain number of stock to be carried within a certain time as an encouragement for the leaseholder to develop his lease. In some areas this requirement has been unfortunate in that the minimum number specified has been in excess of the capacity of the lease or the lessee has endeavoured to carry the required number of stock without providing sufficient other developments such as water to carry the number safely.

Other lease conditions enable the administering authority to cause the number of stock to be reduced if overstocking is in evidence and to cause the lease to be forfeited if it appears that continued stocking will permanently damage the lease.

Rent remissions may be granted in some cases to leaseholders for carrying out some types of work, such as soil conservation.

Lessees generally have no timber or mineral rights and no cultivation rights unless the latter are specifically granted. To prevent speculation and sales at inflated prices when seasons are good and prices high, some states exercise a form of price control on the transfer of leases.

Pastoral Use

Sheep, for wool production, dominate the southern states, particularly New South Wales and southern Queensland, where 87 per cent of the rangeland sheep occur. Cattle are favoured in the north, where 74 per cent are found in Northern Territory and Queensland.

An important restriction to sheep grazing in South Australia and to some extent in Northern Territory is the presence of the dingo and wild dog which attack lambs and adult sheep. In New South Wales and southern South Australia a dog-proof fence is maintained along the New South Wales-Queensland border, New South Wales-South Australian border and from New South Wales towards the Great Australian Bight in South Australia. Beyond the fence sheep need shepherding or are replaced by cattle. Areas in South Australia north of the fence are grazed by cattle, although the country is considered more suited to sheep.

Sheep have been used in Northern Territory but in nearly every case have given way to cattle because of the difficulties of maintaining flocks in the hot climate, on harsh pastures, and in the presence of dingoes. Christian and Stewart (1954) claimed that an important reason for the lack of success with sheep on the Barkly Tableland is the low proportion of highly nutritive forbs present in the pastures during the winter compared with the sheep-carrying Mitchell grass pastures in Queensland.

Management of both the sheep and cattle enterprises has remained simple. Development of the southern sheep areas has progressed further than the extensive cattle enterprises in the north. In general, fencing has proceeded so that sheep holdings have a dozen or more paddocks and most of the pastures on the leases are commanded by water. The cattle leases vary from an open range system where fencing does not exist to well fenced and developed holdings. The open range system is mostly confined to the Kimberley area in Western Australia and parts of Northern Territory and North Queensland.

Where fencing is absent or paddocks are very large, and where bores are used for water supply, some grazing control can be obtained in dry periods by turning off water from the bore. Cattle will normally graze only five miles from water and will confine themselves to pasturage within this radius. Watering points are frequently spaced beyond the grazing range, so that the herd is reasonably confined except in wet periods when cattle will utilise temporary surface water.

There is little or no conscious attempt to manage pastures. In the Barkly Tableland area a form of unconscious spelling occurs. During the rainy season stock utilise surface water but during the dry season they water from bores. Consequently the areas around the bores are able to build up during the growing season.

Pastoralists, generally, are skilled animal husbandmen but lack an appreciation of the importance of the pasture and soil resource. Stock movements and manipulations are inevitably based on the condition of the animals and rarely on the state of the pasture. This situation reflects the inadequate knowledge that exists of the true value of the vegetation components and of the fact that rains at different times of the year can lead to markedly different responses in vegetation, particularly in the arid areas.

Pastoralists almost wholly follow set stocking grazing practice. Grazing rates vary from twenty-five to thirty sheep equivalents per 100 acres to two to three per 100 acres in the driest areas.

Very few landholders practise specialised schemes of grazing such as deferment or rotation systems. Whether increased production results from these systems is doubtful, but there is mounting evidence that such schemes have an important role to play in regenerative procedures.

Cultivation plays a very minor role overall. Isolated instances exist of the establishment of small irrigated lucerne areas, either from bores or rivers. The lucerne is either grazed or baled and fed to selected stock. These schemes are attractive during drought periods but are often neglected otherwise. Some attempts are being made on the better watered parts of New South Wales and Queensland to increase production by clearing timber and scrub and growing a cereal crop with a view to improving the pasture. These schemes are a recent development and their success cannot be fully assessed as yet. Although these developments do not have a large effect on production of the rangelands as a whole they have considerable local and even regional significance. The creation of pockets of improvement can change individual leases from dry sheep areas to breeding areas and can enable a leaseholder to run cattle in conjunction with sheep to control pastures that may develop on cleared ground and be unacceptable to sheep in some stages of their growth cycle.

In brief, the Australian rangeland is almost wholly used to graze sheep and cattle on native pastures under a sedentary, year-long system of management. Although returns per unit area are low the industry is efficient in terms of returns per unit of labour and returns to capital.

THE PRESENT CONDITION OF THE AUSTRALIAN RANGELAND

No quantitative work has been carried out on the condition of the rangelands of Australia, although some detailed assessments have been made of individual leases in some areas. There have been a number of observations on the qualitative degree of deterioration of rangelands but these observations have not endeavoured to indicate the effect that this deterioration has had on either the short- or long-term productivity. That the effect has been substantial can be gauged from a study of regional statistics of sheep and cattle numbers over time.

Cursory inspections of the rangelands can lead to misconceptions of the degree of deterioration that has occurred through erosion. Because of the dominating influence of climate and the effect of variation in weather periodicity on soil movement and vegetation, an exaggerated idea of the amount of deterioration may be obtained on some soils.

Butler (1967) has drawn attention to the difficulties in the concepts 'accelerated' and 'geologic' erosion and shows that erosion as rapid as 'accelerated' can occur through climatic change.

The high rate of geologic erosion is related to sloping and mountainous lands which yield high run-off (particularly during drought periods when the groundcover species are sparse), and to high soil movement due to the presence of crusts, indurated materials and impermeable soil layers. Considerable sheet flow occurs over bare alluvial plains during drought and a considerable movement by wind occurs on sands. The wind movements during droughts are often compensatory, and serious losses from the region do not necessarily occur.

Little evidence exists of significant erosion before the advent of the white man. The sand dune areas, although apparently active, are basically stable. The dunes generally have a solid core of cemented material. They are frequently rich in Aboriginal artifacts, as are the claypans associated with them.

The scalds away from sand dunes occasionally have some artifacts associated and it is possible that some scalds may have predated the white man. However, this is far from established: the majority of scalds are recent in origin and result from the effect of man's activity.

History of Development

The present condition of our rangelands reflects the history of development of the area by the white man. Before his occupation the area was used solely by Aborigines for hunting game. The only apparent influence on vegetation was the occasional harvesting of scrub for construction of shelter and periodic use of fire to assist in driving game or with the object of attracting game to burnt areas (Leichhardt, 1847). In considering the effect of fire Stocker (1966) concluded that it did not appear to have played a significant role in deterioration of vegetation in northern Australia but recommended that management fires should not be used in the low rainfall areas of the Northern Territory as, in the past, the only grazing animals were kangaroos. Under these conditions occasional fires would have little long-term effect on vegetation as it would be almost ungrazed. However, with stocking at present rates, grazing is likely to be concentrated on burnt areas of vegetation and some damage and erosion may result. Winkworth (1967) considers that the Spinifex communities are not susceptible to erosion when periodically burnt.

It is of interest to compare sheep numbers in Australia as a whole with those in arid and semi-arid parts of Australia since reliable records have been maintained. Figures were not available for all areas in the earlier years so comparison is possible only since 1910. In Fig. 2 sheep numbers in Australia have been graphed in 10-year periods and compared with sheep numbers in arid and semi-arid areas of Australia. The latter area does not fully coincide with the defined 'rangelands' area but represents the drier statistical divisions of the Commonwealth. Pasture improvement and other management developments have led to sharp increases in the sheep population of Australia since 1920. In the pastoral areas, however, the numbers of sheep carried have remained virtually static over the same period despite the fact that the actual

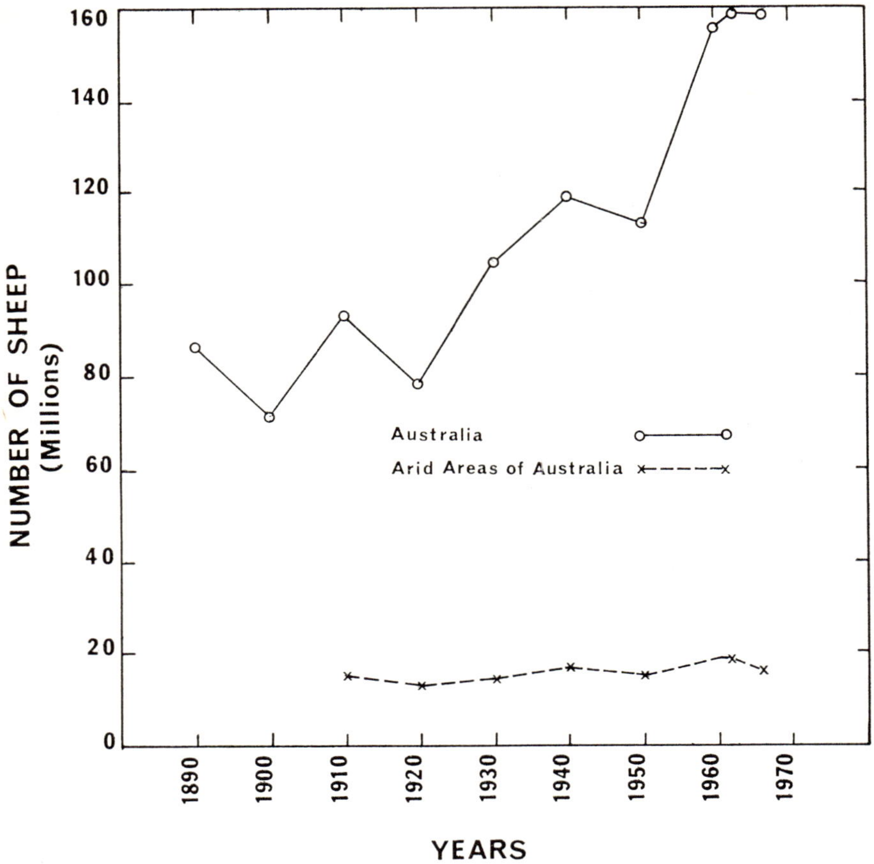

Fig. 2 Total sheep population in Australia and in arid areas of Australia in 10-year periods from 1890 to 1970

area grazed has risen substantially because of increased provision of watering and fencing developments. In effect the number of stock being carried on any parcel of land has fallen during the period in question. If beef cattle figures are included the picture is similar.

The overall picture of stock number variation disguises the detail of individual experiences of the different states and regions. The individual experiences of the different regions have been remarkably similar in that early stock numbers were optimistically high, vegetation was depleted, and the inevitable drought period occurred, with onset of soil erosion. This sequence was accompanied sometimes by a rabbit invasion and general economic slump. The result in each case was heavy financial and stock losses by the pastoralists, and degradation of the soil and pasture resource. This has been followed by restocking which has stabilised at a reduced level.

New South Wales

In the Western Division of New South Wales sheep numbers rose from 6·5 million in 1879 to 15·4 million in 1891. In the period following, the reserve of perennial plants was depleted and soil erosion began. After the major drought of 1901-2, which was associated with severe economic depression, the sheep population fell to 3·5 million. Stock numbers have risen slowly since this time and reached 9-10 million before the drought in 1965, when numbers fell to approximately 6 million. Throughout this period stock numbers have fluctuated considerably, suggesting a dependence on annual rather than perennial species (Beadle, 1948). The collapse of 1901-2 led to the establishment of a Royal Commission to inquire into the conditions of Crown tenants. The result of the inquiry led to the establishment of the Western Lands Commission to administer the lands, and to the present lease tenure in the Western Division.

Beadle (1948) showed that a correlation existed between the rainfall variability and the percentage change in stock numbers—the higher the variability the greater the variation in carrying capacity. This reduction was also associated with a greater loss of timber cover in the higher rainfall variability areas.

During the period in question steady improvements and extension to water supplies were being implemented, using sub-artesian and artesian bores and ground tanks.

Western Australia

The rangelands of Western Australia developed a little later and full occupation of the present pastoral areas did not occur until the 1930s.

Sheep numbers rose in the pastoral areas from 1·6 million to 4·8 million during the period 1900-30. Between 1930 and 1960 the number fell to 3·4 million. The severe drought from 1935 to 1941 led to the decline but, despite a return to normal rainfall sequences, stock numbers have recovered to only approximately 60 per cent of their former levels. The cause has been considered to be reduced production from native vegetation, which is being progressively depleted. The depletion is due to a replacement of the native perennial vegetation by either annuals or unpalatable perennial species (Anon., 1965).

Wilcox (1963) stated that the period of settlement has affected the condition of the pastures. The early settlement before 1890 was along the watercourses of the Murchison plains, and grazing depended on groundwater along the river and on the saltbush of the plain country. The hill country and stony pastures were utilised later. Consequently the major deterioration is in the saltbush country.

In the north Kimberley area grazing has proceeded for only a short period because of the inaccessible nature of the country and the fact that the area is only partially suited to cattle. Leases were let in 1904 but were soon abandoned and have only recently been considered for redevelopment. Again the pastures of the river country are the only areas susceptible to deterioration at present as they are associated with water (Speck *et al.*, 1960).

In the Fitzroy district of Western Australia the graph of sheep numbers shows two peaks, one between 1900 and 1910 and another about 1940. Each peak is followed by a major decline. The first peak is associated with exploitation of the river country and the second with the country away from the river (Perry, 1968).

Queensland

The overall state figures do not show any notable decline in sheep and cattle numbers, but an examination of district statistics suggests that the same pattern has occurred: high and variable stock numbers followed by a fall to a number substantially less than that carried in earlier years. The different periods of development, beginning during the 1860s, in the different parts of the state hide the actual picture in total figures.

Northern Territory

A similar pattern, though more recent, has occurred with cattle numbers in the arid portion of the Northern Territory. Cattle numbers in central Australia gradually increased from the late nineteenth century, when the area was used more or less nomadically by large numbers of cattle, horses, and sheep which depastured around permanent water. In good seasons stock moved away to temporary waters but came back to permanent water in dry periods. A peak number of approximately 360,000 was reached in 1958, after which, due to drought, the number fell to 120,000 in 1964 and possibly lower in 1965. It is not likely that cattle numbers will reach their previous levels for a considerable time.

Degeneration of pastures and erosion have taken place on the areas which have been utilised for a lengthy period (Perry *et al.*, 1962).

South Australia

Ratcliffe (1936) described the conditions which led to deterioration of the South Australian arid lands. The perennial shrubs in general had been seriously depleted by the mid-thirties. Much of arid South Australia began to be occupied from 1850. The severe drought from 1864 to 1869 was said to have destroyed much of the perennial saltbush.

In the early years the number of dams and waters governed the size of flocks, and droughts were 'water droughts'. During a drought stock were concentrated around permanent water and it was common for 10,000 sheep to water from one watering place. Some station owners were said to have deliberately flogged perennial saltbush to obtain better grass. The perennial shrubs in general had been seriously depleted by the mid 1930s.

Rangeland Types

Broadly the rangelands can be considered under the six types (Perry, 1968; Stace *et al.*, 1968):

- Grasslands—mainly Mitchell grass, *Astrebla* spp. on grey-brown cracking clays.
- Shrublands—mainly saltbush (*Atriplex* spp.), bluebush (*Kochia* spp.), and *Chenopodium auricomum* on desert loams, grey-brown and red clays.
- Low woodlands—mainly mulga (*Acacia aneura*), gidgee (*A. georginae*), myall (*A. pendula*), belah (*Casuarina* spp), pine (*Callitris* spp.), mallee (*Eucalyptus* spp.), and rosewood (*Heterodendron oleaefolium*) on red earths, calcareous earths, solonised brown soils and red-brown earths.
- Floodplains and alluvial fans.
- Spinifex grassland, mainly *Triodia* spp. and *Plectrachne schinzii* on deep sands and red earths.
- Mountains and hills—mainly *Acacia* spp. and *Eucalyptus* spp. on lithosols.

Grasslands

The various forms of Mitchell grass pasture have an ability to withstand heavy grazing and the heavy cracking clay soil is not susceptible to degeneration by soil erosion. Mitchell grass pastures generally show excessive degradation only along stock routes and within close proximity to watering points. Country known in the Barkly Tableland area as inferior Mitchell grass (Christian and Stewart, 1954) and comprising heavy clay soil, which is unstable when wet, has suffered from gully erosion as a result of high rainfall intensity, bare soil condition, and proximity to entrenched main streams.

Ebersohn (1963) drew attention to the degeneration of Mitchell grass on river frontages in the Cunnamulla area and in red country where pockets of heavy soil are found. Whilst most of the Mitchell grass stand is in satisfactory condition, the palatable associate species such as Flinders grass (*Iseilema membranacea*) in the pasture appears to have been replaced by unpalatable species such as feathertop (*Aristida latifolia*). This same trend has occurred in the Northern Territory (Perry, 1960) and northern New South Wales. Prolonged heavy grazing leads to the replacement of Mitchell grass by annuals.

James (1960) reported that many graziers in New South Wales maintain their pastures in the neverfail (*Eragrostis setifolia*) disclimax stage by controlled stocking because it forms a more palatable pasture. He considered that over-stocking of the pastures led to the replacement of Mitchell grass and associated palatable perennials by annual grasses and herbs and finally led to a pasture dominated by annual forbs and herbs of low palatability. Weston (1963) reported similar experience on the downs country of north-west Queensland where Mitchell grass country, rated to carry one sheep to four to five acres, was actually being grazed at a rate of one sheep to two acres. Purcell (1963) considered that feathertop was a serious invader of Mitchell grass pastures in central-western Queensland but considered that the reason was mainly seasonal and only modified by grazing practice.

On the Ord River catchment area much of the Mitchell grass and bluegrass (*Dichanthium* spp.) had disappeared from the heavy soils because of continuous, uncontrolled grazing. Protection from grazing has led to a reappearance of these species. On these soils where erosion is not excessive good response is expected from exclosure to stock (Fitzgerald, 1968).

Although some deterioration of the communities has occurred this does not yet appear to be serious and regeneration can be expected with simple management practice.

Shrublands

The shrubland communities are most common in the southern parts of Australia but they extend into southern Queensland and central Australia. The communities in their climax state are dominated by low woody shrubs, chiefly of *Atriplex*, *Kochia*, and *Chenopodium* species. The shrubs have low or moderate palatability but occur with a rich ground flora of annual and perennial grasses, herbs, and forbs. Stock graze chiefly on the ground flora but depend on the shrubs during the normal summer drought periods and extended droughts (Leigh and Mulham, 1966a, b and 1967).

Observations over short periods of heavy grazing suggest that the shrub components can be lost from the community very quickly during dry periods.

It appears that large areas of shrub country were seriously depleted of perennial shrub component early in their occupation.

Ratcliffe (1937) described the conditions that led to the deterioration of the South Australian arid pastoral lands, in particular the chenopodiaceous shrubs. The texture contrast soils often associated with the shrublands have been a significant feature in the deterioration of these lands. The depletion of shrubs is associated with a change in the surface condition of the soil once erosion has begun. The loss of the lighter permeable surface and exposure of the clay subsoil leads to a drastic reduction in the infiltration capacity of the soil and the development of an increasingly arid root environment.

The importance of the shrub stratum, chiefly *Atriplex vesicaria, Kochia planifolia,* and *K. sedifolia,* as a drought reserve and as a soil stabiliser is reflected by the marked variation in carrying capacity of country depleted of bush but not yet scalded. In drought the depleted lands carry virtually no stock whereas in periods of heavy rain carrying capacity exceeds the climax value. Overgrazing of the depleted stands in drought periods has led to serious deterioration by drift and eventual replacement of the palatable annual herbage by unpalatable species, such as *Bassia divaricata* and *B. paradoxa.*

In parts of north-east South Australia the 'gibber' country used to carry scattered *Kochia* spp., but these plants have generally disappeared or been depleted and it is likely that these areas will never return to their original condition.

At Koonamore in South Australia, Hall *et al.* (1964) reported that regeneration of perennial saltbush (*Atriplex vesicaria*) was exceedingly slow on eroded land both with and without exclusion of stock.

In Western Australia, Wilcox (1963) observed that the halophytic shrublands in the Murchison area had been overgrazed early in the development of the area and were susceptible to irreversible damage. Regeneration is slow but response can be obtained on depleted pastures by spelling after autumn or spring rains to obtain germination and then spelling for growth after warm weather rain. Continuous grazing is thought to depress production. Once erosion has proceeded to the scald state, regeneration is virtually impossible without mechanical intervention.

In a study of regeneration of scalded and depleted land at Trida in New South Wales Stannard and Condon (1958) concluded that scalds regenerate very slowly and only after very wet periods. Careful management and spelling is required.

Jones (1966a, b, 1967) observed the mode of reclamation of scalded perennial saltbush land at Hay, N.S.W., under natural conditions and under mechanically assisted conditions. He concluded that 50 per cent of the area which was scalded before 1950 had reclaimed naturally by 1965. Soil changes were associated with reclamation and consequently, although pasture had not returned to the shrub dominant stage, it is believed that the reclamation process is permanent and would be interrupted only by mismanagement. Natural reclamation does not occur haphazardly but requires a hindrance to run-off. Irrespective of slope, free drainage of run-off from a scald maintains the scalded conditions.

Shrublands of central Australia were utilised early in the development of the area and generally are in a severely depleted condition compared with the woodland and grassland types which are generally in fair to good condition.

In New South Wales the shrub communities occur on the stony gilgai or stony desert soils of the north-west and on the texture contrast and heavy clay soils of the Riverine Plain. Only scattered occurrences of shrub-dominated pastures exist in the Upper Darling area but these are generally depleted and are dominated by unpalatable *Bassia* spp. (James, 1960).

Stannard (1963) reported that the bluebush and saltbush pastures in the Central Darling region had been reduced or eliminated through overgrazing and the pastures had been invaded by annual saltbush and the copperburrs *Bassia divaricata* and *B. brachyptera.* In this area he concluded that climate was more potent than stock in causing deterioration but that overstocking favours the spread of erosion and invasion of unpalatable species; in particular overstocking prevents any benefit accruing from the periodic favourable seasons.

The Riverine Plain in the south-west of New South Wales has considerable areas dominated by the perennial saltbush (*Atriplex vesicaria*) and much of this pasture has deteriorated through overstocking. Large areas are invaded by unpalatable *Bassia* spp. and other annuals, substantial areas have deteriorated to the ultimate scald state (Condon and Stannard, 1957a, b, c), while other areas have regenerated from the scald state by natural means to small annual herbs and annual saltbushes (Jones, 1966a).

Knowles (1949) described the processes associated with the degradation of the perennial saltbush and the bluebush (*Kochia pyramidata*) country. The saltbush pastures show an increase in unpalatable species and ephemerals and eventual loss of the perennial saltbush to form a scald. The bluebush, which is less palatable, shows a loss of the associated palatable grasses, drift of the sandy surface begins, and gradually the bluebush is depleted until the whole area supports only ephemerals on drifting sand. This may further deteriorate to scald and small associated sand dunes.

Beadle (1948) claimed that 90 per cent of the perennial saltbush which once occurred in the east Darling of New South Wales has disappeared through overgrazing, particularly in times of drought. The major loss of edible shrubs had been reported as following the severe 1901-2 drought (Anon., 1901).

Overgrazing and subsequent removal of vegetation have led to the disappearance of the dominant species on large areas of shrub steppe in southern Australia. Much of the pasture is now either grassland, annual dominant, dominated by unpalatable species, or is bare of any vegetation. The carrying capacity is often markedly reduced or is subject to marked short-term variations. Erosion forms are scald with or without drift.

Prospects for improving pastures on the semi-arid shrublands are not attractive economically except in specialised situations where increased run-off can be utilised such as on scalds. Any restorative program needs to be accompanied by a lengthy period of controlled grazing.

Low Woodlands

The low woodlands type covers a wide range of vegetation associations throughout the summer and winter rainfall zones and through both the arid and semi-arid lands.

The small tree species in the woodlands are often palatable and have value not only as stabilising plants but as a fodder resource, particularly during drought periods. These species, the most important of which is mulga, are particularly important to cattle, which can graze the plants to a reasonable height. In sheep lands, limited browsing only is possible and a small amount of leaf fall is available. Lopping or cutting of the trees is practised during droughts in some areas to provide survival rations for sheep.

The main value as pastoral lands comes from the ground flora of short grasses and forbs and some palatable shrubs. Mulga has died out in some areas. Various suggestions as to cause have been made but the real picture is not clear. It appears that contributing causes are drought and overgrazing causing loss of undershrub and grass, reduced infiltration of water into the soil, sheet erosion, and general desiccation of the local environment. In some cases senescence of the mulga stands is also contributing. Beadle (1959) has suggested that a major reduction in soil organic matter in surface soils following erosion has reduced regeneration. In other cases mulga has regrown excessively and become so thick that it competes with grass. These occurrences are associated with prolonged moist periods. Ebersohn (1963) reported similar experience in Queensland.

James (1960) reported that a good deal of mulga was killed by excessive lopping during past droughts but has regenerated on the 'hard red' soils when moisture conditions have been favourable.

According to Wilcox (1960b), deterioration of the mulga country is due largely to the reduced mulga shrubs and grasses rather than to damage to the mulga trees themselves.

In Western Australia the mulga understorey species have been described and investigated by Wilcox (1960a). On the deeper soils the associations of wanderrie grasses (which consist of a group of species including *Danthonia bipartita, Neurachne mitchelliana, Eragrostis setifolia, Eriachne helmsii, Aristida contorta, Ptilotus* spp. and others), saltbush and bluebush have been preferentially and continuously grazed since early in the development of Western Australia. Consequently degeneration has occurred through drought and uncontrolled grazing. The wide palatability range of the species has led to replacement of the most palatable species with less palatable species such as *Aristida* spp. and *Eragrostis* spp.

Virtually every pasture in the mulga zone exhibits some form of degradation, ranging from a replacement of perennials by annuals, palatable species by unpalatable species, and in extreme cases to water erosion and severe scalding.

To restore these lands protection from grazing for a period of six months can lead to the re-establishment of *Danthonia bipartita* and *Eragrostis lanipes*. The shrub species are more difficult to replace. Where erosion has proceeded to an exposed hard pan it is almost irreversible.

In the Wiluna-Meekatharra area of Western Australia Litchfield and Mabbutt (1962) have described the hard pan which underlies many of the soils. The consequences of erosion are serious when hard pan is shallow. Mulga intergroves and wanderrie banks normally have shallow red earth over hard pan. Sheet flow occurs naturally over the red earth but when the pasture deteriorates erosion begins and the hard pan may be exposed. Regeneration in these cases is difficult.

In central Australia, short grass-forb pastures occur under a canopy of mulga on extensive areas of red earths and sometimes on calcareous earths and texture contrast soils. The red earths and calcareous earths are reasonably stable and recovery after drought is good. The texture contrast soils, however, have been seriously damaged where excessive grazing has occurred, leading to extensive wind and water sheeting and rilling of slopes. As run-off increases the flats or run-on areas become gullied so that the excess run-off which normally 'irrigates' the flats is confined to narrow entrenched channels.

In the pristine condition the pastures on the red earths and texture contrast soils appear to have occurred as a mosaic of small grassed areas separated by small bare areas. Water runs on to the grassed areas during and after heavy rains. Overgrazing leads to an increase in size of the bare areas, along with removal of the surface soil by wind and water sheeting. As erosion proceeds, the bare areas link up and the 'islands' of grass become elevated above the adjacent surface so that they no longer receive run-on. This marks the beginning of a vicious circle in which increasing aridity causes further reduction in size of the grassed areas. Wind moves the fine sand off the bare areas on to the grassed islands so that these are taken over by ephemerals which leave the islands bare and drifting after every short-term drought.

In south-western New South Wales extensive areas of low woodland occur on sandy solonised brown soils. These are dominated by belah (*Casuarina cristata*) and rosewood (*Heterodendron oleifolium*) in open and dense stands. Pastures are composed of *Enneapogon* spp. Bassias. Although there is evidence of drift in the distant past, most of this area is in good condition with little damage to soils or pastures. Where bluebush (*Kochia pyramidata*) occurred as an understorey, this has frequently been removed.

In parts of northern New South Wales and Queensland serious deterioration of woodland areas is occurring through invasion by woody shrubs such as budda (*Eremophila mitchellii*), turpentine (*E. sturtii*), hop bushes (*Dodonaea* spp.), *Cassia* spp., pine and poplar box (*E. populnea*). This invasion may occur as an increase in density of natural stands or as invasion of cleared and semi-cleared lands. In New South Wales semi-arid woodlands are also being invaded by the galvanised burr (*Bassia birchii*).

These forms of encroachment have caused a serious reduction in the grazing capacity of these lands, sometimes as much as 50 per cent. If the grazier tries to carry stock at the usual level the pastures, depleted by increasing competition from timber and scrub, become chronically overgrazed. Whenever this happens, degeneration of pastures and soil erosion by wind and water sheeting follow the scrub invasion.

This encroachment of woody shrubs appears to have resulted from the prolonged period of above-average rainfall in the early 1950s. A previous invasion took place in the Cobar district of New South Wales after a similar period of above-average rainfall in the 1890s. As early as 1892 Dixon said that settlement in the Cobar district had increased the pine (*Callitris verrucosa*)[1] to such an extent that less stock could be carried there. The original condition of this country was a woodland of cypress pine and poplar box *Callitris collumellaris-Eucalyptus populnea* with scattered trees of mulga, yarran (*Acacia homalophylla*), and budda with shrub underlayers of *Cassia* spp., *Dodonaea* spp., *Eremophila* spp., *Apophyllum anomalum*, and others. Ground flora was *Stipa* spp. and *Aristida* spp. Overgrazing and ringbarking had led to a marked reduction in the palatable flora, and a marked increase in unpalatable components and severe sheet erosion of sloping lands.

The Royal Commission (Anon., 1901) stressed the lack of bushfires following pastoral occupation as an important factor in allowing scrub and timber regeneration to grow unhindered and reach maturity.

Cunningham (1967) has described promising experiments which show that the eroded ridge lands can be restored to productivity with short contour furrows. Further work is in progress in which the run-on areas or flats between ridges are being developed for fodder cropping, using spreader banks (Newman, 1963) for the dual purpose of preventing erosion while the areas are under cultivation and spreading and holding the frequent flows from the badly eroded ridges. Similar techniques are being used in Queensland (Cull, 1963).

The condition of the woodland communities is variable and probably relates to the characteristics of the soil as they affect erodibility. Highly erodible soils, such as level texture contrast soils subject to scalding and undulating hard brown acid soils and texture contrast soils subject to water sheeting, are invariably seriously degenerated. This degeneration has taken place as a result of serious overgrazing in the early occupation of the land.

Degeneration on less erodible soils, such as the softer brown acid soils and red earths, fluctuates according to seasonal conditions and degree of use during drought periods. Minor windsheeting is common throughout on the

[1] Probably *C. collumellaris* as *C. verrucosa* is rare in this district and confined to very specific habitats.

firmer soils. Recovery after severe use is generally good, but continued overgrazing can lead to serious degeneration of pastures and soils. Shallow surface drift must be expected on sandy soils during drought but this is mostly compensatory. Continued over-use leads to severe drift with sandy hummocks and small dunes, and serious degeneration of pastures to ephemerals.

Floodplains and Alluvial Fans

Within this rangeland type may be included the outwash plains adjacent to mountain ranges and the flood-out plains alongside the beds of ephemeral streams in central Australia, the so-called 'channel country' of south-west Queensland, and the broad plains of fine-textured alluvium associated with the larger river systems in New South Wales and Queensland.

In central Australia the soils of the outwash and flood-out plains are variable, with short grass-forb pastures occurring under a sparse low woodland of ironwood (*Acacia estrophiolata*), whitewood (*Atalaya hemiglauca*), and other species. Coarse-textured sands are reasonably stable although drifty during drought conditions. Fine sands are susceptible to severe drift, while texture contrast soils are subject to scalding with over-use. Both the latter soil types have been seriously damaged during the early phases of pastoral occupation when large numbers of stock were constantly forced to retreat on to the longer-lasting waterholes during drought.

The channel country of south-west Queensland comprises the flood-out plains forming the middle reaches of several very large ephemeral streams draining to Lake Eyre. Wide-spreading floods occur in about two years in three, resulting in a flush of annual herbage of good fodder value. The soils of the flooded areas are heavy, grey, deeply-cracking clays and resistant to erosion. However, serious erosion has taken place on light textured soils on the fringes of the channel country.

The fine-textured alluviums of the river floodplains in New South Wales and southern Queensland carry a tall woodland mostly of coolabah (*Eucalyptus microtheca*) or swamp box (*E. largiflorens*). Pastures are dominated by perennial summer-growing grasses (*Eragrostis setifolia*, *Panicum* spp., *Dichanthium* spp., etc.) with plentiful forbs in season. Saltbushes (*Atriplex* spp.) and bassias are prominent components of the pastures in winter rainfall areas. Soils are largely heavy grey clays resistant to wind erosion, with insufficient slope for water erosion. Overgrazing has led, in some areas, to removal of perennial grasses and bush and replacement with ephemeral forbs or spiny *Bassia* spp.

Spinifex Grassland

This occurs extensively on sand plains and dune fields, and on the sandier red earth soils, throughout the drier parts of Western Australia, Queensland,

and Northern Territory, including central Australia. The heavier soils carry the more palatable species of spinifex and associated species. The grassland occurs also on arid hills and on any well drained siliceous habitat.

The range type as a whole, although not very productive, is in good condition. Utilisation is not generally high, and, in fact, it is often used only when adjacent to more productive lands and as a drought reserve (Humphreys, 1963).

Main use comes from associated species, but some increase in productivity by burning before summer rains has been reported from Queensland and Western Australia (Suijdendorp, 1958).

The only damage reported on these pastures is occasional destruction by heavy stock concentrations on small areas and some loss of the softer spinifex species (Nunn, 1956).

The sandplain soils and dunes are generally not mobile, having sufficient coarse sand to prevent undue soil removal by wind. The crests of the dunes are bare drifting sand but there is little encroachment on to dune flanks. Serious erosion may occur where dunes occasionally overlap harder or more productive country on the downwind side of the dune system. Sand activated by excessive grazing and trampling is carried across the plain, stripping vegetation by sand-blast or burying it in shallow or deep drifts.

Mountains and Hills

Much of central Australia and parts of Western Australia and western New South Wales are occupied by stony, rocky and steep hills. Many such areas have no grazing value, supporting only spinifex and stunted mulga or being composed largely of rock and stone with no soil. Other areas are broken by narrow valleys or lowlands of country accessible to stock and with good pastures.

These areas are normally occupied only because of proximity to adjacent more highly productive country. Water is normally not provided and stock graze the areas during periods following rain when temporary surface water is available. Wilcox (1963) reported some loss of shrub component in the hill country in Western Australia.

Minor deterioration can occur by local overgrazing and timber cutting for firewood near settled areas. Some early deterioration occurred at Broken Hill but grazing restriction and control of timber removal has restored the position. Deterioration is still evident on old mining settlement sites which have been abandoned after closure of the mining venture. When erosion has become significant, regeneration has been insignificant.

Because of skeletal soils and general difficulty of access by stock, there is little danger of erosion on steeper hills. In New South Wales the hills have much edible *Bassia* spp. which is little utilised during good seasons but provides a valuable reserve of forage in drought times.

The condition of the lowland areas is generally good. Damage on these lands would be critical, however, in that regeneration is very difficult in the environment and the additional run-off with associated gullying would lead to damage on the valuable outflow areas.

CURRENT TRENDS IN LAND CONDITION

It is very difficult to assess the true trend in land conditions in the arid areas of Australia because of the extreme climatic variability and the fact that short-term weather trends can lead to short-term patterns of vegetative response which can be misinterpreted. For example, in parts of New South Wales there was a feeling during the run of above-average seasons in the 1950s that living areas were too large and that leaseholders should be allowed to sell part of their holdings. However, the subsequent drought reversed this opinion and there has been a movement towards larger living areas.

Wilcox (1963) claims that the trend on all pasture lands in the mulga zone in Western Australia is still toward deterioration.

The position in New South Wales is probably now fairly stable. The major deterioration in early years still largely exists, but observations over twenty years on exclosed plots and adjacent areas suggest that erosion which becomes apparent during each drought period is generally the aggravation of pre-existing erosion which had partially regenerated during above-average seasons. In the Hay area the evidence presented by Jones (1966a) and observations during the recent drought suggest that the trend in this area may be to improved condition. In the experience of the authors, covering more than twenty years, the condition of western New South Wales, as it emerged from the recent drought, was very much better than it was after the 1944-5 drought. The improvement can be attributed to lack of rabbits and a general improvement in the amount of vegetation as a result of a long run of better than average seasons. This improvement is partly reflected in an increase in stock population of about 17 per cent compared with the period before 1944. A marked increase in fencing and watering points consequent upon closer settlement and high wool prices in the early 1950s has also contributed to this increase.

Trend in condition of arid lands is very much dependent on the erodibility of the soil. Short-term droughts of three to twelve months duration are of frequent occurrence. Most arid zone pastures will carry stock for twelve months into a drought, providing there is a reasonable component of perennials in the pasture. It is when stock numbers become excessively high during the unusual run of good seasons that greatest damage is done to pastures and soils during subsequent drought conditions.

Highly erodible soil types, often the most productive, will suffer severely and may never recover. Less erodible soils will return to productivity, the rate and degree of recovery depending on the severity of the abuse.

As an indicator of the pristine condition of the soils and pastures in a semi-arid woodland area, and the rapid rate of decline consequent upon the excessive grazing which seems to have been common to the early occupation of all arid pastoral areas, it is interesting to look at the evidence given before the Royal Commission (Anon., 1901) by James Cotton, Stock Inspector of Cobar.

> In the years 1880 and 1881 before the Cobar district was stocked and when it was being improved the country was covered with a heavy growth of natural grasses—kangaroo grass, star grass, blue grass, mulga and other grasses. The Western half of the district abounded with salt and cotton bush together with the grasses mentioned. The ground was soft, spongey and very absorbent. One inch of rain then in spring or autumn produced a luxurious growth of fresh green grass. The great difficulty seemed to be want of water. The country abounded also in numerous edible shrubs and bushes, and pine scrubs and other noxious scrubs were not noticeable. With the exception of a few gilgai holes the district was absolutely waterless. There has been a gradual deterioration of the country caused by stock which has transformed the land from its original soft, spongey, absorbent nature to a hard, clayey, smooth surface (more especially on the ridges) which instead of absorbing the rain, runs it off in a sheet as fast as it falls carrying with it the surface mould, seeds of all kinds of plants, sheep manure, sand, etc., to enrich the lower lying country and plant it with pine, box and other noxious scrubs.

There is a dangerous concept held by many landholders, and others associated with the pastoral industry in arid lands, that the arid country always recuperates after a drought. It is not easy, when confronted by a sea of waving green grass, to recall what the situation was two or ten years previously.

There is plenty of evidence throughout Australia that once texture contrast soils subject to scalding, or undulating country subject to water-sheeting, become seriously eroded they will not regenerate naturally although they may respond temporarily during good seasons. Other types of country may deteriorate much more slowly, because of gradual depletion of the soil or degeneration of the pastures as palatable species are replaced by unpalatable or perennial species disappear and are replaced by ephemerals.

There is still some highly productive but highly erodible country which will degenerate rapidly if mismanaged. It is necessary for the land user to recognise this country and to find ways of managing it so that it will not deteriorate.

The majority of Australia's arid pastoral lands which are still productive are reasonably stable, but they will degenerate if abused. To prevent deterioration it is necessary to find the level at which they can be stocked without damage.

Both administrative bodies and graziers are becoming increasingly aware of the need to adjust stocking rates to the real productive capacity of the land. The lower grazing rates adopted in all areas after the initial build up to peak numbers and subsequent collapse have undoubtedly played a major role in reducing the rate of deterioration.

An attempt has been made to assist administrative bodies in New South Wales and central Australia to estimate more closely a safe grazing capacity for leases in arid and semi-arid lands (Condon, 1968).

A study of stocking rates in relation to average annual rainfall in western New South Wales has shown an exponential relationship in which the grazing capacity of arid zone pastures increases at an increasing rate with increase in average annual rainfall. Rating scales have been established for factors which influence grazing capacity, giving a bonus for those aspects which increase grazing capacity and a penalty for those which adversely influence grazing capacity.

Rating scales have been compiled for such factors as soil type, topography (as it affects run-off or run-on), tree density, drought forage, pasture composition, percentage summer or winter rainfall, and condition as governed by the degree of erosion. The rating scales for all factors are related to that aspect which occurs on a standard type of country of known grazing capacity. The value of this method is that it can make immediate allowance for factors such as increasing timber density or declining condition due to increasing erosion or deteriorating pastures.

Future trends in condition in sheep-raising areas may be influenced by the lower net returns resulting from the gradual but constant increase in the costs of production without matching increases in wool prices. What was regarded as a living area in the late forties is no longer so regarded. Landholders who cannot increase the productivity of their existing holdings need more land.

The rate of net returns for cattle or sheep enterprises will fluctuate from time to time. Land administration policy needs to be flexible enough to allow the necessary changes to take place.

Other developments which have local significance should be recorded in so far as they affect the soil-vegetation resource. The majority of attempts to modify and improve the productivity of the environment on a large scale are confined to the moister parts of the rangelands.

In New South Wales and Queensland mulga woodlands, run-off is being collected from ridges and spread over flats to provide extra grass or sown fodder. Oats and summer forage such as sudax and lucerne have been grown successfully on small areas. This trend is being adopted to overcome loss of production because of invasion by unpalatable shrubs. The flats sometimes are dominated by dense eucalypt suckers which need to be cleared before development for water-spreading. Water-spreading enables small areas to produce good quality forage which can be grown in sufficient quantity for the grazier to carry breeding stock instead of dry stock only. Such developments also provide the opportunity for relieving the grazing pressure on deteriorating pastures.

Some development of gidgee lands is being carried out in Queensland, while mallee, gidgee, and the cypress pine-eucalypt woodland is being developed in New South Wales. This development involves thinning or clearing the timber, sowing one or several cereal crops, followed by either natural or sown pasture. In Queensland it has been estimated that graziers have increased the area of land cultivated from 82,711 acres in 1959 to 516,263 acres in 1968.[2] In New South Wales the total area sown is somewhat less, and is being developed under the strict supervision of the Western Lands Commission. Conditions which apply to clearing and cultivation limit the size of individual cultivation blocks, require specific belts of timber to be retained, provide for installation of soil conservation measures where necessary, restrict cultivation on specific soils, and specify conditions relating to crop rotation and pasture sowing.

Some thinning of timber to assist native pasture growth is also proceeding, again under conditions aimed at preventing erosion.

Mechanical restorative work is carried out over most of the arid and semi-arid areas, involving ploughing, checkerboard furrowing, contour furrowing, waterponding and related treatments, but these activities have so far been principally exploratory and represent only a small percentage of the potential areas requiring this type of treatment.

A large-scale reclamation project is being carried out on the Ord River catchment area (Fitzgerald 1967, 1968). The area of 17,800 square miles has an estimated 1,450 square miles subject to varying degrees of erosion. Erosion is generally confined to the fine-textured calcareous soils adjacent to the river systems and has been caused by continuous, uncontrolled grazing accompanied by fire and drought. The original grassland and grass savannah has deteriorated seriously and the area is now badly gullied and severely scalded. The aim of the restoration program, which began in 1960, is to re-establish the perennial vegetation. The approach has been to control grazing by fencing and to re-establish vegetation by mechanical treatments with some seeding of introduced species. The response has been related to the degree of erosion, being poorest where scalding is severe.

Extensive mechanical treatment is likely to be used only where the motivation extends beyond pastoral interests. Costs of treatment are relatively high but landholders are likely to adopt them more in the future as they are shown to be economically feasible.

[2] Figures supplied by Department of Primary Industries, Queensland.

SUMMARY

The pattern of events in land use of the Australian arid rangelands in the various states has been similar and is characterised by an early build up of high stock numbers which fell rapidly because of drought and/or rabbit invasions and economic depressions. Stock numbers have subsequently stabilised to a level considerably lower than previous peaks.

Evidence exists that heavy stocking in periods of above-normal rainfall leads to serious damage when drought follows and stock cannot be rapidly moved from the area. Recovery depends on the degree of abuse and the soil and pasture characteristics.

Deterioration has invariably occurred in the early phases of settlement. It is disturbing that the same pattern has occurred in the more recent land development as occurred in the early developments. Lands which have been settled for a long period and which are highly susceptible to erosion damage have often suffered almost irreparable damage. Examples of this are the shrublands of southern and north-western New South Wales, South Australia, and central Australia. The shrub-invaded woodlands of the Cobar Peneplain and some of the mulga lands of the Murchison area in Western Australia have also been seriously degenerated.

In more stable areas deterioration is confined to some replacement of edible and perennial species by inedible or annual species. There is evidence to suggest that informed grazing management practice can restore many of these situations to a satisfactory condition.

Some land developments on the fringe of the pastoral zone offer the possibility of considerable improvement in land condition.

The general findings for the best use of rangeland pastures is utilisation based on expected production during drought periods, with periods of spelling to enable seed setting and regeneration of palatable species.

Government authorities are seeking reliable estimates of carrying capacity on which to base their administration.

To restore present lands in depleted condition and to prevent a general downward trend, rangeland management systems need to be reconsidered and carrying capacities aligned to long-term production capability.

APPENDIX

In an attempt to assess the extent of deterioration of Australian rangelands, the following table has been compiled from experience in some of the areas and from an interpretation of condition as described in the literature of other areas.

The table refers only to the arid lands which are used for grazing and endeavours to relate present pastoral value to the expected pristine condition.

PERCENTAGE OF RANGELANDS TYPES IN PARTICULAR DEGENERATION CLASSES

	The present pastoral value in relation to expected pristine condition			
	Little or no degeneration (80-100%)	Minor degeneration (60-80%)	Moderate degeneration (40-60%)	Severe degeneration (< 40%)
1. Grasslands	30	55	15	negligible
2. Shrublands	10	25	40	25
3. Low woodlands	40	30	20	10
4. Flood plains and alluvial pans	40	30	20	10
5. Spinifex grassland	70	20	5	5
6. Mountains and hills	50	20	20	10

REFERENCES

Anon. (1901). *Royal Commission to inquire into the conditions of the Crown Tenants, Western Division of New South Wales.* Govt. Printer. Sydney.

Anon. (1965). *Sheep in the mulga zone—a handbook for pastoralists.* Dept Agric. West. Aust.

Beadle, N. C. W. (1948). The vegetation and pastures of Western New South Wales, with special reference to soil erosion. Govt. Printer. Sydney.

——— (1959). Some aspects of ecological research in semi-arid Australia. In Keast, Crocker, Christian (eds.), *Biogeography and Ecology in Australia.* W. Junk. The Hague, Netherlands.

Butler, B. E. (1967). Soil periodicity in relation to landform development in south eastern Australia. In J. N. Jennings and J. A. Mabbutt (eds.), *Landform studies from Australia and New Guinea.*

Christian, C. S., and Stewart, G. A. (1954). The land-use groups of the Barkly region. *CSIRO Aust. Land Res. Ser.* No. 3: 150-80.

Condon, R. W. (1968). Estimation of grazing capacity on arid grazing lands. In G. A. Stewart (ed.), *Land Evaluation.* Macmillan of Australia. Melbourne.

——— and Stannard, M. E. (1957a). Erosion in Western New South Wales. Part II. Forms of wind erosion. *J. Soil Conserv. Serv. N.S.W.* **13**: 17-26.

——— and ——— (1957b). Erosion in Western New South Wales. Part III. Forms of water erosion. *J. Soil Conserv. Serv. N.S.W.* **13**: 68-74.

——— and ——— (1957c). Erosion in Western New South Wales. Part IV. Classification and survey of erosion. *J. Soil Conserv. Serv. N.S.W.* **13**: 122-28.

Cull, J. K. (1963). Some agricultural and pastoral problems of central western Queensland. *Proc. Qd Dept. Agric. and Stock, W. Qd Agrost. Conf.* Paper 9 (mimeo).

Cunningham, G. M. (1967). Furrowing aids revegetation at Cobar despite the worst drought on record. *J. Soil Conserv. Serv. N.S.W.* **23**: 192-202.

Dixon, S. (1892). The effects of settlement and pastoral occupation in Australia upon the indigenous vegetation. *Trans. Roy. Soc. S. Aust.* **15**: 195-206.

Ebersohn, J. P. (1963). Mulga scrub lands of the south west. *Proc. Qd Dept. Agric. and Stock, W. Qd Agrost. Conf.* Paper 2 (mimeo).

Fitzgerald, K. (1967). The Ord River catchment regeneration project. *J. Agric. West. Aust.* **8**: 446-52.

——— (1968). The Ord River regeneration project. 2. Dealing with the problem. *J. Agric. West. Aust.* **9**: 90-5.

Hall, A. A., Specht, R. L. and Eardley, C. M. (1964). Regeneration of the vegetation on Koonamore vegetation reserve 1926-62. *Aust. J. Bot.* **12**: 205-64.

Humphreys, L. R. (1963). The spinifex communities of Western Queensland. *Proc. Qd Dep. Agric. and Stock, W. Qd Agrost. Conf.* Paper 9 (mimeo).

James, J. W. (1960). Erosion survey of the Paroo-Upper Darling region, Part III. Vegetation and soils. *J. Soil Conserv. Serv. N.S.W.* **16**: 185-206.

Jones, R. M. (1966a). Scald reclamation studies in the Hay District. Part I. Natural reclamation of Scalds. *J. Soil Conserv. Serv. N.S.W.* **22**: 147-60.

——— (1966b). Scald reclamation studies in the Hay District. Part II. Reclamation by ploughing. *J. Soil Conserv. Serv. N.S.W.* **22**: 213-30.

——— (1967). Scald reclamation studies in the Hay District. Part III. Reclamation by ponding banks. *J. Soil Conserv. Serv. N.S.W.* **23**: 57-71.

Knowles, G. H. A. (1949). Soil erosion and soil conservation in the Hay District. *J. Soil Conserv. Serv. N.S.W.* **5**: 171-7.

Leichhardt, F. W. L. (1847). *Journal of an overland expedition in Australia from Moreton Bay to Port Essington a distance of upwards of 3000 miles during the years of 1844-45.* T. W. Boone. London. (Reproduced by the South Australian Libraries Board.)

Leigh, J. H., and Mulham, W. E. (1966a). Selection of diet by sheep grazing semi-arid pastures of the riverine plain. 1. A bladder saltbush (*Atriplex vesicaria*)-cotton bush (*Kochia aphylla*) community. *Aust. J. exp. Agric. Anim. Husb.* **6**: 460-7.

——— and ——— (1966b). Selection of diet by sheep grazing semi-arid pastures of the riverine plain. 2. A cotton bush (*Kochia aphylla*)-grassland (*Stipa variabilis-Danthonia caespitosa*) community. *Aust. J. exp. Agric. Anim. Husb.* **6**: 468-74.

——— and ——— (1967). Selection of diet by sheep grazing semi-arid pastures of the riverine plain. 3. A bladder saltbush (*Atriplex vesicaria*)-pigface (*Disphyma australe*) community. *Aust. J. exp. Agric. Anim. Husb.* **7**: 421-5.

Litchfield, W. H., and Mabbutt, J. A. (1962). Hardpan soils of semi-arid Western Australia. *J. Soil Sci.* **13**: 148-59.

Meigs, P. (1953). World distribution of arid- and semi-arid homoclimates. In *Reviews of Research on Arid Zone Hydrology.* UNESCO. Paris.

Newman, J. C. (1963). Waterspreading on marginal arable areas. *J. Soil Conserv. Serv. N.S.W.* **19**: 49-58.

Nunn, W. M. (1956). Pastoral research. 'Munda' field day demonstrates real progress in the Port Hedland area. *J. Agric. West. Aust.* **5**: 357-64.

Perry, R. A. (1960). Pasture lands of the Northern Territory, Australia. *CSIRO Aust. Land Res. Ser.* No. 5.

——— (1967). The need for rangelands research in Australia. *Proc. ecol. Soc. Aust.* **2**: 1-14.

——— (1968). CSIRO Rangelands Research Programme. CSIRO Aust. (mimeo).

———, Mabbutt, J. A., Litchfield, W. A., Quinlan, T., Lazarides, M., Jones, N. O., Slayter, R. O., Stewart, G. A. (1962). General report on lands of the Alice Springs area, Northern Territory 1956-57. *CSIRO Aust. Land Res. Ser.* No. 6.

Purcell, D. L. (1963). Gidgee scrub lands. *Proc. Qd Dep. Agric. and Stock, W. Qd Agrost. Conf.* Paper 3.

Ratcliffe, F. N. (1936). Soil drift in arid pastoral areas of South Australia. *Pamph. Coun. scient. ind. Res. Aust.* No. 64.

——— (1937). Further observations on soil erosion and sand drift, with special reference to south western Queensland. *Pamph. Coun. scient. ind. Res. Aust.* No. 70.

Speck, N. H., Bradley, Janice, Lazarides, M., Patterson, R. A., Slatyer, R. O., Stewart, G. A. and Twidale, C. R. (1960). The lands and pastoral resources of the North Kimberley Area, W.A. *CSIRO Aust. Land Res. Ser.* No. 4.

Stace, H. C. T., Hubble, G. D., Brewer, R., Northcote, K. H., Sleeman, J. R., Mulcahy, M. J., Hallsworth, E. G. (1968). *A Handbook of Australian Soils*. Glenside. Rellim Technical Publications.

Stannard, M. E. (1963). Erosion survey of the central East Darling region. IV. Erosion. *J. Soil Conserv. Serv. N.S.W.* **19**: 69-80.

——— and Condon, R. W. (1958). Studies on Trida regeneration area. *J. Soil Conserv. Serv. N.S.W.* **14**: 159-76.

Stocker, G. C. (1966). Effects of fires on vegetation in the Northern Territory. *Aust. For. J.* **30**: 223-30.

Suijdendorp, H. (1958). Problems of the north west of Western Australia. *Proc. Aust. Agrost. Conf.* Armidale. CSIRO. 1: 19-1/19-4.

Weston, E. J. (1963). The mitchell grass association of north west Queensland. *Proc. Qd Dept of Agric. and Stock, W. Qd Agrost. Conf.* Paper 7 (mimeo).

Wilcox, D. G. (1960a). Studies in the mulga pastoral zone. 1. The grazing of wandarrie grass associations. *J. Agric. West. Aust.* **1**: 475-9.

——— (1960b). Studies in the mulga pastoral zone. 2. Some aspects of the value of the mulga scrub. *J. Agric. West. Aust.* **1**: 581-6.

——— (1963). The pastoral industry of the Wiluna-Meekatharra area. In Lands of the Wiluna-Meekatharra Area, Western Australia. *CSIRO Aust. Land Res. Ser.* No. 7.

——— (1965). The grazing of saltbush. *Bull. agric. Dep. West. Aust.* 3329.

Winkworth, R. E. (1967). The composition of several arid spinifex grasslands of central Australia in relation to rainfall, soil water relations, and nutrients. *Aust. J. Bot.* **15**: 107-30.

8

THE PASTORALISTS' VIEWPOINT ON RANGELANDS

R. CASKEY*

FUNDAMENTALLY the graziers' objective in the arid zone is the highest possible net income without damage to his capital asset. In addition we expect an adequate education (generally of a higher standard than we had ourselves) for our children, a reasonable standard of living, the provision of such amenities as are available to us, and to pass on our properties to our sons in a better condition than we found them. These objectives are very like those of people in other industries but by reason of our isolation are harder to attain.

We realise that we may not have the same amenities available to people in more populated areas, but we do expect our pastoral operation to provide reasonable comforts for our families.

I think I should make it clear that management policies are complex. Many factors need to be considered, some of which are not constant, and any policy adopted will need to be elastic. I refer in particular to land tenure, condition of the land, condition of the stock, the available feed, financial considerations, market price and its relationship to transport costs, taxation, and the physical limitation that may be imposed by labour conditions.

LAND TENURE

The arid zone has various types of land tenure which vary from state to state, and even within the state. There are so many variable factors within the arid zone that I will confine my comments to conditions in that part of which I have most knowledge—the Western Division of New South Wales.

The pastoral holdings, with a few minor exceptions, were originally large. A period followed when portions of the original holdings were withdrawn for closer settlement, and here a serious mistake was made. The holdings were too small for economic stability. Efforts were then made to correct the situation by withdrawing further portions from larger holdings in an exchange for a longer term base for the remainder.

The concept of a Home Maintenance Area evolved where the lease was issued in perpetuity. The value of this type of lease is that it enables finance to be obtained for development, restocking after drought, and living expenses.

* Grazier, Byrnedale Station, via Broken Hill, N.S.W.

It also enables the lessee to plan improvements with the knowledge that he will derive benefit from them. An incentive is provided to preserve the land.

The fixed-term lease has serious drawbacks. It can, and in many cases did, lead to land exploitation by overstocking, particularly towards the end of the lease. There was also a reluctance to continue with improvements and provide adequate maintenance of the existing improvements. Soil erosion is, I believe, the most serious problem that we face, and inappropriate land tenures have contributed greatly to the overall erosion problem. Having first arrived at what area of land is an economic unit I favour a Lease in Perpetuity as a sound principle in land conservation.

Past history has shown that the type of land tenure must provide for sufficient size to establish an economic unit. With changing conditions it is apparent that many holdings are still too small. The arid zone is no place for small holdings or individuals with inadequate finance.

MANAGEMENT

The first requirements for management are the provision of water supplies and fencing. Some areas are devoid of suitable underground water, so surface water must be provided. Where underground water is available at an economic depth and is of acceptable quality, it is usual to have both surface and underground supplies.

Several questions related to water supply in arid areas need to be answered through research. There is need for investigation into the impact that salt has on the health and production of sheep and cattle, particularly where the fodder has a high salt content. It may be that our present use of underground water with a high salt content is at fault and we should be providing and reticulating surface water.

There is no general agreement on how far apart our watering places should be. Surely there must be an optimum distance related to vegetation and land types. Pastoralists' opinions vary from between a radius of two and four miles for sheep. I think we need an accelerated program of research on these problems. Accurate information is a necessity for efficient management and land conservation. Far too much time has been lost.

Paddock Size

Paddock size varies from 1,000 acres to 40,000 acres depending on the personal beliefs of the manager and available finance.

Little effort has been made in the past to separate different land types and vegetation communities into separate paddocks. We have no reliable knowledge on the efficiency of paddocks of different size. It could be that our paddocks are too big and, in some measure, by their size alone, accelerate erosion.

Uneven distribution of grazing could be brought about by the habits of sheep, who prefer to graze into the wind. As our winds are predominantly westerly and southerly, over-grazing occurs in the western and southern sides of a paddock. Smaller paddocks may result in a more even grazing pattern.

Provision of water should also be related to sheep habits. There is some evidence which suggests that watering points should not be on the western and southern sides of a paddock.

An investigation into the grazing habits of sheep and cattle is urgently needed. It may alter our management practices, particularly in relation to paddock size, flock size within the paddock, fencing patterns, and placement of watering points.

Shearing

Shearing occurs almost every month of the year, but it is generally concentrated in the months of February, March and April, or July, August, and September. Those who shear in the early months of the year claim a better choice of shearing labour, longer period of daylight, a sufficient length of wool to protect the sheep from cold in the winter months, less stress on the ewe because lambing takes place in the winter months when she has less fleece to carry, and lambs that have grown enough wool to more than repay shearing costs.

Those who shear in the months from July to September claim less stress on man and beast because of pleasanter weather conditions; the sheep are not carrying a full fleece of wool during the summer months, they are not so susceptible to fly strike in the spring months when flies are active, autumn lambs can be shorn at an early age to remove the lamb tip from the wool and reduce fly strike, and spring lambs more than repay the cost of shearing.

Other factors also influence shearing times. Dust contamination is more likely in the summer months, when it is advisable to have as little wool as possible on the sheep. There may be means whereby dust contamination could be reduced. I believe this subject is worthy of investigation. Vegetable fault in the wool may be derived from winter-growing annual plants which mature in the early spring, or summer-growing annuals which mature in the autumn. Individual properties vary on which is the most significant. It is apparent that a good many factors must be taken into consideration before a decision on shearing time is made, and the choice will be a compromise.

Lambing

There are two schools of thought on lambing time. Autumn lambing and spring lambing have both given excellent results. Experiments conducted some years ago indicated that the Merino sheep had its highest fertility period in the cooler months and that mating should take place at that time, but there have been some rather remarkable contradictions to that general recommendation. A further look at this is necessary.

In the last decade the improvement in lambing percentages of the Merino has been astounding. Today 100 per cent lambing and more is not uncommon where not so many years ago it was rare. The reasons for this are far from clear. Seasonal conditions obviously play an important part but there is more than that.

Double lambings are not practised as a general rule. Where they are, condition of the ewe makes her very vulnerable to drought. Double lambings are used after a prolonged drought to build up numbers and most graziers, at some time or another, may attempt the practice. Percentage of rams mated may vary from 1 to 3 per cent of the flock according to the policy of management. It would be of considerable help if we knew the maximum percentage required in relation to given conditions. Rainfall patterns have a considerable influence on mating times. Green feed at lambing time is desirable. Within the arid zone there are areas where the rainfall is predominantly summer, and in other areas predominantly winter, and although it is unreliable the chances of having green feed at a given time are related to rainfall incidence.

Sale and Purchase of Sheep

Surplus sheep on arid properties fall into three major categories: wethers, cast-for-age ewes, and culled ewes.

Wethers may be sold as fat lambs for slaughter at ages from four to eight months or for restocking to woolgrowers in the higher rainfall areas. Some graziers retain the wethers until they are two years old as wool producers. Others sell them at the first opportunity, believing that the ewe is the more profitable animal. There are areas of country which are not best suited for lambing ewes, and in these wethers are carried as a matter of policy.

The age at which cast-for-age ewes are disposed of varies from five to eight years. When the ewe flock has been reduced by drought the ewe may be retained for a further one or two years so that lamb numbers may be increased, but under average conditions the policy of individual management is reasonably consistent. On the basis of lamb production, deaths, and wool production there must surely be an age where a ewe is no longer an economic proposition. What is this age?

Culled ewes are usually disposed of at from twelve months to two years of age. The reason for culling is to discard undesirable types, and the rate of culling may vary from 10 to 30 per cent. There are many graziers who do not cull because they cannot breed enough replacement ewes or they are not convinced that they gain any material benefit.

Sheep are purchased for two principal reasons: to replace numbers after a drought or to take advantage of a lush season. In the second instance, they may be retained until they are shorn and then sold, or sold after they have been fattened and a reasonable profit can be obtained.

Stock Health

The losses from disease in the low rainfall area are not a serious problem. There are, from time to time, spectacular losses from plant poisoning or dietary deficiencies, and no doubt losses could be reduced by more knowledge and better management. Perhaps the most important factor in our losses is blow fly strike. These losses can be reduced by shearing, crutching, and jetting at a strategic time. The performance of the Mules operation on the female portion of the flock is desirable. This operation is by no means general. The benefits have been proved beyond doubt and it is hard to understand why it is not generally practised.

Relationship Between Sheep and Cattle

During recent years there has been an increase in the number of graziers who graze cattle in conjunction with sheep. There is, of course, a large area in the arid zone where cattle alone are grazed. The increase in production costs, combined with a lower return from wool, has caused many pastoralists to try to increase their income by running cattle. I would expect that this trend will extend where the type of country is suitable. In some cases cattle are purchased to take advantage of excess feed and are sold after fattening.

It is apparent that on some types of country a limited number of cattle do not compete seriously with sheep. To what extent they do compete is not clear.

Drought Policies

The general drought policy is to reduce numbers progressively to a point where the nucleus of the breeding flock is retained until their condition and the condition of the land makes it imperative that they be agisted, sold, or provided with purchased fodder. Where they are fed the cost is high and the practice risky.

It is usual to sell the wethers first, followed by the older ewes and the lambs under one year of age. Where the ewes are lambing at a time when it is necessary to reduce numbers the ewe hoggets must be removed.

Should the price for the older ewes be so low as to provide little or no return, financial commitments may dictate the sale of the young sheep. The policy of what to sell depends on the personal beliefs of the manager and his financial position. The objective is to preserve a breeding flock with wool-growing ability and as far as is possible prevent erosion.

When cattle are grazed in conjunction with sheep the objective would be to retain the best of the young cows.

Labour

Labour is perhaps the greatest part of our costs, not only in the amount paid to the employee but in provision for the various other responsibilities—insurance, long service leave, accommodation, sick leave, and holiday pay.

The pressure of costs has forced us to reduce labour to the minimum. I expect this trend to continue. We will rely mainly on what we can provide ourselves and what help we may get from our neighbours. In some cases contractors will be employed for short periods for particular jobs formerly done by permanent employees.

Erosion

Erosion is perhaps the greatest problem of the low rainfall area. It is difficult to lay the blame on the individual. He had, in the first instance, nothing to guide him on the capabilities of the country, and in acquiring experience some damage was done. The recuperative power of the country can so easily deceive, but there is a point beyond which recovery is very slow and in some cases not possible.

In many cases the need to overstock was a sheer economic necessity, to which unsound land policies and insufficient finance contributed. In others there was no way of removing the excess stock when conditions deteriorated. Better roads and road transport have helped enormously in this respect.

Rabbit Damage

The influence that the rabbit had on vegetation was disastrous, not only in what it ate itself, but also in that it forced the domestic animal back on to perennial vegetation much sooner than would have been the case. In many instances the increased pressure was more than the vegetation could withstand.

The effect the rabbit had on the regeneration of some of our most valuable perennials was severe, and in conjunction with stock was responsible for the almost complete disappearance of some plants. I hesitate to think what would happen if we have a return of the rabbit to the extent we had previously.

Myxomatosis has given us breathing space. Some remarkable examples of regeneration can be seen but I feel we should not place too much reliance on this position continuing.

Stock Movement

The four-wheel-drive vehicle and the motor bike have largely replaced the horse in stock mustering and inspection. They are time saving and labour saving, and represent two important factors in reducing costs.

Road transport has an important impact on the movement of sale stock to a market point and to agistment areas. It is expensive but efficient, and with the provision of better roads has enabled us to take advantage of available agistment which was previously impossible as the stock routes were closed for long distances in a drought area. Losses by road transport are usually small.

Transport on the hoof is usually a good deal cheaper than road or rail transport, but shortage of skilled drovers and livestock weight loss when the stock routes are not in good condition have resulted in this method not being as extensively used as it once was. The time taken on a journey is also important. Losses are at times heavy, mostly from plant poisoning or careless droving.

The initial cost of rail transport is usually lower than road transport, but as stock have to be taken to the railway, usually by road transport, double handling occurs and it is not always certain that trucks will be available when the stock arrive. Rail losses are not excessive except when the stock are in poor condition. Then they may be heavy, as proper attention is not possible to the extent that it is on road transport.

The carriage of wool is an important function of rail transport, but increasing competition from road transport has diminished the quantity that was previously carried by rail. There are several reasons for this: competitive price, the necessity for the grazier to load his own wool at unattended railway sidings, road transport that will pick up wool at the shed door and deliver to the wool broker's store without the grower having the responsibility to procure rail trucks which are not always available, and the lack of an acceptable client relationship with the railways. The railways nevertheless provide a very valuable and necessary service.

Transportation costs are very important in our cost structure, and they have in many cases a determining effect on where we sell or buy stock or purchase materials or goods.

I will not pretend that my remarks have covered the whole field of management. The practices are so diverse and elastic that it is an almost impossible task. There is so much variety that it strongly suggests that some of our management policies are unsound.

FINANCE

At some time almost all pastoralists need finance which they cannot provide from their own resources. In the main these funds are obtained from the banks or from woolbroking houses. The degree to which finance is available depends on the policy of the government in relation to the economy, the asset backing for the loan, and the lender's faith in the borrower's ability to meet his commitments.

In recent years the woolbroking houses have not been able to provide the necessary amount of finance from their own resources. Although they are still very important in providing loans, banks have become increasingly important. The system whereby the woolbrokers provide a large part of the loan is a very sound one. Through their officers they have a close personal relationship with their clients and a sound knowledge of their ability and property. On this basis they were prepared to advance money to a greater extent than the banks were, and often this was the financial saviour of the pastoralist.

Unlimited finance is not the answer to our financial needs, of course. It must always be possible for the borrower to meet his interest bill and repay the principal over a reasonable period of years. Obviously the interest rate will have an important bearing on the pastoral enterprise.

I believe that our objective should be to retain the productive capacity of the land; financial stringency often puts this in jeopardy.

I think there is a good case for the provision of means to enable the pastoralists in an arid zone to obtain long-term loans at a lower rate of interest than is now available.

TAXATION

There are few things that have a greater influence on management practices than taxation. To a large extent it dictates the amount of money which can be spent on major items of plant purchase, repairs, or improvements in a particular year. It is, of course, sound practice to do these jobs in a year of high income when the taxation rate is correspondingly high. The extent to which this can be done is dependent on the rate of depreciation or deduction for taxation purposes. Where the need for repairs, improvements, or plant purchase is urgent, the grazier may need to borrow money to meet his taxation bill. It can be appreciated that management policies in these matters can be erratic, by reasons of fluctuating incomes and seasons. A much sounder system would be possible if taxation laws enabled the arid zone grazier to accumulate a financial reserve which would, in part, iron out these fluctuations.

The effect of taxation does at times influence the grazier in his stock selling policy. When seasonal conditions are deteriorating he may be tempted to retain stock so that they may be sold in the next financial year in anticipation of a lower income. The effect that this policy may have on the land could be very undesirable. A credit balance is a good insurance against overstocking. I would not pretend to know how taxation laws can be altered so more flexible operations are possible. It is a job for those more expert, but I think our legislators need a different approach to taxation of arid zone properties.

CONCLUSIONS

I believe that grazing in the arid zone will always be a risky business, but it need not be as risky as it is.

We need a much better understanding of the vegetation of the arid zone. In the past we have relied on experience. I would be the last to underrate its value, but we should not be forced to rely upon it to the extent that we do. It is, after all, largely the sum total of past mistakes.

In summary, I think we can state seven fundamental requirements for continued success of the pastoral industry in the arid zone. They are:

An investigation with the object of determining what area of land is required under differing conditions to provide an economic unit.

A greatly expanded knowledge of the capabilities of the vegetation on different land types and rainfall patterns, a basic requirement in erosion control.

Increased understanding of the grazing habits of sheep and cattle. This is an urgent necessity if we are to practise efficient land management.

A better system of taxation, which will allow the arid zone grazier to build up a financial reserve.

A means of obtaining long term finance at a lower rate of interest than is now possible.

The provision of more and better roads and other transport which can

(a) lower production costs;
(b) bring agistment areas within reach;
(c) provide a means to dispose of excess stock in times of drought.

A co-ordinated approach to our problems with the objectives of retaining the productive capacity of the land.

We should remember in dealing with our problems that we are also dealing with people, who have an inherent desire to be where they are, doing what they are doing, and that they will make many sacrifices to do just that.

I do not believe that Australia, with an arid zone as large as it has, can afford to neglect it. In the past it has made a considerable contribution to our economy, and, given a better knowledge of our problems and a more enlightened approach from our legislators, I am confident that it will continue to do so.

9

THE ROLE OF TRANSPORT IN THE UTILISATION OF ARID LAND RESOURCES

W. J. RICKETTS*

THIS CHAPTER looks at the role of transport in the utilisation of Australia's arid land resources: it comments on transport as a factor of production in the arid regions and draws attention to some aspects of the evaluation techniques used to assess the merit of public investment in additional transport facilities in the arid regions. Its theme is that new transport development together with freight rate policies applied to existing transport systems will continue to be, for some years, among the principal means available to governments to assist improved resource utilisation in the arid land regions.

It is implicit in the chapter that improved resource utilisation in the arid areas will result, generally, in a lessening of the severity of the impact of climatic fluctuations on stock numbers and output and (over time) raise the level of production;[1] that government measures on new transport investment and freight rates may be justified on economic and other grounds from the national point of view; and that the response actions of producers will be economically sound.

RECENT TRANSPORT DEVELOPMENTS

Major advances in the transport field have been achieved both within the arid land regions and between these regions and their marketing outlets[2] during the past decade: new roads have been constructed, other roads have been improved, and the railway network has been upgraded. Decisions have been made in the past year or so for further works to be put in course (see Appen-

* Department of National Development. The author acknowledges the contribution of officers of the Northern Division and in particular of Mr A. G. Bennett in the preparation of this paper.

1 Improved resource utilisation, however, does not necessarily result in increased production in the short term. It may be more efficient to hold or even reduce stocking rates in some pastoral areas because of the pressures on native pastures; this could result in extending the productive life of resources.

2 For beef cattle, such marketing outlets would include meatworks, saleyards, and fattening centres in the non-arid areas.

dix). Governments and authorities are constantly examining prospects and proposals for further transport projects.[3]

Advances have also occurred in the private sector. There have been numerous studies of investment and operating costs on pastoral properties and of developments on the part of transport operators, including those by Allen (1960), Westerman (1966), Keating (1967), Moulden and Jenkins (1968), Bureau of Agricultural Economics (1958, 1959, 1960).

The investment in pastoral property improvement should enable producers to withstand the pressures of climatic fluctuations better and assist them towards changes in the organisation of their production and marketing practices aimed at ultimately raising output. Rapid expansion has occurred in the tourist industry, and developments in the mineral industry have been spectacular. In the transport field, private operators have designed and brought into use larger and more efficient vehicles, including articulated road trains. Cattle producers have shown a marked preference for road transport vehicles in the place of droving.[4]

The investigations that precede government decisions to invest in such transport facilities as beef roads and railway improvement include the prospective productivity of the zones of influence (with and without the transport facility) and the market outlooks for their products. These investigations also need to raise for government consideration aspects of development policy not amenable to quantitative analysis. Economic evaluation inevitably involves estimation and prediction to returns and costs extending far into the future and it raises questions of time preferences and alternative uses of capital. Papers on the application of benefit-cost technique have been prepared by Patterson (1961), van Holst Pellekaan (1964) and Commonwealth Treasury (1966).

Despite the estimation and prediction entailed, benefit-cost analysis is now recognised as a useful tool in determining priorities for major government investment and has been used in a number of instances for evaluation of agricultural and pastoral projects (e.g. B.A.E., 1964a, 1964b, 1965).

Existing transport facilities and the freight rates being charged are important factors, too, in the decisions on resource utilisation made by rural pro-

[3] Transport projects affecting the arid regions of Australia include: the east-west railway standardisation and improvement project; the use of nuclear explosives for a new port at Cape Keraudren, W.A.; a feasibility study for re-development of the port of Darwin; road development within the Northern Territory; land communication between Alice Springs and Port Augusta; railway and port development in Queensland associated primarily with coal production and export; and examinations leading to a new Commonwealth Aid Roads Program.

[4] The Department of National Development data on road transport of cattle in the more remote areas of Australia over the past ten to fifteen years indicate that road transport had expanded from a very low level of activity less than ten years ago to a point where producers prefer this method.

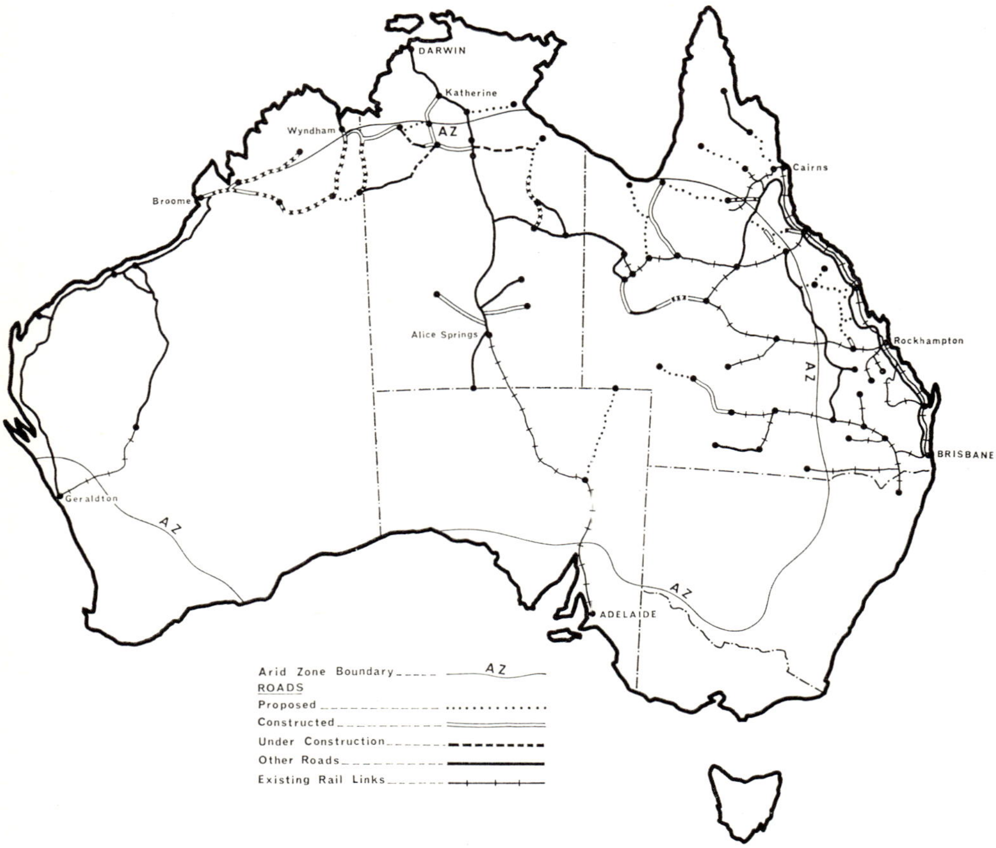

Fig. 1 Beef roads in Australia's arid lands (roads partly constructed and under construction shown thus: xxx)

ducers in the private section. Since the major trunk transport systems in Australia are government-operated and rating policies are often influenced by other than financial criteria, these rates can influence the resource allocation patterns of producers and private operators.

The arid regions (Fig. 1) cover an immense area, taking in large areas of all of the states other than Victoria and Tasmania. Notwithstanding wide divergences in many important features, there is a common land use characteristic—the dominance of pastoral production. Although there are areas of agriculture, these are restricted mainly to places where irrigation is feasible and potentially viable economically and to some wheat-producing areas in the south. Mining—of extreme economic importance in the arid regions—occupies a very limited area compared with the pastoral industry.

TRANSPORT AND THE PASTORAL INDUSTRY

A good transport system is essential for efficient resource utilisation anywhere; it is crucial in the arid regions where management strategy may involve the movement of large numbers of stock over considerable distances sometimes at short notice.

Some characteristics of the pastoral industry illustrate this point.

- The productivity of the arid zones is low, so that property areas are large and stocking capacity low, and so transport and communication are basic needs.
- The potential of arid lands for rural production is limited by low rainfall. Thus pastoral production will continue to be in the form of extensive range management and producers must pursue management systems which will preserve natural pastures. In general, opportunities to raise safe stocking rates significantly are limited.
- Low branding rates, insufficient culling, poor growth rates, and high mortality rates affect production and this is accentuated by inability to market young cattle and by problems of integration with fattening areas if transport is inadequate. As a result, turn-off in some areas is as low as 10 per cent of the herd. Much higher rates (up to 20 per cent and more) are obtained in similar areas where producers have invested in property improvement, have made organisational changes, and have transport available to them.
- The relatively slow development in some of the arid areas is partly attributable to the difficulties and high costs of transport, for the properties need inward movement of substantial quantities of heavy equipment such as fencing materials, building materials for yards, sheds, dips and accommodation, equipment for water facilities such as piping, troughs, windmills, pumps and engines, materials for dipping cattle, and fuel. Further, the existing transport network and its limitations are such that producers frequently have to obtain goods by more than one mode of transport involving a number of trans-shipment points.
- Because of the low productivity ceiling and the great distances involved, unit costs of transport are a relatively large proportion of pastoralists' production costs. Accordingly, producers must seek reasonably efficient transport systems with low unit charges in a situation where transport operators incur relatively heavy operating and maintenance costs because of low-standard roads and access tracks, limited availability of spare parts, and severe climatic conditions. Moreover, transport vehicles generally have unused capacity since the movement

of cattle coupled with seasonality of operation is predominantly one way. (Under state transport regulations, long articulated cattle trains are generally not permitted on the roads in more populous areas.)

The cattle industry in the arid areas, then, is based largely on a system of grazing unimproved pastures on large holdings, on many of which there is still scope for further improvement by fencing and waters and by organisational changes. In general, productivity per acre or per head is low, and there are transport problems.

It is suggested that greater stability and increased production can be achieved, with existing technology, in the following respects:

- General property improvements and some changes in management methods to make optimum use of existing native pastures and reduce wastage levels.
- Breeding, selection and (where practicable) introduction of cattle better suited to the environment.
- A greater degree of specialisation into breeding, or breeding and fattening, enterprises and integration with fattening regions in non-arid areas.
- Adoption of organisational practices designed to iron out, so far as is practicable, the effects of climatic fluctuations but at the same time to take the fullest advantage of 'opportunity' country.[5]

Transport is involved in each of these—in the provision (largely by governments) of transport facilities, in operators charging acceptable freight rates, and in producers reacting by carrying out property improvements and management adjustments which make use of the transport facilities.

TRANSPORT NEEDS

The Committee of Investigation into Transportation Costs in Northern Australia (1965b) reported that 'pastoralists have to pay up to 30 per cent of gross proceeds to move their cattle out to marketing or processing centres' (see Fig. 2). It also reported on the magnitude of the direct transport costs for inwards stores. Surveys of the economic and financial position of cattle producers carried out by the Bureau of Agricultural Economics provide further particulars of transport and marketing costs incurred by pastoralists.

[5] I.e. country of the kind found in south-western Queensland and north-eastern South Australia, which, from time to time following rains either locally or in the headwaters of its rivers and channels, possesses an abundance of good native pasture suitable for fattening store cattle brought in from other areas to take advantage of this relatively temporary situation.

Fig. 2 Percentage of gross proceeds taken by transport costs—northern Australia (Cattle sold to nearest meatworks—September 1964)

Predictions have been put forward from time to time of the future beef cattle turn-off that will result from property improvement and industry re-organisation, given favourable markets and adequate transport facilities (e.g. Queensland Royal Commission, 1946; Bureau of Investigation, 1947; Kelly, 1959, 1966). Without attempting to predict future turn-off it is possible to calculate the effect of particular improvements.

There would be an increase in the demand for cattle transport facilities if one or both of the following situations eventuate:

- A significant reduction in the present high wastage rates. With no change in stocking rates (and thus in herd numbers or in the pressure on native pastures) but with modest improvements in branding, culling, mortality, and marketing age, the volume of turn-off requiring

L

transport facilities could rise from an estimated present figure of 550,000 to approximately 700,000 annually. This would require transport facilities, acceptable freight rates, satisfactory market prices, and producers would need to make moderately positive adjustments to their production and marketing policies. In all, this situation could be regarded as short-term, i.e. capable of being achieved fairly soon after, say, road construction.

- A reorganisation of the cattle industry resulting in more integration between production areas. In the arid and semi-arid areas suitable for the purpose, breeding and young stock (together with some steers) would impose no more pressure on native pastures but would be a source of supply to fattening regions from which there is now and is likely to be for some considerable time in the future a very strong demand for store cattle. In this situation, coupled with the 'immediate' type of improvement described above resulting from better branding rates etc., the estimated present turn-off figure of 550,000 could rise to over one million annually subject to certain assumptions. It might also be argued that an organisational change of this kind would tend to ensure added stability in an environment characterised by severe seasonal fluctuations.

It is imperative, however, to keep in mind that the construction of a new road or railway cannot in itself improve the basic stocking capacity of the country within its zone of influence; the impact is essentially of an organisational kind.[6]

The broad pattern of cattle movements in Australia is illustrated in Fig. 3, which shows the volume of movements, direction of flow, and methods of transport used in 1964.

The great bulk of the movement is to the meatworks: if anything, this feature has been accentuated further since 1964. Store cattle movements in the Kimberleys are few but may increase, locally, with the development of irrigated agriculture at the Ord. In eastern Australia the flow is heavily eastwards to meatworks at or adjacent to the coast. Store cattle movements have been inadequate to satisfy the demand for stores in the regions suited for fattening such as along the coast, in the brigalow area, in the south-east, and (in good seasons) the Channel Country. In the Northern Territory there has been a pronounced increase in road transport to the railhead at Alice Springs. In New South Wales, Victoria, and the southern areas of South Australia, the increases in local beef cattle numbers, and in particular in breeding stock numbers, point

[6] One feature is that a producer in a drought-prone area might be able to stock closer to capacity, in the knowledge that efficient transport was readily available to move stock out rapidly if need arises.

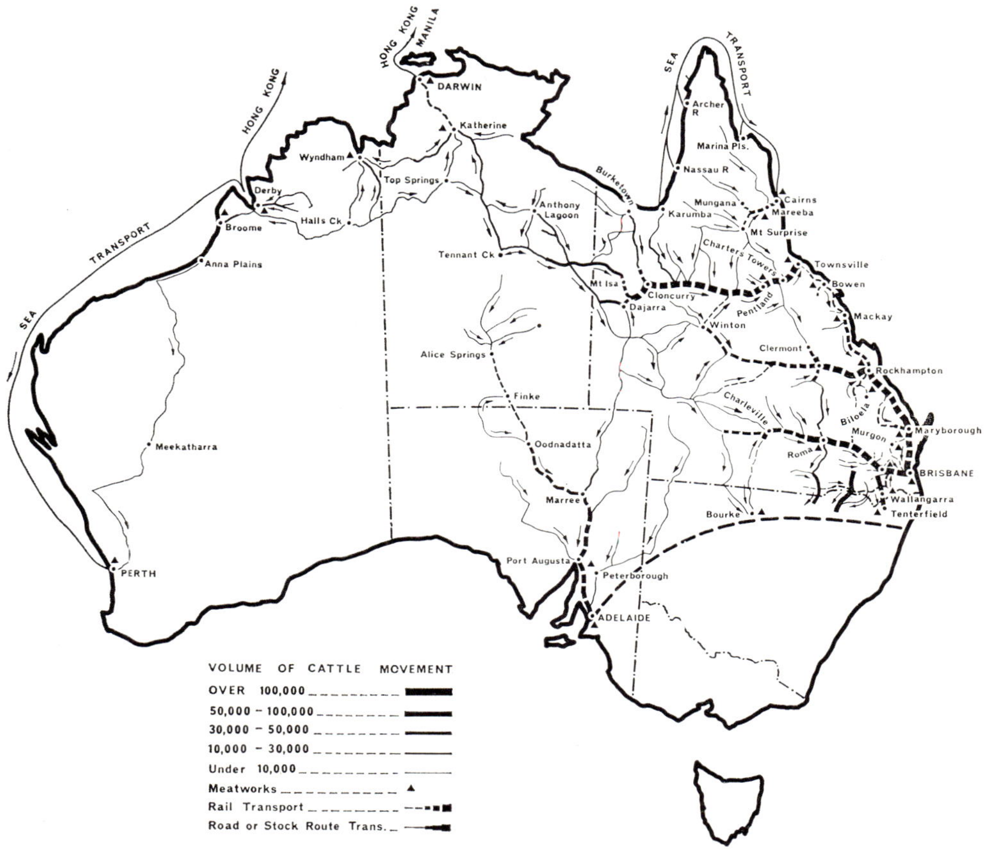

Fig. 3 Principal cattle movements—northern Australia, 1964

up the inability of producers in those states to obtain sufficient store cattle from the arid areas of those states and further afield.

There are several serious practical obstacles, other than transport, to integrating cattle production regions. Briefly summarised, these are:

- Quarantine restrictions impose barriers some of which cannot be passed, some of which can be, but only by procedures involving time and costs.
- Fatteners' attitudes are highly relevant. While attracted by the possibilities of integration, they may be inhibited by the risks involved: in the large financial outlay required, in estimating the market (at time

of purchase and predicting it when the fat cattle will be turned-off), in gauging the seasonal and pastoral conditions during the intended fattening period; and there sometimes is a problem of contracting with distant suppliers. Nonetheless, there is a strong demand for store cattle.

- Climatic conditions fluctuate and an area generally suited to breeding and the sale of stores may be suited, on occasions, to marketing fat cattle.

THE EFFECT OF IMPROVED TRANSPORT

The absence of transport facilities in a remote area has restricted producers to droving cattle to market and has inhibited property development. Given ready access to transport, property improvements and organisational changes can be expected.

The immediate gains are likely to be:

- Turn-off of cattle over a longer period of the year to meet the needs of producers in fattening areas who buy stores; turn-off of small mobs of cattle mustered after the main mobs have been moved; turn-off of old cows now left to die on properties; turn-off of cattle at a younger age, thus increasing the overall turn-off rate.
- Progressive investment in property and organisational adjustments to husbandry methods and marketing practices and policies.

Reactions of Producers

Availability of transport facilities will not, in itself, lead to increased productivity from arid lands: it will make possible and provide an inducement for changes on the part of producers to many aspects of property management. The reactions of producers normally will depend on their assessment of whether changes which have become practicable will in fact be profitable. They will also bear in mind other factors outside the scope of this paper, such as land tenure conditions.

It is not difficult to illustrate the apparent attractiveness of some changes involving the pastoralists' use of transport facilities. Some instances, in assumed situations, are given below:

- While the immediate cash outlay for droving is relatively small in comparison with other transport charges, the total costs to the pastoralist and the industry are actually much heavier. As an illustration, the haulier's charges for road-transporting the turn-off of 600 fat cattle over a distance of 200 miles would be approximately $4,200 (at 70 cents per K wagon/mile). If it were necessary to drove the animals, losses of the following order could commonly occur—

Deaths in transit (say 1 per cent) of beasts with an export beef value of	$500
Weight losses (594 at 50 lb.) with an export beef value of	$3,900
	$4,400

In this illustration the road haulier's charges would be exceeded by the losses averted.

- The added value of beef from old cows that are unable to walk long distances can be substantial. As an illustration, on a breeding and fattening property in a remote area with a herd of 5,000 and a present turn-off rate of 11-14 per cent, the added value could be assumed as about 160 animals involving some 30 tons of beef worth around $10,000 to the export trade.
- In the event of a change-over to breeding entirely for sale of young stores, turn-off can be assumed to rise from around 11-14 per cent to as much as 25-30 per cent of herd numbers. As an illustration, a turn-off of 11 per cent from a herd of 5,000 would result in an output of 550 head and beef production of approximately 120 tons with an export value of about $45,000. If there were a change as complete as that mentioned, output would rise possibly to some 1,500 animals and at store prices could involve a return of some $65,000.

Illustrations of the relative profitability of fattening and breeding and fattening are also given in a number of economic studies (e.g. van Holst Pellekaan and Robinson, 1964).

To ensure that changes in management systems are economically sound in the longer run, particularly with respect to large-scale expenditures on property improvements, producers may be expected to refer increasingly to methods of investment analysis, including types of 'benefit-cost' analysis at the property level. Investment policy is one of the most crucial areas in managerial decision-making. The problem confronting a grazier in the arid areas is whether the increased stability and productivity resulting from additional on-property investment in waters, fences, yards, improved breeding stock, additional labour, and plant would, over a period of years, justify in financial terms the program of development whose prospects were enhanced by the provision of improved transport facilities.

Clearly no decision-making on the investment value of effective transport systems is directly required of the individual grazier. Rather, the provision of improved transport facilities and the determination by transport operators of acceptable freight rates have the effect of modifying the economic climate within which he will be making his longer-run investment decisions.

Governments can play an important additional role with regard to freight rate structures. Through freight rate policies, governments are in a position to provide incentives for increased use of these facilities and thus influence the resource allocation patterns of producers and private operators.

Three examples of government action aimed at reducing the impact of freight costs in the fairly recent past are:

- To encourage the use of superphosphate, which is very severely affected by transport costs, the Commonwealth Government pays a per-ton subsidy to enable primary producers in the Northern Territory to obtain their superphosphates in Darwin at Townsville prices. In the period of the subsidy, the use of superphosphate in the Territory has increased more than threefold.
- To encourage herd improvement, the Commonwealth Government pays a subsidy, based on the value of the animals, on the introduction of good breeding stock into the Northern Territory. It subsidises railway, ship, and road charges, and droving and handling charges between the place of purchase and the destination of the animals in the Territory. This policy has contributed towards much greater imports of breeding stock into the Territory.[7]
- To assist producers in their drought strategies, the Commonwealth and states from time to time subsidise rail freights for moving starving stock to and from agistment pastures and for importing drought fodder into the affected areas. State governments also subsidise freight charges on cattle imported for post-drought restocking.

As Campbell (1966) noted, freight concessions for the transport of cattle (and fodder) in stress periods play a significant part in the drought strategies of producers by:

- Helping to prop up the current viability of the pastoralists concerned.
- Encouraging further investment in remote areas.
- Reducing mortality rates and thus increasing production levels.

[7] The Loder Committee Report suggested that the practice of subsidising freight on good breeding stock might well be applied in principle to the north of Western Australia and the far north and west of Queensland.

EVALUATION OF TRANSPORT SYSTEMS FROM THE NATIONAL VIEWPOINT

A government is faced with a wider range of investment issues than a grazier; it has to look to a multiplicity of objectives, many of them being far less tangible than the profit-type aims of the investor in the private sector. In the context of investment in transport facilities for pastoral areas in Australia, benefit-cost analysis has been used largely by the Commonwealth Government to assist in ordering priorities within a national program of beef road construction.

Despite the intrinsic merit of various requests, only a limited number can be accommodated within the funds available. The merit of expanding transport facilities in arid lands has therefore to be weighed against the economic and other merit of increased expenditure on defence, education, water conservation, health, etc. Having accorded a priority to transport facilities in arid lands and earmarked a portion of the national budget for this purpose, the government can then be further assisted by evaluation analysis in reaching its decisions on the claims of various regions for shares of the funds voted, on the priorities for roads within particular regions, and on the standard and rate of construction justified for particular roads.

The identification of benefits and costs associated with investment proposals in the public sector is not always straightforward, since the relatively simple criterion of profitability is not sufficient: the effects of the proposed investment on activity in the economy as a whole have also to be taken into account. For example, the construction of a road through some arid region may:

- Provide an aid to stability in stock numbers and production; accord producers the advantage of using 'opportunity' country on occasions; stimulate a positive and viable increase in beef production in that area.
- Permit tourists to visit a previously inaccessible point of interest; reduce the real costs of the transport of other products such as minerals; confer income stability on a remote settlement by providing a market for petrol, refreshments, and Aboriginal artifacts; or, on the other hand, reduce the traffic and profitability of a section of railway.

A further aspect which can cloud the issue of defining relevant benefits and costs is that some may have definite regional significance and yet have no net national significance. For example, two roads in different regions may give rise to similar external or intangible effects by way of increased tourist traffic. From the point of view of the respective regions such an increase in tourist traffic is important and valuable; from a national point of view, the relevant consideration is whether a net increase in tourist traffic is generated (e.g. from overseas) resulting from either road proposal; which road is used and which region is visited by these tourists is immaterial. Where regional development—for some reason of policy—is an over-riding objective, however, it is essential to incorporate such matters into the analysis.

Benefits and costs must be quantified as well as identified. This is usually done in monetary terms and the market mechanism normally provides a guide by pricing them. Since imperfections in the market such as monopoly, marketing controls, and subsidies are common, market prices may have to be adjusted before they are used.

Sometimes a commodity is not traded in a commercial market and no price is available as a guide to its value. For example, a benefit in a form of reduced road fatalities resulting from better road alignment, or a cost in terms of reduced aesthetic value of a national park marred by major road works, are intangibles having no market price and their incorporation in a quantitative analysis would require that monetary values be imputed. On the other hand, if left to the policy-interpretation stage, cognisance may be given in qualitative terms.

Investment in transport facilities, like many other investments in the public sector, are long term and considerations of risk and uncertainty are of particular importance. Estimates of both benefits and costs can go awry through changes in technology, the demand for goods and services and supplies of factors of production. A number of methods, of varying sophistication, have been suggested for taking into account risk and uncertainty. In practice, the commonly adopted procedures are to outline the possible contingencies for consideration by the decision-maker along with other evidence; to estimate benefits and costs under a number of sets of specified conditions; or to adopt conservative estimates of benefits and include contingency allowances in costs.

The final ratio in benefit-cost analysis should not be the sole determinant of public investment decisions. Other criteria are not only desirable but also necessary to give a balanced picture of proposed projects, including transport facilities in arid lands. In this regard the Vernon Committee (1965) concluded:

> The form of the analysis should not be allowed to endow it with an aura of accuracy not warranted by its underlying assumptions on such matters as likely technological developments and markets. Moreover the initial derivation of an unfavourable cost-benefit ratio for a project should not necessarily be regarded as a final condemnation. The community may sometimes choose to develop resources for a number of reasons, other than purely economic, even when cost-benefit figures are unfavourable. Projects may be embarked on for reasons of their social desirability, or for defence purposes, including such long-term defence purposes as are thought by some to be served by settling Australia's 'empty spaces', or for reasons of 'State sentiment'.

APPENDIX

Although there had been Commonwealth assistance to the beef industry to the extent of \$4·3m. for upgrading of roads between 1949 and 1954, it was in 1961 that the Commonwealth initiated the 'Beef Roads Program' by approving grants of \$10m. and \$1m. to Queensland and Western Australia respectively. In addition a program estimated to cost \$9·14m. was announced for construction of beef roads in the Northern Territory.

Further assistance was provided subsequently, and at 30 June 1968 the expenditure that had been incurred on beef roads in the previous seven years was:

	Commonwealth Funds	State Funds	Total
	\$	\$	\$
Queensland	24,241,000	2,309,000	26,550,000
Western Australia	9,801,000	8,978,000	18,779,000
Northern Territory	14,216,000	—	14,216,000
Total	48,258,000	11,287,000	
Grand Total			59,545,000

Altogether thirty-two roads have been approved as 'beef roads' and the following table sets out the present (February 1969) status of each:

Roads Completed to Sealed Standard

Georgetown-Kennedy Highway	Qld	92 miles
Boulia-Dajarra	,,	92 ,,
Julia Creek-Normanton	,,	270 ,,
Quilpie-Windorah	,,	153 ,,
Katherine-Willeroo-Top Springs	N.T.	180 ,,

Roads Completed to Gravel Standard

Stuart Highway-Yuendumu	N.T.	168 miles
Stuart Highway-Plenty River	,,	128 ,,

Roads In Course of Construction

Mount Isa-Dajarra	Qld	96 miles
Currawilla-Windorah	,,	105 ,,
Winton-Boulia	,,	231 ,,
Mareeba-Laura	,,	167 ,,
The Lynd-Charters Towers	,,	160 ,,
The Battery-Townsville	,,	72 ,,
Mt Douglas-Collinsville	,,	120 ,,
Dingo-Mt Flora		149 ,,
Oxford Downs-Mackay	,,	63 ,,

Dunmarra-W.A. Border	N.T.	364	miles
Anthony Lagoon-Borroloola	,,	162	,,
Top Springs-Wave Hill	,,	100	,,
Daly Waters-Cape Crawford	,,	167	,,
Barkly Highway-Anthony Lagoon	,,	140	,,
Duncan Highway	W.A.	335	,,
Great Northern Highway (Broome-Wyndham)	,,	713	,,
Derby-Gibb River	,,	232	,,

Roads Not Started

Cloncurry-Burketown	Qld	270	miles
Normanton-Georgetown	,,	190	,,
Mungana-Highbury	,,	115	,,
Mt Coolon-Nebo	,,	110	,,
May Downs Road	,,	35	,,
Willeroo-Timber Creek	N.T.	98	,,
Mataranka-Roper Bar	,,	108	,,
Birdsville-Marree	S.A.	306	,,

At 30 June 1968 a total of $44·8m. of Commonwealth funds was available for spending on roads under construction and roads approved as beef roads in Queensland, Western Australia, and South Australia over the following six years. In addition it was planned to spend $15·8m. over about five years on beef roads in the Northern Territory.

Although not under any agreement to finance roads themselves, the States have indicated their intention to spend from their own resources over the seven year period 1967-74 as follows: Queensland, $7·004m.; Western Australia, $6·619m. and South Australia, $2·0m.

Railways

Prominent among improvements in the railway system have been the rehabilitation of the Mount Isa line and the standardisation project to provide a uniform gauge railway across Australia.

Completed in 1966, the Mount Isa project involved reconstruction of 603 miles of track, from Townsville to Mount Isa, and 161 miles from Collinsville to Stuart. The work was carried out at a cost of $52m. to facilitate carrying of products and to handle the increased output of Mount Isa mines eastward and coal from Collinsville to Mount Isa.

The standardisation project was commenced in 1962 to provide an unbroken 4ft 8½in. gauge over 2,440 miles from Sydney to Perth. The proposal was also linked with the establishment of an iron and steel industry at Kwinana, the new railway facilities being used to transport iron ore from Koolyanobbing.

The project, which is well advanced, will cost more than $200m. and will enable passenger trains to complete the trans-continental journey in slightly over sixty hours.

Standard gauge trains have been operating since late 1967 on the Western Australian section carrying grain and iron ore.

The entire project including new rolling stock is expected to be operating in 1970.

REFERENCES

Allen, W. T. (1960). The road transport of beef cattle in the Northern Territory: A case study. *Q. Rev. agric. Econ.* **13**: 164-72.

Bureau of Agricultural Economics (1952). *The beef cattle industry in Northern Australia.* Canberra, A.C.T.

——— (1958). *The economics of road transport of beef cattle, Western Australian pastoral areas.* Canberra, A.C.T.

——— (1959). *The economics of road transport of beef cattle, Northern Territory and Queensland channel country.* Canberra, A.C.T.

——— (1960). *The economics of the development of road transport of beef cattle, far north of South Australia.* Canberra, A.C.T.

——— (1964a). *The economics of brigalow land development in the Fitzroy Basin, Queensland.* Canberra, A.C.T.

——— (1964b). *Economic analysis of road development in the Northern Territory buffalo area (Beatrice Hill to Oenpelli Mission).* Canberra, A.C.T.

——— (1965). *Extension of the comprehensive water supply scheme (agricultural areas of Western Australia): An economic evaluation.* Canberra, A.C.T.

Bureau of Investigation (1947). *The channel country of south-west Queensland with special reference to Coopers Creek.* Department of Public Lands. Brisbane, Qld. Technical Bulletin No. 1.

Campbell, K. O. (1966). Problems of adaptation of Pastoral Businesses in the Arid Zone. *The Australian Journal of Agricultural Economics.* **10**: 14-26.

Commonwealth of Australia (1965a). *Report of the Committee of Economic Enquiry* (Vernon Committee).

Commonwealth of Australia (1965b). *Report of the Committee of Investigation into Transportation Costs in Northern Australia.*

Commonwealth Treasury (1966). *Investment Analysis, Supplement to Treasury Information Bulletin.*

Keating, C. (1967). The Queensland beef cattle industry. *Q. Rev. agric. Econ.* **20**: 192-203.

Kelly, J. H. (1959). *The beef cattle industry in the Leichhardt-Gilbert region of Queensland.* Bureau of Agricultural Economics. Canberra, A.C.T.

——— (1966). *Struggle for the north.* Australasian Book Society, Sydney.

Moulden, J. O. and Jenkins, E. L. (1968). Road transport of beef cattle in Northern Australia. *Q. Rev. agric. Econ.* **21**: 87-100.

Patterson, R. A. (1961). Application of benefit-cost analysis to development investment in cattle roads. *Q. Rev. agric. Econ.* **14**: 160-72.

Royal Commission appointed to inquire into proposals for the establishment of additional abattoirs or meatworks and matters pertaining to the welfare of the livestock and meat industry (1946) *Report.* Brisbane, Qld.

van Holst Pellekaan, J. W. (1964). The application of benefit-cost analysis to the evaluation of brigalow land development. *Q. Rev. agric. Econ.* **17**: 14-23.

van Holst Pellekaan, J. W. and Robinson, V. (1964). Property development in the brigalow area of the Fitzroy Basin, Queensland. *Q. Rev. agric. Econ.* **17**: 154-71.

Westerman, P. A. (1966). The Kimberley beef cattle industry. *Q. Rev. agric. Econ.* **19**: 206-16.

Vernon Report, *see* Commonwealth of Australia (1965).

10

CONSERVATION AND RECREATION IN ARID AUSTRALIA

A. B. COSTIN* AND J. G. MOSLEY†

CONSERVATION OF the arid zone—in the sense of management of its resources to ensure long-term benefits for man—is an underlying theme of this symposium. Major aspects have been covered in earlier chapters. The topics we shall consider are:

Nature conservation
Aboriginal sites and reserves
Recreation and tourism
Research problems in the conservation of grazing lands
Alternative land uses and uncommitted lands

NATURE CONSERVATION

There is increasing acceptance, both in Australia and internationally, as through the International Biological Program, that representative reserves of the main plant and animal communities should be established and conserved throughout the country. Such reserves have value as biological standards and reference areas, as breeding reservoirs for more mobile species, as gene pools for future plant and animal improvement programs, as sources of biologically active compounds, and for recreation.

Considerable progress has been made in the last few years in most Australian states and territories towards the establishment and extension of national parks and flora and fauna reserves. Figure 1, based on Mosley's recent (1968) survey of *National Parks and Equivalent Reserves in Australia*, shows the distribution of reserves in the arid zone. Table 1 gives further information.

These reserves, some thirteen million acres in total, represent about one

* CSIRO Division of Plant Industry, Canberra.
† Australian Conservation Foundation, Canberra.

Advice and criticism are gratefully acknowledged from Messrs J. H. Calaby, CSIRO Division of Wildlife Research, Canberra; D. Howell, School of Civil Engineering, University of New South Wales, Kensington; J. P. M. Long, Department of the Interior, Canberra; and D. J. Mulvaney, Department of Pre-history, Research School of Pacific Studies, Australian National University, Canberra.

per cent of the Australian arid zone. More than nine-tenths is contributed by the Tanami Desert Sanctuary and the Simpson Desert National Park, with huge areas of sandhill and sandplain country.

Most of the reserves have been chosen largely on a scenic basis or simply because they were available and, although some major habitat types are well represented, the coverage is by no means complete. For example, the wide-

TABLE 1

PARKS AND RESERVES IN ARID AUSTRALIA AT 30 JUNE 1968*

State	Name	Area (acres)
Queensland	Simpson Desert National Park	1,248,000
New South Wales	Kinchega National Park	96,700†
	Mootwingee Historic Site	1,200
South Australia	Simpson Desert National Park	1,708,800
	Elliot Price Wilderness National Park	160,000
	Yumbarra National Park (in part)	262,000
Western Australia	Barrow Island Reserve	50,000
	Dales Gorge Reserve	58,400
	Cape Range National Park	31,171
	Bernier and Dorre Islands	26,000
	A28220 (Carnarvon)	1,293
	A24392 (Fortescue River)	1,090
	A19145 (Carnarvon)	130
	A19144 (Carnarvon)	120
Northern Territory	Tanami Desert Sanctuary	9,273,600
	Ayers Rock-Mt Olga National Park	311,680
	Palm Valley Flora and Fauna Reserve	113,280
	Ormiston Gorge and Pound	11,520
	Devil's Marbles Reserve	4,519
	Trephina Gorge Scenic Reserve	4,378
	Emily and Jessie Gap	1,718
	Green Valley Scenic Reserve	1,239
	Alice Springs Telegraphic Station National Park	1,096
	Glen Helen Gorge National Park	954
	Simpsons Gap National Park	640
	Henbury Meteorites National Park	40
	Corroboree Scenic Reserve	18
	Heavitree Gap Police Station	1
	John Flynn's Memorial	1
	John Flynn's Grave	—
	Central Mount Stuart Memorial	—
	Attack Creek Memorial	—

*Listed are national parks, fauna and flora reserves, fauna sanctuaries, historic reserves, etc., with relatively secure tenure. The reserves included have either been reserved or may be revoked only by Parliament, or are under the control of a special reserve authority.

†At December 1968.

spread and important Mitchell grass communities are not extensively protected in any reserve, and at least one community—of old man saltbush (*Atriplex nummularia*)—has now virtually disappeared as the result of grazing. With the latter exception, however, sufficiently large areas of the major plant communities are still available, at least in a semi-natural condition, for the deficiencies in the present reserve system to be made good. Most states are now working in this direction, assisted by the scientific surveys of vgetation resources and conservation requirements, as carried out by the Australian Academy of Science (1969), Specht (1961), Frankenberg (1967), Hayden (1969) and others. For instance, the Western Australian Sub-Committee (1962) of the Academy Committee has made recommendations covering more than ten million acres of arid lands in various parts of the state. Despite the large amount of habitat modification which has occurred in the arid zone, there is

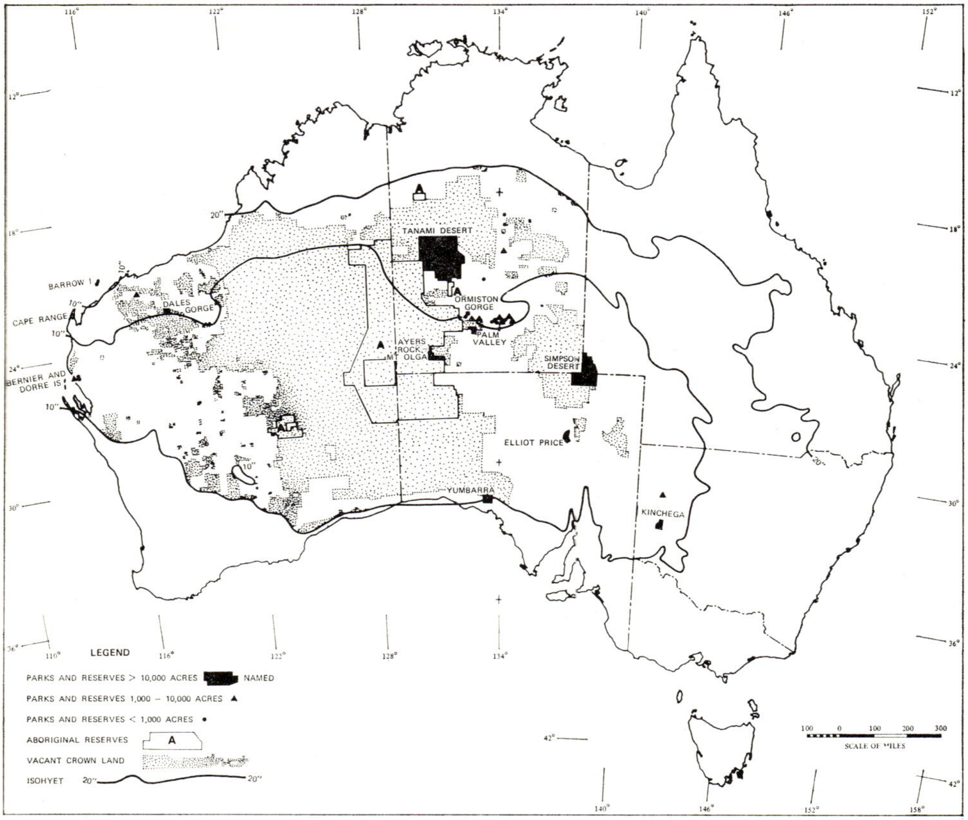

Fig. 1 Australia: Parks and Reserves in the Arid Zone

still time for an arid zone parks and reserves system in Australia second to none.

The situation with regard to smaller areas and minor communities of special scientific concern (e.g. Palm Valley), and individual species of flora and fauna, is not so clear. There has been widespread and, in places, severe modification of the shrub and herbaceous components of the vegetation as the result of grazing, although, as far as is known, actual losses of species of the arid zone flora have not occurred. Some species of fauna have become increasingly rare during the period of grazing by livestock or rabbits. However, cause-and-effect relationships are not always clear. For example, the night parrot (*Geopsittacus occidentalis*), known from specimens from inland Western Australia, Central Australia, and South Australia, has not been collected since the 1880s. Lesueur's rat kangaroo (*Bettongia lesueur*), once common right across the continent, is now known only from Bernier and Dorre islands in Western Australia. A burrowing species, this rat kangaroo may have been ousted by the rabbit (even where there were no domestic livestock) through direct competition for burrows and burrowing sites (Calaby, pers. comm.). By contrast, the red kangaroo (*Megaleia rufa*) and the euro (*Macropus robustus*) have been favoured in places by livestock grazing, through changes in the vegetation from woody to herbaceous species (Frith, 1964; Ealey, 1967), and perhaps by the development of stock waters. Some species appear to have very restricted occurrences, possibly requiring precise conservation measures, which may not be feasible. For instance, the Eyrean grass wren (*Amytornis goyderi*) is known only from one small area near Lake Eyre, having been collected there in 1874 and not seen again until 1961 (Calaby, 1969). The conservation status and requirements of many species and communities must remain uncertain until more detailed biological surveys and ecological research have been carried out. The CSIRO Division of Wildlife Research has plans to obtain some of this information for arid zone fauna. There should be a corresponding effort with respect to the flora and vegetation.

Linked with the need for an adequate reserves system, biological surveys, and ecological research is the requirement for permanent reference areas where the condition and trend of plant communities and soils can be recorded over long periods of time; otherwise it is not possible to attribute the normally wide and often sudden changes in arid vegetation and soils to the appropriate factors of climatic variation, native fauna, livestock—as the case may be. Such information is essential for successful management, whether the resource be wildlife or livestock. The only long-term reference areas for which such data have been collected are at Koonamore (saltbush in South Australia: Hall, Specht and Eardley, 1964), and at Gilruth Plains (Mitchell grass in Queensland: R. Roe, unpublished data), although a start is now being made at a few other centres, notably in the Riverina. Because of the long-term nature of these investigations, questions of methodology, repeatability by changing workers, and availability of original data are particularly important. There is a strong case for a national set of arid zone vegetation standards, under the continuing investigation of a single research group working collaboratively with state conservation agencies.

ABORIGINAL SITES AND RESERVES

The future of the large Aboriginal reserves in the arid zone (Fig. 1) is obscure, especially in the contexts of nature conservation and tourism. These reserves are controlled by the Aboriginal welfare authorities of the appropriate state or territory, the policies of which are generally well integrated and conservation oriented. The Central Australian Aboriginal Reserves in the Northern Territory, South Australia, and Western Australia have an area of some 100,000 square miles and an Aboriginal population of over 2,000. Access to the reserves requires a special permit; scientific collecting is permitted in some, but not in others.

In view of the fact that these reserves contain some of the most scenic and best preserved of the arid lands (e.g. the Petermann Ranges), it is not surprising that there is a growing demand to visit them, and the number of tourists, both authorised and unauthorised, is increasing. Within the next few years the permit system may have to be abandoned as ineffective. The welfare authorities, believing that it is no longer possible for the Aborigines to exist by hunting, have plans to bring cattle into some areas, as a source of food for the Aborigines rather than as a commercial venture. Such developments make it increasingly difficult to view the Aboriginal reserves in isolation. More realistically, the reserves could be visualised as having a key position as part of a huge Central Australian Park, in which the interests of the Aborigines, nature conservation, tourism and recreation are carefully integrated, as outlined in the Academy of Science (1969) report. Management of such a reserve to the international standards it would merit would require special arrangements for Commonwealth participation—as this symposium is also demonstrating in other fields.

Both within the reserves and at hundreds of other localities in the arid zone, there are problems of conserving cave and rock engravings, paintings, stone arrangements, surface scatters of implements, and other Aboriginal antiquities. As Mulvaney (1968) sees it: 'In Australia we must assume custody of over 30,000 years accumulation of antiquities, dotted over a continent whose European population is sparse, [and] whose ethos is both pragmatic and hedonistic.' As well as the more obvious problems of conscious and unthinking vandalism, there are also those of deterioration of sites by natural weathering. There is no simple solution. The protective legislation which now exists in most states should be backed up as far as possible by local rangers, preferably Aborigines where these are available, or by the landholders in the case of occupied land. Laboratory and field experiments in techniques of conservation are an urgent and totally neglected requirement. At certain sites, expert 'touching-up' of deteriorating paintings is possible, preferably by local Aboriginal custodians aware of the special significance of the paintings in question. In the latter context, Long (pers. comm.) believes that Aborigines should also play a key role in the survey of the large, virtually unmapped areas of the Aboriginal reserves, fauna reserves, and national parks, so that features can be identified and named appropriately before their special significance is forgotten.

RECREATION AND TOURISM

According to the estimates of the Harris, Kerr and Forster Report (1965), the number of overseas visitors to Australia will approximately double, from about one-quarter to one-half million, within the next five years, and visitor expenditures will increase from one hundred to more than two hundred million dollars. Tourism is now the largest item in world trade, greater even than the entire international trade in wheat (Piesse, 1968). In Australia, international tourism is the eighth highest earner of foreign exchange. These figures, which do not include those in respect of Australians themselves, give some idea of the rate at which tourist pressures are increasing.

Tourist pressures in arid Australia are increasing similarly, especially in the Centre, which is now a 'must' for most international tourists. It has been estimated that the total number of visitors to the Northern Territory rose from 28,000, spending four million dollars (exclusive of fares to and from the Territory), in 1964 to 41,000, spending thirteen million dollars, in 1966. The majority of these visitors were tourists who spent at least part of their time in the Centre.

For some tourists, the condition of the soils and vegetation is not an important consideration—in fact, the strong red colours of eroded, arid country might heighten its appeal. The most important tourist qualities of the Australian arid zone are probably its remoteness, vastness, and spaciousness. With more than two million square miles of arid country, it might be thought that there was no danger of these qualities being impaired. However, since the tourist also seeks areas of spectacular scenery and easy access, tourist interest is, in fact, concentrated on particular areas such as the Centre and along roads. For example, visitors to Ayers Rock increased from 4,332 in 1960 to 18,963 in 1966, an increase of well over 300 per cent (Piesse, 1968). In such areas, tourist development is now proceeding rapidly, and it is a matter of urgency that the essential qualities of the arid zone are preserved through careful planning and control of tourist operations. For instance, the facilities in Ayers Rock-Mt Olga National Park should have been located well away from Ayers Rock itself; it is to be hoped that the mistakes of the past will not be repeated in the developments now under consideration. Sympathetic treatment is also necessary for historic sites, especially those associated with early explorers such as Sturt, and Burke and Wills, who have an important place in Australian history.

The Harris, Kerr and Forster and the Academy Reports stress the tourist and conservation values of the Centre, and the possibilities of integrating them in the huge Central Australian Park referred to earlier. The Reports suggest further that Aboriginal arts and crafts could be revived and presented as an important part of the tourist industry; and they see the development and management of such a park to international standards as requiring increasing co-operation between the states, and Commonwealth participation.

RESEARCH PROBLEMS IN THE CONSERVATION OF GRAZING LANDS

Even if 10 per cent of the arid zone is reserved for nature conservation, recreation and tourism, the major part—about one million square miles—will continue to be used for grazing, and there is almost as much unoccupied land, of which part could be developed for livestock (Fig. 1). We shall now examine some of the main conservation problems of these areas. The problems selected include: the recognition of key species for grazing and for site stabilisation; the ecology of these key species; and the management of water supplies.

Key Species

For nature conservation purposes, most, if not all, species of plants have value. But in sparse grazing systems this is far from the case. The work of Leigh and Mulham (e.g. 1966) has shown that rangelands grazing, in which overall livestock numbers are low and the livestock are more or less unrestricted, is extremely selective, often involving species that are inconspicuous in the community. For example, in a bladder saltbush-cotton bush pasture in the Riverina, it was found that minor components in the pasture made up to 96 per cent of forage eaten by sheep in spring and up to 72 per cent in summer. One of the first requirements for developing stable systems of livestock husbandry, therefore, is to ascertain which species from the grazing viewpoint should be preserved. So far, these investigations on diet preferences have been carried out for Merino sheep only on grassland, saltbush, and cotton bush pastures in the Riverina. It is important that these studies should be extended—for cattle and marsupials as well as for sheep—into the other main communities of the arid zone.

It is also becoming clear that the species selected by the grazing animal may depend for their existence on micro-site conditions produced or maintained by the more perennial, often woody, and perhaps unpalatable species in the community. Long-term management, therefore, also depends on maintaining an adequate cover of the stabilising species. What is an adequate amount of cover depends on several factors, including the type of vegetation and soil and the nature of the stress conditions against which protection of the soil surface is required. Normally, this is protection from erosion from the strong (rather than average) winds likely to be experienced in a given area during the driest period of the year. The general problem is to define the space-between-plants/height-of-plants relationship of the stabilising species at which erodability in the spaces between the stabilising species is effectively zero. Marshall, in the CSIRO Riverina Laboratory, has been working along these lines initially with vegetation models in a wind-tunnel, then with plant communities in the field. For cotton bush, he finds the critical distance/height relationship is 5:2. It is of great importance that similar relationships be obtained for the other main pasture types of the arid zone, so that 'cover standards' or 'minimum acceptance levels' can be defined as a basis of management. This is an area of research which lends itself particularly well to co-operative Commonwealth/state activity from basic research to field testing and practical application.

Ecology of Key Species

The conservation of the grazing value of arid lands depends in part on recognising for the main pasture types the key grazing and stabilising species. The next step is to understand the ecology of these species to the extent that they can be maintained, increased, or decreased by virtually the only management tool available—the grazing animal, in places perhaps assisted by the occasional use of fire. The water and nutrient relations of key species, and their reaction to defoliation, are priority problems requiring much greater research effort.

Thanks to the work of Slatyer and others, considerable progress has been made in the theory of plant-water relations of dryland species, which can be used in the solution of applied problems. It is not generally appreciated, however, that nutrients may also limit production. According to the work of Charley and Cowling (1968), soil nitrogen and phosphorus may often be limiting factors in the regeneration of arid zone plants, and Main (1968) suggests that productivity of some species of fauna may be limited by nitrogen supply. Charley and Cowling emphasise the extent to which the working nutrient capital of arid soils is concentrated in the surface, with the consequent risk of serious nutrient losses and long-term reduction in productivity, even though soil erosion, as such, might appear to be slight. In so far as the application of limiting nutrients is likely to be uneconomic except in the most productive parts of the arid zone, it is important to know the magnitude of nutrient inputs, losses, and cycling in the main communities, and the extent to which these can be affected by management. Studies of defoliation should define the critical conditions for survival in the seedling, adult, and reproductive phases of the key species under investigation. Hodgkinson's work at the Riverina Laboratory is a step in the right direction.

Management of Water Supplies

Development of additional stock waters in arid lands is often the quickest means of bringing more land into grazing production. Unfortunately, such developments in the form of large tanks and bores are relatively expensive and are fixed to a particular site. Over the years a pattern of concentric overuse towards the watering-point develops, until in 'well-watered' areas, such as parts of western New South Wales, the whole country can only be described as a large scald.

The problem here is to develop low-cost, effectively-moveable watering systems whereby given areas of country can be grazed or protected from grazing at will. One approach is to interconnect bores or tanks with polythene piping from which stock water is turned on or off when, and at whatever intervals, required. Howell, at the Fowlers Gap Research Station of the University of New South Wales, is working along these lines. New thinking in water engineering is nowhere needed more than in the arid zone.

ALTERNATIVE LAND USES AND UNCOMMITTED LANDS

The arid lands are at a disadvantage compared with most of the more humid areas in lacking versatility of land use. Sparse grazing is likely to remain the main means of harvesting plant material economically over large areas, despite the fact that only a small percentage of the available dry matter is utilised. Cropping, if any, will be restricted to sites where water is locally abundant. The main possibility for diversification at present appears to be within the grazing industry, along the lines of substituting or combining meat- with wool-production, substituting cattle for sheep, and harvesting marsupial products. There would appear to be sufficient interest in these alternatives for the necessary research to be carried out.

Another alternative, for which a more far-sighted approach is required, is the utilisation of extensive areas of woody vegetation in the arid zone for timber, chips, or other primary wood products. Although present economic prospects for such activities are poor, it is a possibility we should not ignore, and the necessary data on primary productivity, harvesting, and regeneration procedures, timber and fibre properties and their utilisation should be collected now. Professor Harold Young, of the School of Forest Resources of the

University of Maine, estimates that the biomass and productivity of some of the Australian semi-arid scrub-forests are comparable with those of northern American forests being utilised economically on a long-term basis. The extensive cypress pine forests and scrubs of the Cobar area—some half million acres in extent—would be a logical resource to investigate in this regard.

There are also large areas of land for which there is at present no economic use; these are mainly where stock waters have not been developed (as on the Nullarbor Plain) and/or where the quality of the pasture is regarded as too low, as in the huge spinifex areas of Western Australia. Such areas (Fig. 1) comprise almost half of the arid zone. Our experience with arid zone grazing to date is that in most areas and for most plant communities the initial increase in livestock is followed by a crash both in livestock numbers and in pasture condition, without subsequent recovery of carrying capacity to more than about half of the original level (Perry, 1967). Therefore, we should think hard and carefully about developing additional country. A prerequisite for such development should be our ability to devise stable management systems for areas at present in use. Until we can do this, the uncommitted lands should be kept in reserve.

A general rider on most land use in the arid zone, in view of the uncertainty of the future, should be to retain flexibility. Implications of this policy would be the retention of the leasehold system of land tenure (in preference to alienation), amalgamation rather than subdivision of leases to obtain more flexible operating units, and a minimum of capital investment in avenues likely to perpetuate existing systems of use.

REFERENCES

Australian Academy of Science (1962). *Report of Western Australian Sub-Committee, National Parks Committee.* Australian Academy of Science, Canberra.

——— (1969). *National parks and equivalent reserves in Australia.* Australian Academy of Science, Canberra.

Calaby, J. H. (1969). The vertebrates. In J. le Gay Brereton, D. Whitelock and L. J. Webb (eds.), *The last of lands.* Jacaranda Press, Brisbane (in press).

Charley, J. L. and Cowling, S. W. (1968). Changes in soil nutrient status resulting from overgrazing and their consequences in plant communities of semi-arid areas. *Proc. Ecol. Soc. Aust.* **3**: 28.

Ealey, E. H. M. (1967). Ecology of the euro, *Macropus robustus* (Gould), in north-western Australia. I. The environment and changes in euro and sheep populations. *CSIRO Wildlife Res.* **12**: 9.

Frankenberg, J. (1967). Nature conservation survey of Victoria. Vic. Nat. Parks Assoc. (mimeo).

Frith, H. J. (1964). Mobility of the red kangaroo, *Megaleia rufa*. *CSIRO Wildlife Res.* **9**: 1.

Hall, E. A. A., Specht, R. L. and Eardley, C. M. (1964). Regeneration of the vegetation on Koonamore Vegetation Reserve, 1926-1962. *Aust. J. Botany* **12**: 205.

Harris, Kerr, Forster and Company (1965). *Australia's travel and tourist industry 1965*. Stanton Robbins & Co. Inc., New York.

Hayden, E. (1969). Survey of plant communities of New South Wales. Nat. Parks and Wildlife Service N.S.W. (in preparation).

Leigh, J. H. and Mulham, W. E. (1966). Selection of diet by sheep grazing semi-arid pastures on the Riverine Plain. I. A bladder saltbush (*Atriplex vesicaria*)-cotton bush (*Kochia aphylla*) community. *Aust. J. Exp. Agric. and Anim. Husb.* **6**: 460.

Main, A. R. (1963). Physiology in the management of kangaroos and wallabies. *Proc. Ecol. Soc. Aust.* **3**: 96.

Mosley, J. G. (1968). *National parks and equivalent reserves in Australia. Guide to legislation, administration and areas.* Aust. Conserv. Found. Spec. Publ. No. 2, Canberra.

Mulvaney, D. J. (1968). Human factors in the deterioration and destruction of antiquities and their remedy. *Proc. Conf. Aust. Inst. Aboriginal Studies, Canberra, May 1968.*

Perry, R. A. (1967). The need for rangelands research in Australia. *Proc. Ecol. Soc. Aust.* **2**: 1.

Piesse, R. D. (1968). Tourism, aboriginal antiquities and public education. *Proc. Conf. Aust. Inst. Aboriginal Studies, Canberra, May 1968.*

Specht, R. L. (1961). Flora conservation in South Australia. I. The preservation of plant formations and associations recorded in South Australia. *Trans. Roy. Soc. S. Aust.* **85**: 177.

11

THE IMPACT OF CAPITAL INTENSIVE INDUSTRIES ON ARID AUSTRALIA

M. J. WEBB*

BECAUSE OF the extraordinary concentration of the population of Australia on the humid temperate and sub-tropical coastal margins, the economic development of the arid interior, the so-called 'dead heart', must depend upon decisions by those who have little direct knowledge and even less understanding of its problems. The task is complicated by the concentration of political and financial power in the state and Commonwealth capitals.

Formerly, this was perhaps not as important as it is today. The pastoral industry, for long the only effective method of land utilisation, represented in Australia, as it does the world over, one of the most primitive forms of commercial land use. The demand for capital investment was low, the number of persons employed *directly* in the industry few, and because most of its products were sold overseas, any urbanising influence of the pastoral industry was restricted almost entirely to increasing the population at one or other of the state capitals. The major development of service towns and centres in the pastoral districts took place forty or more years ago. The lack of subsequent development in these centres, and the way in which so many retain the manners and architectural styles of the horse and buggy era confirms the pastoral industry's limited possibilities for intensive development.

Thus, before one can discuss the impact of capital intensive industries upon arid and semi-arid Australia, reference has to be made to the conditions under which capital intensive industries would and should be located in this remote region. The fact that this type of environment is a continuous tract embracing over 74 per cent of the total land surface of Australia and, with the exception of Victoria, extends into every mainland state makes the problem of considerable state and Commonwealth importance. Thus, the pressures to develop central Australia have become as much political as economic and the criteria used to justify development include factors, such as

* Department of Geography, University of Western Australia.

ACKNOWLEDGMENTS. During the preparation of this paper direct enquiries were made of all major Commonwealth agencies and of the appropriate departments of the State Governments of New South Wales, Queensland, South Australia, and Western Australia. The author wishes to acknowledge the help received from these quarters.

defence, not always susceptible to a proper economic evaluation. Furthermore, the attitudes of the state and Commonwealth governments and of private individuals toward our semi-arid and arid interior are seldom free from emotionalism engendered either by support for, or opposition to, the use of scarce capital resources in Australia's interior and northern regions. From the days of Brady's (1914) *Australia Unlimited* to those of Davidson's (1966) *Northern Myth* we have had ardent proponents for both development and non-development, but none was as effective as the stimulus of the last war.

This stimulus amounts in practice to the development of the north for demographic rather than economic reasons. Herein lies a distinction which is often not considered when merits of alternative investments are discussed. In addition, the areas defined by most post-war schemes for northern development have been restricted to only a portion of what is after all a naturally undivided zone.

Thus, the pastoral industry, from the demographic standpoint, is remarkably ineffective. However, the seeming weakness of this industry is an illusion from the economic standpoint: the pastoral industry has for decades produced export earnings out of all proportion to its importance in population. It is this signal fact which has separated the economy of the large metropolitan centres from that of the pastoral outback. The metropolitan centres under tariff protection have waxed under national policies which are more overtly demographic than economic. Consequently, from the 1860s onward, and beginning in Victoria, the economic policy in Australia has been torn between the protectionism of the major metropolitan centres and the free trading of the pastoral and cereal-growing country districts.

That the free trading issue as such was alive until recently is evidenced in Western Australia by the secession movement of the 1930s, which made free trading the movement's economic plank. This was at a time when the major states of Victoria and New South Wales were totally committed to protectionism.

The schism undoubtedly continues in other ways and is evident in the differences which lie between the Liberal and Country parties of the present-day Commonwealth government. We should know that a return to free trading could be one of the outcomes of a massive dependence upon the export of minerals.

The essential differences between these two policies are important because they are especially relevant to the understanding of the prospects for investment in capital intensive industries in semi-arid and arid Australia. The export-orientated industries, and for our purposes those located in central Australia, will tend to see population growth as a product of economic development. In other words, economic criteria will be important and the play of the market through the pushes and pulls of international trade will be the most important factor in the decision-making process. By contrast, those concerned with economic development behind tariff and customs barriers are concerned principally with economic development as a by-product of population growth. That is, the most important criterion is the extent to which economic development can provide employment for a population whose rate of growth is conditioned in very large measure by an effective immigration policy.

Because of the vastly different population densities involved, it is obvious that the cost of capital facilities must be very much higher in semi-arid and arid Australia than in metropolitan Australia on a *per capita* basis, particularly when the facility, such as a road or telegraph line, is linear. For example Western Australia, with approximately 53,000 miles of proclaimed or declared roads and a population of less than one million, has nearly four times the mileage of Victoria although less than a third of its population. Such examples emphasise that remoteness, distance, and absence of population, rather than aridity, are the important criteria when initially judging the capital investment possibilities of interior Australia.

In sum, the differences between metropolitan and interior Australia are exaggerated because of the clear-cut political, social, and economic advantages which the former possess over the latter. Thus, those who are interested in relating a rainfall gradient to a capital investment gradient will discover that the latter from the coast inland falls steeply after the first few miles. The concentration of Australia's capital wealth in the metropolitan areas is probably even more than the concentration of population, since a high proportion of rural population makes little use of capital equipment.

Thus, in considering the impact and the development of capital intensive industries in low rainfall areas, we not only have to contend with remoteness but with the possibility that capital will be less effectively used under these conditions. Furthermore, development can proceed according to either economic or demographic criteria.

The position is further complicated by the fragility of the biological envelope in these semi-arid and arid zones. The biological shocks which the pastoral industry experiences from time to time are specific reactions to a high rainfall variability; the general absence of agriculture is evidence of low rainfall. The irreversibility of so many of the biological and edaphic processes under semi-arid and arid conditions is a major reason for imposing special types of control over investment and development in interior Australia.

THE IMPACT OF MINING

The initial impact of mining upon central Australia was frequently to open it to subsequent occupation by pastoralists. Where pastoralism was already established, albeit precariously, very often early mining enterprise provided a market for animals for food and transport. This was especially true of the Western Australian pastoral industry, where the constant tracking of prospectors across the Pre-Cambrian Shield, principally in search of gold, helped to spread settlement from Norseman in the south to Halls Creek in the north. The way in which the Halls Creek gold rush helped the Durack family to establish their East Kimberley station at Ivanhoe is well known, but there are other examples. Conversely, many original mineral discoveries were made by members of the pastoral industry, and although the well-known episode of the discovery of the lead-silver-zinc lode at Broken Hill by Charlie Rasp was the most dramatic, it was only one of a whole series.

The legacy of this stage was a road system which continues to be the backbone of the network in Queensland and Western Australia, and many of the smaller settlements. Indeed, it is doubtful whether settlements such as Marble Bar and Nullagine would have survived without the continued presence of fossickers and prospectors in the district.

On the other hand, the pattern of mineral discovery was also strongly influenced by the pre-existing roads. Until recently the mineral map was, in fact, a road map using different symbols. The dependence of prospecting on roads has decreased with the increasing scale of the operation and the greater use of aeroplanes and helicopters, but, it is not until the mineral resources are exploited that the transport system is improved.

What is perhaps more significant, at least in the early stages, is the impact of aridity and isolation upon mining, rather than the other way about. Semi-arid and arid Australia is so large a contiguous area that it generates a 'mass effect' in which the values and even the character of the individual parameters are modified, and new ones created. This is especially true with regard to isolation, which as a consequence, is magnified by the thinness of the road network, the poor standards of road maintenance (especially as water is difficult to get), the lack of ancillary services and supporting social and economic services and structures. The cumulative effect of these factors is frequently enhanced by psychological repercussions brought about by family separation, culture shock, and disorientation; these are also commonly reinforced by real or imagined physiological difficulties. The combined effects of these and other factors upon the attitudes of management, workers, and their families to permanent settlement is not known, but must be related to high employment turnover rates in the mining industry.

Consequently, the way in which the mineral industry has developed in arid Australia contrasts with its growth in the more settled areas where these and other factors are not operative. The total lack of facilities inland means either a make-do-and-mend operation with a small labour force highly individualistic in character and often operating either on its own account or on a share basis, such as we have in the opal mining field at Coober Pedy or in the small-scale tin and manganese operations in the Marble Bar district, or, by complete contrast, large-scale operation which involves the planning and installation of the total requirement of a mining operation, including housing and social services, shopping and entertainment facilities for its labour force and its dependants.

Arid Australia impinges upon mining, then, since the marginality of the enterprise in arid Australia will be fixed, as compared with other sites, by the difference between the gains made possible by the richness of the ore body and the increased costs put upon extraction by these environmental 'surcharges'.

The outcome will thus be a very much bigger commitment on the part of the larger organisation than would otherwise occur, provided sufficiently large mineral sources are found to justify these big investments. This is precisely what has occurred in the Ashburton-Pilbara district of Western Australia. The provision of schools, shopping and entertainment, and other social facilities to serve the townships at the mine and port at either end of the railways owned and operated by the mining companies could thus become the primer for a whole series of secondary developments of wider importance.

However, as with other centres, such as Mt Isa and Mary Kathleen, the mining settlements of Dampier, Mt Tom Price, Mt Newman, and Mt Goldsworthy are company towns somewhat divorced from the rest of the economy. Nearby Port Hedland is an open town and will become a focal point for the newly emerging region. However, such was the rapidity of development in what is generally thought to be a highly competitive ore market that Port Hedland willy-nilly became focal point for the region, its local authority thrust, thoroughly unprepared, into the major role for the district.

The smaller mining organisations down to the one-man operations are principally in gold, tin, manganese, and opals.

Their more plastic response to the environment considerably reduces the capital requirements per man. For example, the goldminers are very often doing no more than either reworking or extending shafts dug at some earlier date and using facilities very often originally installed in more palmy days; this is especially true of centres around Marble Bar.

The first impact mining has upon the arid zone is through the development of transport and the emergence of an air of buoyancy, even optimism, in the region as a whole. Thus, the pastoral industry and the established towns are inevitably drawn into the system simply because of increased movement and activity. Pastoral leaseholders also reappraise property values, and mining and construction companies begin to purchase pastoral leaseholds. The Hamersley Iron Co. for example, has purchased the pastoral leasehold which immediately surrounds their port of Dampier, and the Utah Construction Co.

has bought, for altruistic reasons, a pastoral property which is unrelated to its mining operations.

The subsequent impact of mining operations is generally localised, and excavations such as we have at Mt Tom Price or Mt Isa, though possessing a certain primeval grandeur, are essentially destructive. In proportion to the total area of central Australia the amount involved is small. However, this is misleading. Most mining enterprises are either on or close to the main avenues of communication. Many mining sites, such as those in the Ashburton-Pilbara region, are also in scenically interesting areas and many are of considerable scientific interest.

The report of the Western Australian Sub-Committee of the Australian Academy of Science on National Parks and Nature Reserves recommended the creation of a million and a half acre reserve in the Hamersley Range area. It also drew attention to the importance of the Dampier Archipelago, the islands of which are visible from the new iron ore port of Dampier. In neither case have reserves been declared. The shortage of water has meant the extraction of water from stream beds and more recently one of the most remarkable stretches of water in the Hamersley Range, known locally as Mill Stream, has been tapped for water.

Researches have begun into schemes for a large-scale integrated water storage system covering most of the major streams in the Ashburton-Pilbara area and an overseas company has been commissioned to study the development of the region as a whole. On the other hand, Dampier has installed two stills for fresh water.

Air and water pollution occurs at the time of loading of ore at the ports. That at Dampier is so great as to be seen from miles away and the dust could affect the marine biology of an area well known for its marine life. The Port Hedland pollution is less than it once was; nonetheless it has reduced the quality of living of the town and surrounds.

There are only two areas within semi-arid and arid Australia in which oil and natural gas are in the process of exploitation, Gidgealpa, South Australia, and Barrow Island, Western Australia. Again it is not possible at this stage to detail the nature of any threat to the region. Barrow Island is, however, an A Class Reserve for fauna preservation and thus raises the question of the priority: preservation or exploitation? In Western Australia mining has the prior right over preservation; indeed, because reserves happen to be Crown property they are more easily entered for the purpose of exploration than most other forms of tenure. Over Barrow Island an agreement was reached between the petroleum company and the state government in an effort to preserve this mammal-rich island. However, the land is not vested in any particular authority and the frequent coming and going of ships and aircraft is bound to lead to the introduction of harmful alien rodents.

Moreover, the realisation of the offshore potential for oil and gas in Australia will increase the risk of large-scale oil pollution. The importance of this could be overlooked because of isolation and lack of regard.

The long-term impact of mining will be upon the development of secondary industries: refining, smelting, power production, and even manufacturing. These are already part of the plans for the Western Australian iron ore program and in the long term the impact of this type of capital intensive industry could become widespread. The ore reserves are so large and so rich that direct conversion from ore to metal is a possible development. Should this occur, the great Western Australian reserves could become even more valuable than they are now. The attraction for secondary industrial development would be enhanced, with population growth prospects a great deal more than they are now.

The most optimistic forecasts are that as many as 125,000 persons could be living in the Ashburton-Pilbara region by 1985. At Broken Hill, Mt Isa, and Kalgoorlie, which is seeing a new lease of life with the exploitation of nickel from Kambalda, the real population growth followed secondary developments.

At Kalgoorlie and Broken Hill the pressure of population on the immediate neighbourhood has encouraged the establishment, although not entirely successfully, of vegetation regeneration belts around parts of the urban tract. The provision of water for recreational uses is vital to the mental and physical welfare of the inhabitants of arid zone settlements.

THE IMPACT OF IRRIGATION

By contrast with the widespread exploitation of the mineral resources of Australia's semi-arid and arid zones, the development of capital intensive irrigation schemes is more localised and most of the developed resources may be regarded as forward edges of the existing agricultural districts and thus marginal to the semi-arid and arid zones. This is especially true of the Murray-Darling-Murrumbidgee Basin irrigation works. In part this may be explained by the tendency of agricultural development to expand peripherally rather than through isolated centres; in part through the proximity of these areas to the major water catchments of the Great Dividing Range; in part because of the high concentration of population in New South Wales and Victoria: of 3,000,000 acres of irrigated land in Australia, 2,400,000 acres lie within their borders. Western Australia, with the largest area of arid land, has about 85,000 acres irrigated.

Away from the south-east section, with the exception of the Ord scheme and the Carnarvon (Gascoyne River) area, all developments are either small or experimental. For example, the Northern Territory's *Water Supplies*

Development Ordinance 1960-1963 gives to the Water Resources Branch power to grant financial and technical assistance for the development and improvement of water supplies on pastoral and agricultural leases. Most of the irrigation potential is either just within or on the borders of the humid north in the Daly River plains and on the Adelaide River dam site. The only truly arid zone investigation appears to be the Ti-Tree Basin area, 120 miles north of Alice Springs. Yet of 5,478 bores registered and abandoned in the Territory 314 were for agricultural purposes, irrigating approximately 1,500 acres.

In Western Australia the Carnarvon (Gascoyne) irrigation area is a remarkable example of local enterprise based upon a very simple technology. Thus far, in the absence of any large capital investment, it is labour intensive rather than capital intensive; but it could be representative of a whole range of future schemes in areas dependent upon either bores or shallow well water supplies. Nonetheless, further development depends upon the provision of storage in the upper basin of the Gascoyne. The Carnarvon irrigation scheme is important in providing fresh vegetables and fruit for the growing mining settlements of the arid zone. Schemes at Alice Springs are comparable. Elsewhere many pastoral stations have demonstrated the feasibility of growing vegetables under extreme conditions for local consumption.

However, the Ord scheme has demonstrated that the problems of developing large-scale cropping in conditions unfamiliar to both scientist and farmer are numerous. The political-cum-economic arguments for and against the scheme were not entirely separable from the general argument concerning so-called Northern Development, but the economic arguments of the type advanced by Davidson (1966) concerning alternative returns on investments in the more humid and accessible southern areas, though economically relevant, ignore the demographic undertones to the scheme. Unlike the exploitation of mineral resources, which are concerned with the expansion of highly localised reserves, the development of water resources for agricultural purposes in semi-arid and arid areas at a distance from the main centres always raises the possibility that the increased production could be obtained through either extending the existing agricultural boundary or intensifying existing agricultural practices in better located dry or wet farming areas.

Thus the bio-economic base of this type of scheme is not secure, unless the project is seen as a part of a demographic policy, in this case as a part of the populate-the-north movement.

The Ord scheme must be regarded as the most successful to date of any of the oasis-type developments. Its local impact has been less than was hoped for because the various cropping systems adopted have not been related to the local pastoral industry. Nor has the experimental work at Camballin, on the Western Australian Fitzroy, been directed principally toward the integration of pastoralism and agriculture. These contrast with the Tipperary projects, where the growing of sorghum grain could become part of a raising-fattening program involving the export of fresh and canned beef. The Northern Division of the Department of National Development reported that the Tipperary Land Corporation had assets and cash equivalent to $8,000,000.

One side-effect which is likely to become more important is the regeneration of catchment areas to reduce the rate of reservoir silting. In the Ord scheme, for example, 1,400 square miles out of a total of almost 18,000 square miles in the catchment comprise a 'most serious erosion hazard' and most of this has been resumed, cleared of stock, and reseeded. As other basins come under irrigation control the pastoral industry will be affected increasingly by regeneration, restocking, and recapitalising and fencing programs which will extend the impact of the irrigation scheme far beyond the immediate confines of the agricultural section of the scheme.

The impact of the great Murray-Darling-Murrumbidgee irrigation scheme upon the semi-arid environment is well known. Some of the problems highlight the need for integrated supra-state commissions capable of resolving, on a continuing basis, all the regional problems; some, the growing competition between power needs and irrigation needs; others concern soil conservation in areas of high evaporation. Nonetheless, these areas have far fewer problems, because development can be related to nearby major centres of population and a reliable source of water. Consequently, they may be regarded as an extension of the existing agricultural zone.

THE IMPACT OF TOURISM

The general increase in the tempo of life brought about by the exploitation of mineral and water resources in the semi-arid and arid areas of Australia and the quickening of public interest in these areas has led to a rapid rise in tourism. The round-Australia route, or the trip to Alice Springs and on to Darwin has become the regular, even if rugged, haunt of the tourist. The advent of well organised air or air-road tours has made holiday travel through the interior no more difficult than in other parts of Australia except in time and cost. For example, a survey of visitors to the Northern Territory showed that nearly 80 per cent of the sample stayed for more than seven days, and almost the same percentage spent more than $100 in getting to the area, although nearly half the sample spent less than $100 while actually in the Territory.

The impact of tourism is both direct and indirect. It is direct by way of encouraging investment in transportation services, including roads and railways, hotel, motel, and other forms of accommodation, and in the provision of shops and other facilities. The indirect benefits lie principally in the spread of knowledge about the central themes and values of central Australia. Whether this information, gained from a fast moving aeroplane or motor vehicle, and without expert guidance, is of real value is open to question.

Other indirect benefits which may flow from tourism involve the use of, and the demand for, local products and facilities. But this is not always so. In Western Australia, for example, the needs of tourists are catered for almost entirely from Perth and a significant proportion of the labour force in the dry areas is in transit. Consequently the impact is less than expected.

The most important direct impact is the construction and the upgrading of hotel, motel, and caravan and camping facilities in the major centres of attraction or at intermediate points. However, because tourism is a seasonal enterprise, these facilities tend to be located in areas where a base load demand for some or all of the facilities exists. For the time being the base load and tourist demands are largely complementary. Mining developments, such as those at Mt Isa, provide facilities which tourists use because Mt Isa is itself worth seeing;[1] and because of the base demand for hotel accommodation etc. by the mining community Mt Isa is also a place where good facilities exist. It must therefore be admitted that much of what the tourist wants to see at present in arid Australia is man made. Not all of this is modern: centres such as Broome attract tourists because of their quaintness. Many old mining and pastoral centres in Queensland and Western Australia also attract tourists and their preservation will become increasingly important.

The distance between centres is so great that land travel is time consuming and costly. Consequently, a high proportion of tourists do not use hotel facilities but camp along the roadside or use caravan parks. The camping-out approach to tourism is also encouraged by so-called 'safari' tours, some of which combine road and air travel in order to reduce total travel time. The extreme example of this use of air and road lies in tours designed for overseas investors who are flown from one investment possibility to another with short road tours at each setdown; a 7-day tour of this kind may cost over $1,000.

The main physical limitation upon tourists at present is the lack of sealed roads. The consequent risk of wear and tear and of serious breakdowns, plus the low average speeds which result, reinforce, in the minds of potential tourists, the terrors of outback travel. The distance-time element has really to be experienced and unless one turns a holiday of two or three weeks into a marathon, journeys in excess of two thousand miles by road are very difficult for the average family. Usage is further restricted, at least for the northern parts, by a summer rainy season.

Alice Springs is the product of a series of fortuitous circumstances: war-time publicity, the terminus of the unfinished transcontinental railroad, comparative proximity to Ayers Rock and Mt Olga. Much has been made of its

[1] 10,000 tourists visited Mt Isa Mines in 1966.

assets, and Alice Springs is the premier tourist centre for arid Australia. Few centres in Australia are as widely known internationally or nationally. Although not fully developed, the construction of accommodation at Ayers Rock is part of a trend which will inevitably involve many sub-centres and tourist routes connected to Alice Springs. Nothing comparable has occurred elsewhere in central Australia, but Alice Springs will probably set a pattern for the future based upon good accommodation and communications, a wide variety of scenery, contact with Aboriginal culture, and a hint of adventure rounded off by a topping of outback flavouring. The result thus far is impressive: with a population of some 6,000, Alice Springs is by far the largest non-mining settlement in arid Australia.

The development of tourism could be important for the conservation and development of arid Australia, especially the development of national parks and nature reserves. No other single factor could so forcefully affect governments, and although the creation of reserves and national parks is not yet one of the impacts of tourism, no plan for tourism or conservation which does not take the other into account is satisfactory. The four states with the highest percentage of arid lands have also the lowest proportion of their land surface devoted to parks and reserves. It is startling that Tasmania, with less than 17,000,000 acres of territory, has more than 700,000 acres devoted to national parks and reserves, compared with South Australia's 243,000,000 acres and 2,889,000 acres in parks and reserves, and Western Australia's 624,000,000 acres and 3,000,000 acres of parks and reserves. Queensland is marginally better, with 2,400,000 acres out of 427,000,000 acres but still has only 1 per cent of its total land surface protected to some degree or other.

The HKF Report on tourism (Harris *et al.*, 1965), basing its findings upon American experience, unequivocally recommends that the entire tourist and travel industry of Australia should support those working toward the goal of more and better parks and reserves. Thus far, this has not occurred and the attitude of conservationists is not always favourable to tourism. The existing scale of tourist expenditure is probably not fully appreciated. In the Northern Territory, for example, tourists spent more than $14,000,000 in 1966-7—a sum equal to half of the amount spent by the Territory administration.

The possibility of large tourist income from national parks and reserves in arid Australia has not been fully investigated, and the size of reserves in relation to our needs over the next twenty-five years is underestimated. What is also appalling is the extraordinary way in which we do nothing to preserve even the immediate environs of tourist attractions in dry Australia; in utter disregard of the fact that open space in Australia generally exists only because someone or some organisation has not yet found a way of exploiting it. So much of what is admired in arid Australia is there only for this reason, and large areas are, or could be, changed to permanent or quasi-permanent ownership without regard to the needs of conservation and thus of good tourism.

The only reserves which are appropriate in scale and purpose to the needs of arid Australia are the Tanami Desert Reserve and the Simpson Desert Reserve. Although there are reserves within the Alice Springs tourist region, they are so distributed as to indicate no cohesive plan and no conception of the real problems involved. In Western Australia nothing is comparable with either of these reserves, despite the magnificent report of the Western Australian Sub-Committee of the Australian Academy of Science Committee on National Parks.

The impact of tourism upon a National Parks policy for arid Australia will not be felt until it is harnessed to conservation. In existing tourist attractions, such as those in the Pilbara-Ashburton and Kimberley districts of Western Australia and the Flinders Ranges of South Australia, we find little in the way of conservation, even less of tourist development. As for continental wilderness routes, such as the forbidding Canning Stock Route in Western Australia, there are literally no concepts of these as potential reserves which will become tourist and commercial highways.

CONCLUSIONS

It is difficult to set out logically conclusions to a paper which is tentative and experimental. The writer has been conscious throughout of the importance of, and the lack of knowledge about, this important issue. Capital intensive investments tend, by nature, to be point located; but their impact radiates outwards. The rate of development of mining over the past five years has been remarkable. Yet the basic transport network remains and the essential relationship of the parts is very much as it was before. Even the routes from the Pilbara-Ashburton for the new iron ore mines at Mts Tom Price and Newman relate more closely than is generally realised to the routes for the export of asbestos from the, now closed, Wittenoom mines. On the other hand, the quality of living has improved and the permanence of occupation is now far more secure than it was in the past. To some extent we may now regard Broken Hill as less of an accident and more of a signpost to the future; Mt Isa is rapidly becoming a more modern version, and with the development of secondary industries similarly large populations will develop in the Pilbara-Ashburton region.

Since all mining operations are big export earners, there is every reason to believe that miners and pastoralists might see eye to eye on certain issues, especially concerning government investment in arid areas. But even before this becomes a possibility we should consider more closely a situation in which both industries are extractive in the sense that neither, in any real way, contributes to the wider development of the regions from which they obtain their wealth. One of the great locational problems of the twentieth century is that of accumulation at the centre and exploitation at the periphery. Arid Australia is no exception, and unless means are found to divert some of the resources back into the area for conservation, irrigation, and human welfare, then the region will continue to be exploited and poorly managed.

The establishment of regional development funds specifically earmarked for work in Australia's arid lands would enable the development of a consistent investment policy, directed to the best long-term interest of this vital region. Indeed, it would be even more preferable if, by some means, the politician could be weaned from his remarkable, almost limpet-like, attachment to the now outworn and thoroughly inappropriate concept of northern development and directed instead toward the more realistic and viable concept of semi-arid, arid, and tropical Australia. At present arid Australia tends to be regarded as a barrier between the north and the south; it is, in fact, the bridge.

The development of irrigation could be partly related to this special development fund, but what is needed even more is a new approach to irrigation investment in arid Australia. The extraordinary allocation of money for urban road construction represents a blow to outback Australia and to its developing areas. The abandon, and that is perhaps the best word to use, with which money is devoted to roads in urban areas for which there are generally no thoroughly worked out justification, as against the extraordinarily restricted way in which capital loan or capital grants are given for irrigation is something to which insufficient attention has been given. For example, in Western Australia we have two projects; one for two freeway interchanges in a mile of freeway estimated to cost some $63,000,000, the other the Ord Irrigation Scheme second stage, costing perhaps $50,000,000. Very few will have heard of the former; most people will have heard of the latter, mainly because of the amount of politicking necessary to achieve a loan of smaller dimensions than the outright grant for the freeway segment.

The distinction between economic and demographic aims in non-industrial development has not been made and should be made in the future. Oasis-type developments, such as we have in the Ord, cannot be thrust into the economic arena. Similarly, the ownership of land and our approach to land in these vast areas should be reconsidered. With increasing understanding of the crop ecology of certain areas, principally on the fringes of the semi-arid zone, the pressure for capital intensive investment will increase. This is especially important where companies with large capital resources and more go-ahead management enter the field, companies such as the Kaiser Corporation at Mulla Bulla Station in the East Kimberley, or the Tipperary Land Corporation in the Northern Territory. It is clear that failures to match capital to land tenure could lead to long-term difficulties with regard not only to size of holding and capital-land appreciation, should very large leasehold be converted to freehold without proper regard not only to the best systems of land management but to the long-term pressures for sub-division and even greater appreciation of land values. This is especially important where irrigation is concerned, and there is no doubt that current policy in both Queensland and Western Australia is partly a reaction to the extreme difficulty in getting capital for irrigation out of centrally controlled capital funds. As with Western Australian iron ore, arrangements are bound to be made which by not drawing upon these central funds make co-ordination difficult and the eventual outcome uncertain.

The transport revolution which follows the mining boom is more of a change of mode than of route. Nonetheless, traffic flows and tourists appear. At present the great central mass of Australia is crossed by only one effective route, the Darwin-Alice Springs-Oodnadatta line of communications. This may appear admirable, from the conservation standpoint. There is no doubt, however, that conservation and tourism should be regarded as symbiotic. Even the total ban on the shooting of wild animals without a special shooter's licence and the payment of set fees for each animal shot could, for example, provide a steady and useful income for conservation purposes. The cause of conservation probably has its best support from this direction. However, the interest goes wider. The comparatively small impact of tourism at the present stage is largely the product of a lack of reserves and national parks and of a settled policy on their management, control, and investment. Too many people see the present open spaces as permanent features. They are not.

The position is further explained by what must be considered a thoroughly incorrect interpretation of the access which present leaseholders have to their land. For all practical purposes this usage is for the pastoralist one of ownership. Consequently, the area of Crown lands from which reserves and national parks may be selected is unnecessarily reduced to the point where they are either inaccessible or inappropriate. Much of the present leasehold land in the Northern Territory, Queensland, and Western Australia should be surveyed and much of it placed within a special category which could be called, for example, 'protected lands', by which is meant land open to use as at present but managed in such a way as to preserve amenity, flora and fauna, and scenery. The Alice Springs region should be the first to be thus investigated and developed. Many of the best conservationists are to be found among the pastoral population. The possibility of relating conservation to pastoralism could be made a part of a development program for the region. The significance of this would become immediately apparent should an integrated road policy be drawn up for the Alice Springs region to serve the needs of pastoralism, tourism, and conservation. It is worth mentioning in this connection that the estimated earnings from tourism for the Northern Territory as a whole is equal to that of present earnings from the pastoral industry. From this it should be obvious that the so-called beef roads ought to have been looked at from the standpoint of tourist earnings, but, so far as can be gathered, the cost-benefit analysis used following the initiation of the 1964 investigations did not include an item for tourist benefits.

The final and overall conclusion is, of course, that there has been a limited impact to the present. The investment in mining and minerals has risen steeply and it looks as though new funds will maintain a significant flow of intensive capital into the region. The switch away from traditional pastoralism has begun, but the impact of irrigation is not yet really discernible, especially when the largest scheme of all, the Ord, is not integrated into the regional pastoral economy. Irrigation schemes in the south-eastern areas of New South Wales, Victoria, and South Australia are generally to be regarded as extension of the humid zone agriculture, and therefore more closely related to their nearby population centres. In any case, all irrigation schemes are peripheral to the semi-arid zone, and apart from local developments at Alice Springs and Carnarvon, in particular, there are really no true arid zone irrigation developments of any magnitude.

It is unfortunate that the importance of the semi-arid and arid zones as earners of foreign exchange, increased since the great mineral discoveries of the 1950s and 1960s, is frequently glossed over in favour of consumption expenditure in metropolitan areas. The distinction between the criteria used to allocate what is essentially capital investment as between metropolitan and interior Australia, the one basically demographic and the other economic, could doom the latter to a further period of extractive exploitation. The future can only be assured if a constant feed-back of capital into semi-arid and arid Australia through special funds is maintained. The impact of capital intensive industries is still small in those areas where capital is most needed: arid Australia.

REFERENCES

Australian Academy of Science (1965). *National parks and nature reserves in Western Australia.* Aus. Academy Sc., and Nat. Parks Bd. W. A. Perth.

Brady, E. J. (1914). *Australia unlimited.* Robertson. Melbourne.

The case of the people of Western Australia (1934). Government Printer. Perth.

Clarke, G. F., Jones, N. O. and Smith, K. H. (1968). Government members and national development committees tour July 21st-28th 1968 (unpublished).

Condliffe, J. B. (1964). *The development of Australia.* Ure Smith. Sydney.

Davidson, B. R. (1966). *The northern myth.* Melbourne University Press. Melbourne.

Department of the Interior. (1968). *Annual report on the Northern Territory of Australia: 1966-67.* Canberra.

——— (1968). *Major development projects.* Canberra.

Department of Public Works and Water Supply (1968). *Report.* Perth.

Harris, Kerr, Forster and Co. (1965). *Australia's travel and tourist industry, 1965.* Stanton Robbins & Co. Inc. New York.

Kerr, A. (1967). *Australia's north-west.* University of Western Australia Press. Perth.

Macdonald Holmes, J. (1963). *Australia's Open North.* Angus & Robertson. Sydney.

Northern Territories Tourist Board (1968). *Annual Report 1968.* Darwin.

Raggatt, H. G. (1968). *Mountains of ore.* Lansdowne Press. Melbourne.

Shann, E. (1948). *An economic history of Australia.* Georgian House. Melbourne.

12

LAND TENURE SYSTEMS: PAST AND PRESENT

R. L. HEATHCOTE*

> One of the merits of studying the history of legislation is that laws are excellent indications of the abuses which provoked their enactment; they are less trustworthy guides regarding accomplishments because of the gap between intent and enforcement.
>
> —Glacken, 1956:63

THE FRAME of reference for this paper is 'the evolution of various systems of land tenure in various states; the degree to which they have helped or hindered development, management and conservation'. Before attempting to tackle such a broad topic, however, certain key words must be defined and a specific and manageable frame of reference drawn up.

DEFINITIONS

The definitions of 'land tenure' used here are influenced by the set frame of reference, which by implication is land tenure as affecting and effecting the development, management, and conservation of arid land resources. Thus emphasis will be upon the past and present relationships between land tenure and land use in the arid lands of Australia.

'Land tenure' is defined here as 'the customary and codified rights which individuals and groups have to land and the behavior characteristics which directly result from these rights' (Bertrand and Corty, 1962: 7). Under common law, 'land' means 'any ground, soil, or earth whatsoever, whether dry or under water, and includes all buildings on the land, and also everything attached to the land, whether above or under the surface' (Collins, 1949: 11). 'Rights' are defined as those facilities for ownership and, by implication, use of land, which are either conveyed by specific forms of legislation regarding the land or are assumed to be implicit in common law. Finally, in the definition of land tenure it will be useful to bear in mind the two types of general land tenure systems suggested by Toennies, and noted in Bertrand and Corty (1962: 11-12):

1. *Gemeinschafts systems*—those group relationships and rights which develop unconsciously over time as the result of experience of land use.

* Senior Lecturer in Geography, The Flinders University of South Australia.

2. *Gesellschafts systems*—those rights set up deliberately by legislation and policy decisions for the achievement of certain objectives. These two types represent the division, in part, between unofficial group systems and the systems set up by officialdom. This dichotomy is an important characteristic of most land tenure systems and the Australian arid lands are no exception.

THEORETICAL CLASSIFICATION OF LAND TENURE SYSTEMS

Leaving aside for a moment, however, the Toennies dichotomy, it is possible to suggest a general classification of tenures which should help clarify the evolution of the systems found in the arid lands. Five factors seem to be relevant: first, the range of items included in the system—the Content Factor; second, the duration of the system—the Time Factor; third, the possible size of the units of the system—the Size Factor; fourth, the range of ownership permissible in this system—the Personnel Factor; and fifth, the range of intensity of uses permissible in the system—the Use Factor. Each of these factors can be

TABLE 1

THEORETICAL LAND TENURE CLASSIFICATION

Factor	Controls		Description
	Key	Title	
CONTENT (C)	C1	SPECIALIST	Only one item owned + use allowed.
	C2	LIMITED	Several items owned + use allowed.
	C3	UNLIMITED	All items owned + use allowed.
TIME (T)	T1	SHORT	Term of 1 year or less.
	T2	LONG	Term of 1-98 years.
	T3	PERPETUAL	Term of 99 years or more.
SIZE (S)	S1	RESTRICTED	To support only one family.
	S2	LIMITED	To support more than one family but maximum area is fixed.
	S3	UNLIMITED	No limit to area fixed.
PERSONNEL (P)	P1	INDIVIDUAL	Ownership limited to one person.
	P2	INSTITUTIONAL	Ownership limited to corporations or other institutions (public or private).
	P3	GENERAL	No restrictions on ownership.
USE (U)	U1	LIMITED	Complete control of use + products limited.
	U2	SUPERVISED	Controls only to maintain public interests or to meet temporary stresses.
	U3	UNLIMITED	No limits on use of items.

subdivided to show the range of possible controls and the subdivisions are indicated on Table 1 and explained below.

Content Factor (C)

Not all the items or characteristics of land, as defined above, are always included in systems of tenure, hence the need to distinguish between the different scope or extent of the systems. Possible variations of content might be classified as:

C1. *Specialist Content Systems*, where only one item of land is held under the tenure system, e.g. a goldmining right.

C2. *Limited Content Systems*, where several, often complementary, items are held under the tenure system, e.g. a pastoral grazing right to forage and water.

C3. *Unlimited Content Systems*, where there appears to be no limit to the items held under the tenure system, e.g. some forms of freehold tenure.

Time Factor (T)

Not all the tenure systems are intended to be permanent and some recognition is needed of the possible durations of the systems. Thus systems might be classified as:

T1. *Short-term systems*, where the tenure only exists for short time and is recognised as being of only a temporary character. An arbitrary time limit of one year or less would include such systems, e.g. the occupation rights offered as interim tenure at the end of a longer tenure and before a renewed system comes into operation, or as a preliminary to instigation of a new system of tenure.

T2. *Long-term systems*, where the tenure is offered for periods ensuring a measure of security and relative permanency of occupation. The implication of such systems is that the particular tenure and associated use needs to have a long period of operation to be effective, but that at some future date the system may be changed. Arbitrarily the duration of such long-term systems is limited to at least one year but less than 99 years, e.g. most of the pastoral grazing rights in Australia.

T3. *Perpetual systems*, where the tenure is awarded in perpetuity, the implication being that the tenure will give security for the foreseeable future with no practical time limit. Legal definitions of perpetual, i.e. 99 years or more, will be the arbitrary limits adopted here. Examples are found in many of the closer settlement grazing right tenures in Australia.

Size Factor (S)

Not all the tenure systems allow unlimited areas to be held, hence the need to distinguish the various size limits. The implication here is either that certain uses need certain minimum areas to be successfully effected or that policy requires some sharing of the opportunities for use between potential users, hence some maximum limits on areas. Such limits might be classified as:

S1. *Restricted area systems,* where the unit area is limited to that which will support by its productivity only one family, e.g. 'home maintenance' or 'living' areas for agricultural and pastoral rights.

S2. *Limited area systems,* where the unit area is more than adequate for one family but which has some arbitrary upper limit attached to it, e.g. several types of pastoral leases.

S3. *Unlimited area systems,* where the unit area has no set limit to its size, e.g. some pastoral leases.

Personnel Factor (P)

Not all tenure systems allow any institution or individual to be eligible for ownership of the land. The implication here is that certain types of tenures ought to be controlled by certain types of owners. Generally the limitations on ownership are related to the implied needs (e.g. capital) of certain tenures and to policy decisions by governments. The possible variations of eligible personnel can be classified as:

P1. *Individual systems,* where tenures are restricted to individual people, who must exercise the tenure only in their own name, e.g. many of the closer settlement tenures.

P2. *Institutional systems,* where tenures are restricted to institutions, governments, or public organisations acting in their own name or as trustees for the general public, e.g. Crown reserves for any purposes.[1]

P3. *General systems,* where no restrictions on tenure, apart from those referring to the age of legal responsibility of individuals, are made. Any person legally of age, or institution, or public body, therefore, may hold these tenures, e.g. most freehold tenures.

Use Factor (U)

While the range of items, and by implication uses, of land tenure systems are covered by the classifications of the Content Factor, there is need for a further classification of the different restrictions on *intensity* of use, found in many tenure systems. The inclusion of this factor is in recognition of the modern trend towards increasing outside, i.e. non-owner, controls of management of resources through specific references in tenure systems. The restrictions may be classified as:

U1. *Limited use systems,* where the intensity of use of the land is completely controlled by an authority other than the owner. This tenure implies a

[1] 'Crown' is used in this paper both for the rights claimed for the British Crown and the rights as transferred later to the various States and the Commonwealth.

right only to a certain item of the land, and only a certain *amount* of that item, after which further use is prohibited, e.g. a goldmining right where the amount of gold to be dug is limited by a total figure or an annual rate of production.

U2. *Supervised use systems*, where the intensity of use of the land is not specifically or permanently controlled by an authority other than the owner, but the general use of the land *is* so controlled and temporary restraints on use *may* be applied periodically. This system implies that the use may need to be controlled at some future time, but that this control will be only temporary to meet a changing situation, e.g. the implicit controls of overstocking built into pastoral grazing rights.

U3. *Unlimited use systems*, where there are no restrictions on land use, other than those existing in common law as to trespass, noxious activities, and infringements of personal liberties, e.g. early freeholds in Australia.

Thus, with these five major criteria the sequence of land tenures in Australia may be classified and the historical trends in the character of tenures identified.

AUSTRALIAN LAND TENURE IN THE 1960s

Before examining the details of current arid land tenures we must briefly note the scope of tenures throughout the continent. Three main types of tenure are differentiated on legal grounds:

Types of Tenure

Alienated land is held in fee simple as freehold land. In terms of a theoretical classification this would comprise virtually unlimited content (C3 or C2), perpetual duration (T3), limited or unlimited size (S2 or S3), unrestricted personnel (P3), and virtually unlimited intensity of use (U3).

Leased land is held by lease or licence, i.e. in terms less than freehold tenure, and where ultimate ownership rests in the Crown. Classified, this tenure would comprise limited or even specialist content (C2 or C1), durations varying from perpetual to short term (T3, T2, or T1), restrictions in size from complete to unlimited (S3, S2, or S1), restrictions in ownership which varied from general to individual (P3, P2, or P1), and some controls upon intensity of use (U2).

Reserved land is held on behalf of the general public or groups of the general public. In virtually all cases ownership rests in the Crown. Classified, these tenures would include limited or specialist contents (C2 or C1), long-term or perpetual durations (T2 or T3), restricted sizes (S1), institutional (including Crown) ownership (P2), and limited use (U1).

The relative characteristics of these three types of tenure are classified according to the proposed theoretical framework in Fig. 1. The characteristics of the three tenures are classified in terms of the five factors and range of

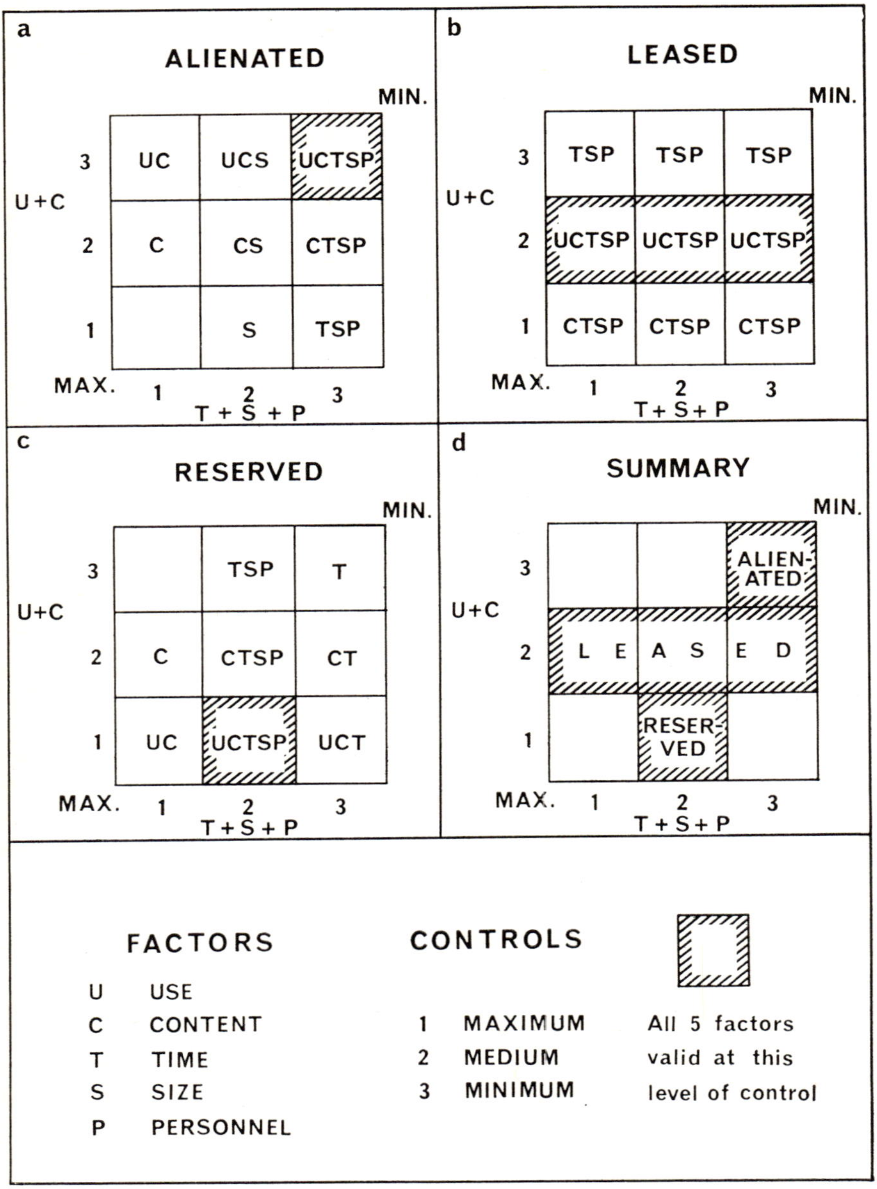

Fig. 1 Theoretical Land Tenure Characteristics

controls and plotted on a three-by-three matrix. Within the matrix there is a concentration of factors and controls at certain locations which are indicated. Locations where all five factors coincide with at least one control might be said to show the core characteristics of the particular tenure plotted, and the matrices enable the relationships between the core characteristics of the three tenures to be seen at a glance in Fig. 1d. Thus the relative freedom from controls of the alienated, as opposed to the reserved land, and the intermediate position of the leased land are immediately apparent.

Distribution of Tenures

The distribution of the three tenure types between the states and territories in Australia as a whole is shown in Table 2 and Fig. 2.

TABLE 2

AUSTRALIAN LAND TENURE 1966

State or Territory	Total acreage (000)	Percentage of Total Area					
		Arid lands‡	Alienated	Alienation in process	Leased or licensed	Reserved	Unoccupied
S.A.*	243,244	87·1	6·6	0·2	61·8	9·4	22·0
W.A.*	624,588	86·3	4·9	2·4	39·4	12·5	40·8
N.T.*	332,979	86·3	0·1	—	57·3	18·3	24·3
QLD*	426,880	62·7	6·2	1·8	85·6	5·8	0·5
N.S.W.*	198,037	51·9	30·9	2·6	57·1	8·1	1·3
VIC.†	56,245	—	56·7	4·1	11·1	15·9	12·2
TAS.*	16,885	—	39·2	1·2	7·7	25·0	26·9
A.C.T.*	600	—	10·6	6·9	4·4	12·4§	23·7
AUSTRALIA	1,899,462	74·0	9·1	1·6	56·5	11·3	21·4

*30 June 1966. †31 December 1965. ‡Rangelands ex Perry. §A.C.T. city tenures.

Sources: Commonwealth and State Yearbooks, Perry 1967.

The first point from the data is the remarkably small proportion of Australia held as freehold or alienated land. Only approximately one acre in every ten is so owned and ownership of the remaining nine acres still rests in the Crown either indirectly as landlord of the leased lands or directly as trustee for the reserves and yet vacant lands. Thus the disposal of complete title to land and its uses has been remarkably ineffective in Australia despite the intentions for such disposal throughout the history of settlement since 1788. However, it must be said that as the Crown does still, directly or indirectly, retain ultimate title to about nine-tenths of Australia, if new tenures or modifications of old systems are needed, now or in the future, the task of compulsory reacquisition of land will be minimised.

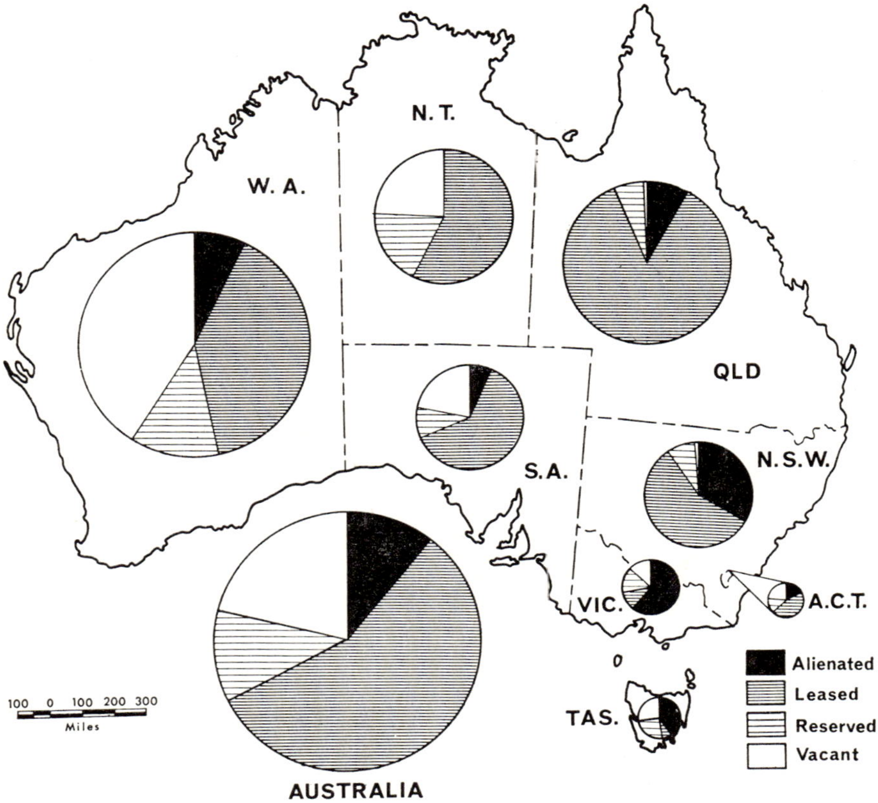

Fig. 2 Australian Land Tenure 1966

Sources: Commonwealth and State Yearbooks.

The leased lands occupy the largest areas of Australia, comprising over half the Commonwealth area and locally even a higher proportion, e.g. 85 per cent in Queensland. For the bulk of the continent, therefore, not only does the ultimate title rest in the Crown, but there may be certain restrictions on occupation and scope of land use.

The reserved lands occupy almost an equivalent area to the alienated lands and, if the vacant lands are included, approximately a third of the continent has either extremely strict controls upon occupation and land use or they are prohibited altogether.

AUSTRALIAN ARID LAND TENURES IN THE 1960s

As shown in Table 2 there appears to be some relationship between arid lands and types of land tenure—put crudely, the greater the proportion of the state

TABLE 3

AUSTRALIAN ARID LAND TENURES 1960s

State or Territory	Alienated	Leased or licensed	Reserved	Unoccupied
	%	%	%	%
S.A.	4·1	58·0	7·5	30·4
W.A.	0·6	47·9	6·8	44·7
N.T.	0·1	59·4	14·9	25·6
QLD	0·9	97·8	1·1	0·2
N.S.W.	0·4	97·5	1·4	0·7
AUSTRALIA	1·0	64·9	7·0	27·1

Sources· Fig. 4.

or territory classed as arid land, the smaller the area alienated and the larger the proportion vacant or under lease or licence. This general relationship is seen more explicitly in Table 3 and Figs. 3 and 4.

Distribution of Tenures

When the types of land tenure within the arid lands are examined, it is immediately apparent that alienated land is not a feature of this portion of Australia. In fact, the small areas which are alienated were in most cases acquired before 1900 in order to consolidate control of land held by pastoralists under other, mainly leasehold, tenures. Thus the bulk of the alienations are of a relict character and do not reflect a tenure currently expanding in area (Heathcote, 1965b; Roberts, 1924:225-300, 425-7).

In contrast the areas of leasehold tenures do represent both a much larger area and a currently effective arid land tenure. Covering over half the arid lands, the bulk of the leaseholds are for pastoral purposes, approximately 17 per cent of them held under tenures which limit their size and the remainder under no size limits.[2] An increasing proportion of the leases are for mining purposes.[3] The arid lands in fact contain the bulk of the leasehold tenures of Australia and if any one tenure can be said to characterise the arid lands it is the leasehold, with the implied restrictions on content, term, and intensity of use, and the ultimate ownership still resting in the Crown.

The arid areas contain a large proportion of reserved land of various

[2] The intensively leased areas (Fig. 4) represent approximately 17 per cent of the total leased arid lands and approximately 11 per cent of the total arid lands.

[3] Because of the rapidly changing pattern of mineral leases and search rights, not all current tenures are shown on Fig. 4. Thus for reasons of clarity the data on the recently published map of Australia—showing 'Petroleum Exploration and Development Titles', Department of National Development, Canberra, December 1968—have not been included.

types; the largest area is reserved for protection of certain items of the land, from the Aboriginal peoples' hunting grounds and tribal areas to flora and fauna reserves, while there are still large areas reserved for communications and transport—the travelling stock and camping reserves—and finally for strategic and military purposes.

The bulk, over two-thirds, of the vacant lands of Australia are to be found in the arid interior. These lands are vacant in the sense of not being currently used, but portions of them are currently being prospected by mineral companies, so their status may be an interim one.

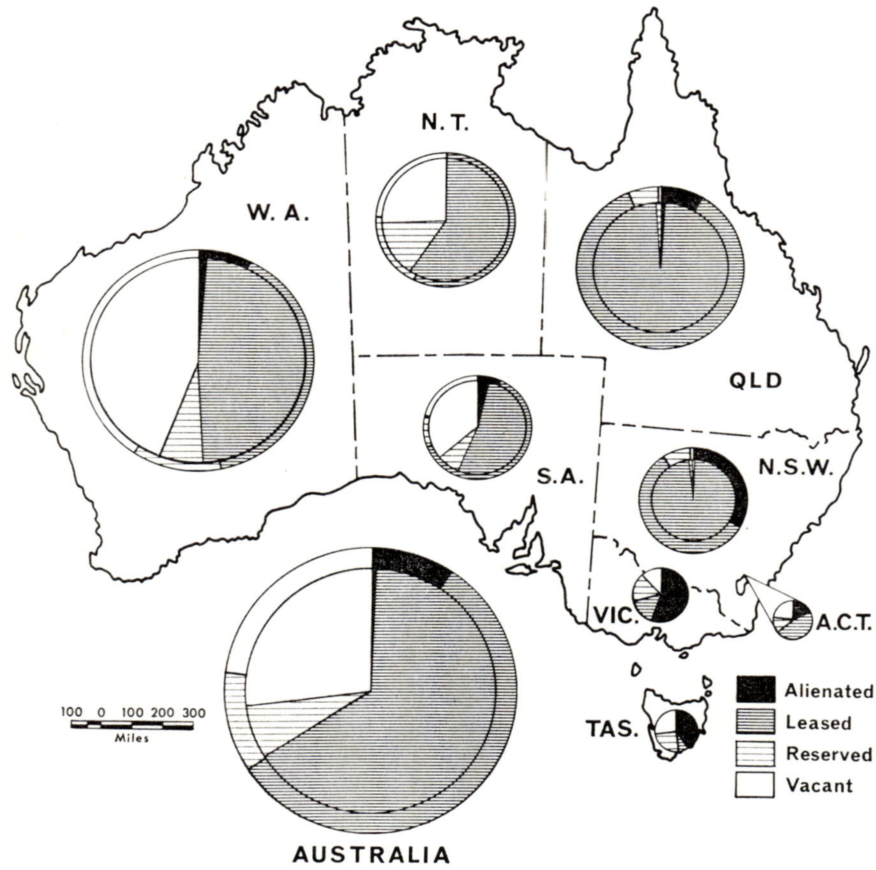

Fig. 3 Australian Arid Land Tenures in the 1960s

Inner circles show proportions of arid lands under various tenures; outer circles show the tenures for the whole state, territory, or Commonwealth.

Sources: As for Figs. 2 and 4

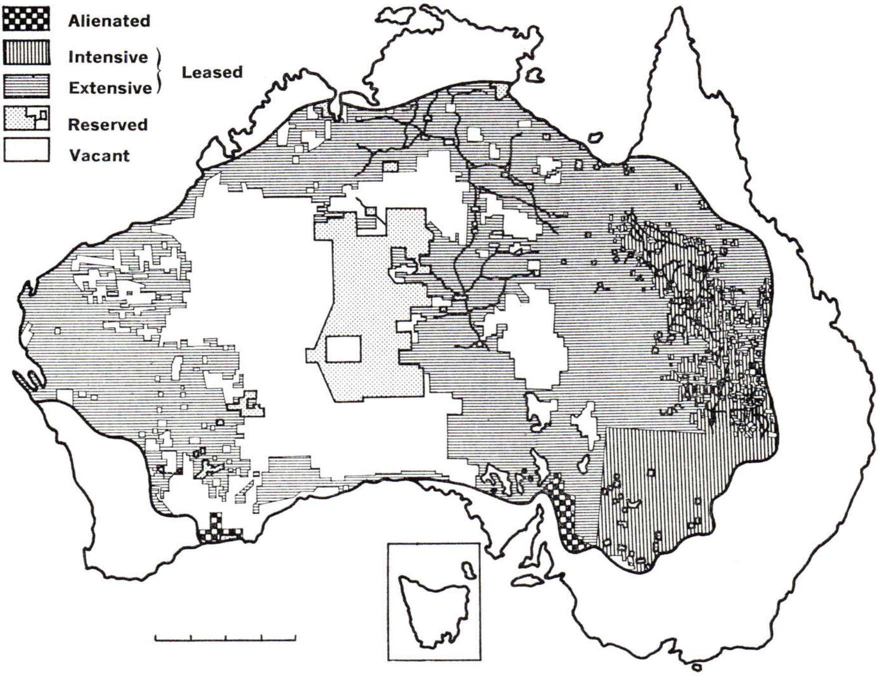

Fig. 4 The Pattern of Australian Arid Land Tenures in the 1960s
Note: This map has been generalised from the original at 1:4,500,000 scale. The intensively leased areas are pastoral leases where the size of individual leases is limited in terms of 'living' areas; the extensively leased areas are pastoral leases where no such limit applies. Mineral rights and leases are not included on this map.

Sources: Map of the Southern Portion of South Australia, June 1967, Surveyor General, Adelaide; Map of the Western Division of New South Wales, n.d. (1967), Soil Conservation Service, Sydney; Pastoral Map of the Northern Territory of Australia, June 1967, Department of National Mapping, Canberra; Map of Queensland showing Land Tenure, June 1962, Department of Public Lands, Brisbane; Map of Western Australia (showing pastoral stations), 1968, Department of Lands and Surveys, Perth (2 sheets); Map of Australia (National Parks and Equivalent Reserves), 1967, Australian Conservation Foundation, Canberra; Map of Australia (Unoccupied Land and Forest Reserves), 1968, Department of National Development, Canberra, (manuscript)

General Influences on the Evolution of Tenure Systems

For the explanations of current arid land tenures we need to look at two kinds of influences, first the general historical context of land settlement and legislation, and second the details of that legislation as evidence of reactions to the arid lands. For the first, we must examine the philosophical, social, and economic contexts for individual and institutional action, while for the second an examination of the sequence of pastoral land tenures will illustrate the detailed reactions to some of the qualities of the arid lands.

Rationalism and Progress

Land settlement in Australia began in the eighteenth century Age of Reason and developed in the nineteenth century Age of Progress. Perry (1966) has shown how the characteristics of the Australian climate were rationalised in 1744 by their latitudinal location well before the details were known, and a similar attitude characterised the views of the continent as a whole. Rationalists, using analogies with the conditions in the Northern Hemisphere, drew a new Mississippi River through the interior (Heathcote, 1965a:20) and argued that the whole continent would be of value, for all things had a place under the beneficent Hand of Providence (Yi-Fu Tuan, 1968). If deserts did exist, then with man's newly demonstrated skills and successes from the Industrial Revolution, and the evidence of natural change through time in Darwinian concepts of evolution, the deserts could be made to bloom and the irrational waste lands brought into rational production. To achieve this transformation all that was required was the investment in the land of labour and capital, and the aim of the governments was merely to pass on the ownership of that land to their subjects for development, usually by the conferring of freehold title by sale or gift.

Settlement

The occupation of land was to develop its resources to an optimum situation where population was relatively densely and stably settled, and the civilisation of the waste lands thus achieved. The emphasis was upon density and stability of land occupation and use. If there were deserts in Australia they were not to be settled in the same way as the deserts of the old world but civilised by sedentary settlement. 'Nothing would be more unfortunate', commented the English Secretary of State for the Colonies to the Governor of New South Wales in 1831, 'than the formation of a race of Men, wandering with their Cattle over the extensive Regions of the Interior, and losing . . . almost all traces of their original Civilization' (HRA, XVI:116). Even now official attitudes to nomadic and shifting (but not necessarily, despite the implication, shiftless) land use are strongly antagonistic, even on a world scale (UNESCO, 1962: 357).

Equality

Opposition to any form of land monopoly has been a basic theme throughout the history of Australian land tenures. Even the free grants of the early years of colonisation were for limited areas, and with the various selection policies and legislations of the latter half of the nineteenth century, not only was land use to be intensified, but also shared among as many users as possible. Hence the tendency in the subsequent twentieth century closer settlement policies was for controls upon aggregation of holdings and continued attempts to subdivide the larger units.

Grafted on to the anti-monopolistic policies has been the concept of the family-owned production unit. References to yeomen settlers and their families, whether farmers or graziers, are sprinkled throughout the debates on the Land Acts from the 1850s onwards, and the success of these Acts was often judged solely on their ability to put individual families upon the land (Campbell, 1967; Roberts, 1924).

Profit

For the man on the land, however strong his concern for and belief in the previous considerations, the main question was the profitability of his occupation. In contrast, the concern for the economics of settlement at an official level did not seem to have appeared until at least the late nineteenth century, with the recognition that the subdivisions for closer settlement would need to be large enough for economic operations. The sequence of changes in the definitions of the subdivisions, the 'home maintenance' or 'living' areas, illustrates the increasing official recognition of the need to secure profitable as well as intensive settlement (Heathcote, 1965a: 74-7). In many cases, however, closer settlement and profitability have not been synonymous (Campbell, 1967:180-2; Lewis, 1963).

Conservation

The concepts of progress, development, and settlement have received setbacks in the twentieth century as a result of increasing scientific concern for the effect of use upon the quality of the land. The concept of conservation dates back to at least Medieval time in Europe (Glacken, 1956), but the ideas have been reinforced by recent ecological studies, particularly in areas opened up to European settlement within the last two hundred years, which have suggested that uninhibited use has led not to the improvement but to the deterioration of the land. As a result there has been increasing legislation in Australia, as elsewhere, to control the use factor in land tenures, particularly with regard to soil erosion. In addition there has been concern for the continued occurrence of catastrophes, such as droughts and floods, and the recognition that not all the influences on land settlement can be rationalised nor seen as leading to the improvement of conditions.

Thus, the general influences upon land tenures might be summarised as:

i. Suggesting a uniform development of the 'waste lands' of Australia.

ii. That development to be by uniform private (freehold) land ownership and unlimited use, preferably by family-owned units.

iii. That development to result in the intensification of land use and progressive intensification of rural population densities by the development and improvement of the latent resources and their productivity.

iv. The above points might be liable to modification in the light of possible conflicts in the policies and the economic and ecological questions posed particularly within the last thirty to forty years.

The details of the contemporary Australian land tenures show that these general influences have had to be considerably modified for the arid lands, and as an example of some of the necessary modifications, as well as illustrating the complexities of tenures for the currently largest land use in the arid lands, the pastoral tenure systems will be examined.

Pastoral Tenure Systems—a case study

Pastoral land use in Australia initially copied British practices and took place on land held under tenure systems imported from Britain. In western Scotland at the time of the First Fleet, cattle and sheep were being herded over the extensive freehold estates of local and absentee landlords or on lands leased for grazing for up to twenty-five years, and then driven south over the 'drove roads' to be fattened in eastern England for slaughter (Mackenzie, 1810). In Norfolk, sheep were grazing and treading the sandy heaths back into a cultivable state, on lands held for twenty-one years as agricultural leases (Young, 1804). Over the rest of England and Wales, villagers had, since earliest records, grazed their livestock on the common meadows and woodlands, and as late as 1873 there were still 1·6 million acres of such commons in use (Royal Commission, 1958: 150-65). Thus in British experience domestic livestock could graze over land held under all forms of tenure, from freehold to commons.

Before the Imperial Waste Lands Act of 1846 and the Orders-in-Council which applied it to the local conditions, Australian pastoral tenures had been merely transfers of British experience. Having acquired all rights over land from the Aborigines, the British Government and local officials allowed grazing on all forms of tenures. Townsfolk were to be provided with grazing lots for their animals from the beginning of settlement (HRA, I: 127) and 'commons' were set up in New South Wales in 1804 'to encourage the grazing of livestock' (HRA, V: 6 and 91), and in South Australia in 1845 (Lewis, 1915-16). Macarthur ran his sheep on the free grant of Camden Park (Ellis, 1955), and leases, provided for in the first governor's instructions, were beginning to appear by the 1820s in the form of 'Tickets of Occupation' which gave the right to graze stock up to two miles from a stockyard. More genuine forms of lease appeared in 1827 when rents of £1 per 100 acres were charged, but these were still within the settled counties and not until 1836, when the system of 'Depasturing Licences' was introduced in New South Wales, was an unlimited area to be grazed beyond the settled Nineteen Counties for £10 per year (Burroughs, 1967: 150; Roberts, 1924: 189). In South Australia leases for 'commonage' on unsold lands were provided for in 1834 and standardised for 3-year terms in 1836 as purely grazing operations (Lewis, 1915-16).

The Imperial Waste Lands Act of 1846, however, was the first indication at official level of the need to modify the general policies of land settlement to cope with the specific problems of the arid lands. The modifications were not complete, but a trend was begun which was to produce the contemporary pattern of tenures. Alongside the general influences, which continued to affect some policies in the arid lands, were ranked new attitudes to the challenge of the arid lands.

To understand the parallel development and continuing conflict between general and specific official land policies and tenures, the specific challenges of the arid lands will be identified and the responses to them, positive or negative, discussed. Space prevents the detailed documentation of the legislations and proper recognition of the variations in detail between the states but these may be consulted in Heathcote (1965a and 1965b), King (1957), and Roberts (1924).

The challenges of the arid lands which faced the pastoralists were four. First was the sparse nature of the natural feed and consequent low productivity per unit area. Second was the challenge of competing users, whether indigenous fauna and the Aboriginal population or alternative contemporary resource uses, such as mining. Third was the challenge of general resource variability over the arid areas in time and space, whether feed or water supplies, and fourth was the challenge of specific and catastrophic variations in the availability of resources.

Low Unit Area Productivity

Before 1846 the initial reaction to this challenge had been to propose increases in the sizes of free land grants, but these had failed from fears of monopoly. The sales of land, once the free grants had been abolished in 1831, did not meet the challenge either, because of limitations on the maximum size of blocks and relatively high reserve prices for the lands. Despite sales on credit and free selections from the 1860s, arid lands were not selected as *bona fide* intensive independent production units, but usually as illegal acquisitions of strategic locations (waterholes and stockyards) by dummy owners to enable continued monopoly for pastoral use of the surrounding leased lands. Only where high capital investment had taken place or was envisaged, as on stud properties or irrigated agricultural land, were sales achieved.

The 1846 Act was the first recognition of the need to diversify tenures to meet the diverse nature of the lands and particularly the low yields from the pastoral use of the interior. As a result of the Orders-in-Council enforcing the Act in Australia, by 1850 New South Wales had been divided into Settled, Intermediate, and Unsettled (mainly arid) lands, with pastoral leases, varying from one, through eight, to fourteen years respectively, but with no limits on the total lands so held; South Australia had lands within the proclaimed hundreds which were to be sold, and lands outside the hundreds (mainly arid) which were to be leased similarly for up to fourteen years; and Western Australia had a parallel division with leases for up to eight years in the arid country. The basic division between extensive leasehold occupation of the arid areas and other tenures in the humid lands was already established and was to be re-emphasised throughout the following century of occupation.

The basic tenure system for the pastoral occupation of arid lands was a compromise by which the problem of disposal of title to the arid lands was sidestepped by disposing of only part of the title for only part of the time. Only the rights to natural waters and feed were initially granted, and in this way only a portion of the whole was used; the returns to official revenues (rents) could therefore be acceptably lower than if the whole was used; the cost to the user (rents again) could be kept to a minimum (usually less than a halfpenny per acre per year), and large areas could be allowed to fall into few hands since the tenure was terminating and reverted to the Crown. The usefulness of this tenure system has been demonstrated for over a century of legal and generally profitable occupation of the arid lands by pastoralists, and it has even been incorporated into the current closer settlement policies (Table 4B). The basic principle of division of title into various leaseholds is today the characteristic response to the main challenge of the arid lands (Fig. 5).

Competing Users

By standards of world history the occupation of Australia's arid lands has been relatively peaceful, but it has not been without conflicts between rival resource users. European livestock faced competition from native herbivores as well as imported game animals, and were victims of native predators and Aboriginal hunters. Controls on 'vermin' (arbitrarily defined from rabbits to eagle hawks) have been written into most leases, and native fauna has been protected only relatively recently and for very limited areas. The acquisition of Aboriginal hunting grounds has only been partly offset by the creation of Aboriginal Reserves in the most unpromising pastoral country, but it is interesting to note that Aboriginal hunting rights have been preserved in South Australian and Northern Territory pastoral leases. Here, at least, complementary use of the grazing lands is envisaged.

Resource Variability

To meet the challenge of scarce and variable resources the initial official reaction was to reserve such resources for general use. The water reserves set aside from the 1840s onwards around waterholes and river frontages, and the reserves around government-dug artesian wells from the 1880s onwards, fall into this category of response. To encourage private search for scarce resources, grants to successful explorers and the specific Artesian Well Leases of New South Wales in the 1880s and currently similar leases available in South Australia have been provided. Finally, to encourage private capital expenditure to cover resource variations, all a lessee's 'improvements' (buildings, tanks, wells, etc.) are generally compensated for on termination of the lease. However, despite this concern for the water resources in the arid lands, there has yet been no attempt to standardise on a continental scale the legal control of water supplies, and the application of the unsatisfactory English Common Law system is still widespread.[4]

To meet the variation of fodder resources, various tenures have been applied. Timber has been excluded from grazing rights, but edible trees can be browsed as long as the tree itself is undamaged, but permission is usually required for them to be cut down. In contrast, the clearance of inedible and therefore useless scrubs has been officially encouraged by special developmental 'scrub leases', as in South Australia 1866-7 and New South Wales after 1884. Finally, the need to move livestock not only to markets but also to agistment areas has been met by the reserves for travelling stock routes (as successors to the Scottish 'drove roads'), which also have offered unscrupulous 'grass pirates' free 'commonage' on occasions.

Catastrophes

At the extreme levels of resource variability, when droughts destroyed pastoral resources, the relief policies have included tenure modifications. One early and continuing method of relief has been the remission of rents until the drought ended; another, the extension of term of the lease to allow drought losses to be recouped by profits in the good seasons which are assumed to follow. In Western Australia the first method is currently applied also to relief from cyclone and flood losses.

The relatively short-term effects of drought, however, are currently overshadowed by fears of the long-term effects of over-use of the pastoral resources. In extreme cases, where the resources appear to have been exhausted, leases have not been renewed and the land has reverted to the Crown,

[4] Water rights and legislation affecting water use vary between states and territories and it is therefore surprising that no consideration has been paid, even in recent studies (Australian Academy of Science, 1964; Langford-Smith and Rutherford, 1966), to the legal background to this vital resource use.

TABLE 4

PASTORAL LAND TENURES IN THE 1960s

A. EXTENSIVE TENURES

Factor	State or Territory			
	QLD*	S.A.†	N.T.†	W.A.†
CONTENT	← – – – – – – – – – – – – Fodder and water	– – – – – – – – – – – – →		
TIME (years)	30	{ 21 on new country 42 standard	50	50
SIZE (acres)	No limit	No limit	3,200,000‡	1,000,000‡
PERSONNEL	← – – – – – – – – – – – – – No restrictions	– – – – – – – – – – – – – →		
USE	Stocking liable to controls. Control of noxious weeds required.	Minimum and maximum stocking limits. Control of noxious weeds.	Stocking limits. Soil erosion survey possible.	Minimum and maximum stocking limits and continuous stocking required. Pasture management required. No agistments allowed.
	← – – – – – – – No agriculture without official consent	– – – – – – – →		

*Pastoral holding †Pastoral lease ‡Maximum allowed

B. INTENSIVE TENURES

Factor	State or Territory	
	N.S.W.*	QLD†
CONTENT	← – – – – – – – – – – – – Fodder and water	– – – – – – – – – – – – →
TIME (years)	Perpetuity	Up to 30
SIZE	Up to 2 home maintenance areas	Living area‡
PERSONNEL	Personal residence on, or within 'working distance' of, for 5 years.	No agents corporations or partnerships of more than 2 persons. Personal residence for 7 years.
USE	Control of noxious weeds and vermin. No agistment without approval.	No agistment for more than 6 months in year. Control of noxious weeds.

*Western land lease, perpetual lease †Grazing selection ‡Maximum, 45,000 acres but most evidence of this concern is in the increasing controls of land management written into current pastoral leases. Control of noxious weeds and limits upon the numbers of stock to be grazed are found or implied in most current pastoral leases (Table 4).

PROBLEMS OF LAND TENURE IN ARID AUSTRALIA

To summarise the main points of this paper the problems of devising a suitable land tenure for the arid lands will be analysed in terms of the theoretical framework put forward earlier.

Content Factor

The most successful arid land tenures seem to have been leasehold systems. Alienation failed because the reserve price initially was too high for merely grazing operations, the potential of the land was untried and uncertain, and the authorities refused to allow unlimited alienation anyhow. Under the leasehold system, land resources are differentiated and tenures awarded to each resource separately, thus enabling a division of land uses and of users (Fig. 5). So far the potential for multiple use has not posed many problems because the area of the arid lands is so vast that overlapping uses are rare and certain uses do not seem mutually exclusive. Thus the maintenance of Aboriginal hunting rights over certain pastoral leases seems to be possible (although no Aboriginal opinion of its success has been canvassed), and grazing can be carried on over parks or recreational reserves in Western Australia for example.

Recently, however, as a result particularly of renewed development of mineral resources, there have been clashes of interests between rival resource users. The clash between pastoral and mining uses seems amenable to settlement by some system of compensation, but already there have been significant clashes and forecasts of future problems between mining and Aboriginal interests. While the Gove Peninsula Case (*The Australian*, 14 December 1968) is, strictly speaking, outside the arid lands and still *sub judice*, it is a type of clash which will be inevitable within the arid lands. The problem is complicated not only in that the content factor of the Aboriginal rights within the Arnhem Land Reserve is not clear (at least at time of writing—December 1968—it was doubtful whether the reserve included the mineral rights), but because there is a conflict of opinion between Commonwealth officials, who are prepared to allow a private mining company to mine a reserved area, and the Aborigines, who believe that all the land rights belong to them and who wish to prevent the mining company's trespass. Here in miniature is the Toennies dichotomy, the conflict between the official-Gesellschafts-system and the group-Gemeinschafts-system. This clash could be

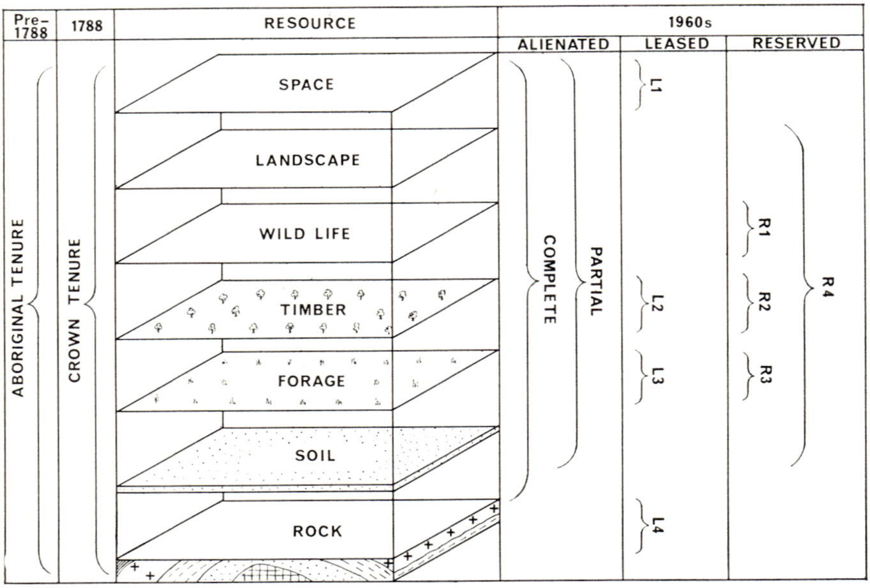

Fig. 5 The Framework for Arid Land Tenures

Note: The range of resources covered by Aboriginal, Crown, and contemporary types of land tenure is indicated. Initially, post-1788, alienation covered the complete range of resources, but since approximately mid-nineteenth century the alienations have excluded rock and mineral rights. Examples of leased tenures: L1—military leases; L2—timber licences; L3—pastoral leases; L4—mineral leases. Examples of reserved tenures: R1—fauna protection; R2—forest reserves; R3—travelling stock routes; R4—national parks.

repeated in any of the Aboriginal Reserves, or in any of the National Park or flora or fauna reserves in the arid lands, the clash being in these cases between the official view on the one hand and, on the other, the Aboriginal or general public's view of what reserves should be.

Differences of content occur not only between factual and fanciful views of tenures, but also between the facts across political boundaries. The contrasts between the pastoral tenures in each state have been drawn up as an example on Table 4 and similar tables for mineral leases, water rights, and the reserved areas could be provided. The question here is whether such variations are in the best interest of the rational development of the arid lands.

Time Factor

The dominance of leasehold tenures means that even in the case of so-called perpetual leases, the tenure is terminating and will revert to the Crown eventually. Most of the leaseholds are timed to run for long enough to achieve the intended land use and to enable the occupation to be effective and profitable. Despite arguments over the optimum length of these tenures, the success of settlement so far, the uncertain latent resources of the arid lands, and the growing concern over possible misuse of the land are strong arguments to retain the leasehold system at least until the full scope of the resources of the arid lands is understood.

Size Factor

The anti-monopolistic controls upon the size of holdings do not seem particularly suitable for the arid lands. Certainly the concept was not designed for the arid lands, but for much richer agricultural lands, and the enormous costs and capital needed for arid lands uses, whether mining or pastoral, make the 'living area' concept difficult to apply without continuous modification. Whether this can be modified to include a capital requirement or not (Lewis, 1963), it is doubtful whether any but the largest production units can survive. Even the largest commercial concerns venture into the arid lands only after careful economic surveys and under detailed safeguards and guarantees of protection from state governments, the Mount Newman and Mount Goldsworthy iron ore agreements being cases in point (Western Australia, 1964a, 1964b). Such giant concerns, however, will always have to face public suspicion, particularly if in the good years their profits appear to be unreasonably high, for the public tends to forget the bad years very quickly (Campbell, 1968: B 7-4).

Personnel Factor

All the available evidence suggests that the single family-owned production unit, largely because of inadequate financial backing, is not the most successful type of land ownership in the arid lands. Official acceptance of company and institutional control of the larger units and even partnership owners for closer settlement pastoral leases, together with relaxation of the residential requirements which marked earlier agricultural closer settlement tenures (Heathcote, 1965: 74-8) all point to a growing recognition of the need for greater flexibility in ownership and settlement in the arid lands. The more stringent selection of, and financial requirements needed by, new settlers prevents all but the wealthiest of family men from applying, and even then success is not guaranteed.

Use Factor

The increasing official controls upon land management found in the new lease tenures pose problems. One aspect of the problem is illustrated in the pastoral leases, all of which contain developmental clauses requiring capital to be spent on improving the stock-carrying capacity of the land and all of which also contain some form of controls upon grazing intensities, to limit if necessary (because of overgrazing) the number of stock carried. The lessee might well complain in such a situation that he is caught between the goad and the knout, pricked on to spend money to increase his stock and then rapped over the knuckles when he becomes embarrassingly successful. Development and conservation may not be complementary policies.

A complication which will also need to be considered in future dialogues between the developers and the conservationists is the general ethic of land use. To return to the pastoral lessees and the controls of overgrazing, several questions need to be posed. For example, why is overgrazing implied to be evil? If because it causes soil erosion, does the evil rest in the deterioration of the environment's ability to reproduce equivalent fodder in the future or because of the destruction of a resource for which use was not included in the lease? If the evil lies in the effect on future fodder supplies, can we be sure how much of the deterioration can be ascribed to overgrazing and how much to natural causes such as drought and geological erosion? If the main cause *is* overgrazing, why is this in itself evil? Is it because one man receives a temporary benefit denied to future generations? If so, presumably the difference between this and mining operations lies in the so-called renewability of the resource. But can we say for certain that the vegetation of the arid lands is a renewable resource? The answer, I suggest, is not yet known, although Campbell (1966: 16) has already suggested the possibility 'that the lands of Central Australia can only be economically used for pastoral purposes as a slow mining proposition. In other words, pasture deterioration might be inevitable if the land is to be used at all.' In this case the ethics of land use are not merely a choice between private and public good, but something much more complicated.

Finally there is the problem that where tenures are so closely tied to use as they are in the arid lands, the Crown has a responsibility to the lessee, if, through no fault of his own, the particular land use for which he obtained the lease becomes impossible. Official drought relief policies are based partly on recognition of this fact and partly on general official attitudes to national catastrophes (Heathcote, 1967). However, the officials have recently argued:

> that if farmers and graziers elect to carry on their businesses in areas which past experience shows to be subject to violent climatic fluctuations they should be prepared to accept such fluctuations as part of ordinary business hazards and not expect the community to bail them out when conditions get tough.
>
> (Campbell, 1966: 15)

Future tenure policies will need to distinguish between drought relief measures and drought subsidies.

CONCLUSION

The future land tenure systems of the Australian arid lands are likely to remain basically leaseholds and reserves, the former to allow flexible use of specific resources and the latter to protect and preserve scarce resources either from all use or for specific public uses. The two basic types, however, are likely to become much more complex in the future, for the scope of resource use has widened considerably within the last twenty years. To the extensive grazing and limited mineral developments of mid-twentieth century have been added the new mineral discoveries, the military and civilian rocketry and aviation uses, the tourist attractions, and the recognition of rare scientific phenomena—from insects and Aboriginal relics to whole landscapes—which need to be preserved. Before the new types of leases and reserves to cope with these new resources are finalised, however, some thought will need to be given to some of the problems noted above. I would draw attention in particular to the question of competing uses in the future in the arid lands. Priorities will need to be decided, the various state and Federal land tenures will need to be, if not unified, at least reconciled, and the experience of other nations with similar problems consulted. In the arid areas of Australia, possibly even more than in the more humid areas, we need to know much more about not only the quality of the available resources, and the goals towards which resource use is to be oriented, but also the means by which those goals can be achieved—the land tenure systems.

REFERENCES

Australian Academy of Science (1964). *Water resources and management*, Melbourne University Press, Melbourne.

Bertrand, A. L. and Corty, F. L. (1962). *Rural land tenure in the United States: a socio-economic approach to problems, programs, and trends*, Louisiana State University Press, Baton Rouge.

Burroughs, P. (1967). *Britain and Australia 1831-1855: a study in Imperial relations and crown lands administration*, Clarendon Press, Oxford.

Campbell, K. O. (1966). Problems of adaptation of pastoral businesses in the Arid Zone. *Australian J. Agric. Econ.*, **10**(1): 14-26.

——— (1967). Land Policy. In D. B. Williams (ed.), *Agriculture in the Australian economy*, Sydney University Press, Sydney, 171-84.

——— (1968). Pastoral management under production uncertainty—problems and strategies. In *Proceedings of bankers' conference on drought*, Reserve Bank of Australia, Sydney, B1-1—B1-13.

Collins, C. M. (1949). *The valuation of property, compensation and land tax*, Law Book Co., Sydney.

Ellis, M. H. (1955). *John Macarthur*, Angus and Robertson, Sydney.

Glacken, C. J. (1956). The origins of conservation policy. *J. Soil Wat. Conserv.*, **11**(2): 63-6.

Heathcote, R. L. (1965a). *Back of Bourke: a study of land appraisal and settlement in semi-arid Australia*, Melbourne University Press, Melbourne.

——— (1965b). Changes in pastoral land tenure and ownership, an example from the Western Division of New South Wales, *Aust. geogrl Stud.*, **3**(1): 1-16.

——— (1967). The effects of past droughts on the national economy. In *Report of the A.N.Z.A.A.S. Symposium on Drought*, Commonwealth of Australia, Bureau of Meteorology, Melbourne, 27-45.

H.R.A. *Historical Records of Australia*, 1st Series, Sydney (volumes indicated).

King, C. J. (1957). *An outline of closer settlement in New South Wales, Part I The sequence of land laws 1788-1956*, Department of Agriculture, Sydney.

Langford-Smith, T. and Rutherford, J. (1966). *Water and land: two case studies in irrigation*, Australian National University Press, Canberra.

Lewis, J. (1915-16). Our early pastoral possessions and the pioneers of settlement in South Australia, *Proc. Roy. Geog. Soc. Australasia,* **17**: 1-83.

Lewis, J. N. (1963). Is the concept of the home maintenance area outmoded? *Aust. J. agric. Econ.* **7**(2): 97-106.

Mackenzie, G. S. (1810). *A general survey of the counties of Ross and Cromarty*, R. Phillips, London.

Perry, R. A. (1967). The need for rangelands research in Australia, *Proc. ecol. Soc. Aust.*, **2**: 1-14.

Perry, T. M. (1963). *Australia's first frontier: the spread of settlement in New South Wales, 1788-1829*, Melbourne University Press, Melbourne.

——— (1966). Climate and settlement in Australia 1700-1930: some theoretical considerations. In J. Andrews (ed.), *Frontiers and men*, Cheshire, Melbourne, 138-154.

Roberts, S. H. (1924). *History of Australian land settlement 1788-1920*, Macmillan of Australia, Melbourne (reissued 1968).

Royal Commission (1958). *Report of Royal Commission on common land 1955-1958*, HMSO, Cmnd 462, London.

UNESCO (1962). Nomadic pastoralism as a method of land use. In *The problems of the Arid Zone: proceedings of the Paris symposium*, Arid Zone Research Programme No. 18, UNESCO, Paris, 357-64.

Western Australia (1964a). *Iron ore (Mount Newman) agreement, 13 Elizabeth II, No. 75*, Government Printer, Perth.

——— (1964b). *Iron ore (Mount Goldsworthy) agreement, 13 Elizabeth II, No. 97*, Government Printer, Perth.

Yi-Fu Tuan (1968). *The hydrologic cycle and the wisdom of God: a theme in geoteleology*, University of Toronto Press, Toronto.

Young, A. (1804). *General view of the agriculture of the County of Norfolk*, B. McMillan, London.

13

ASPECTS OF THE ECONOMIC HISTORY OF THE ARID LAND PASTORAL INDUSTRY

ALAN BARNARD*

THE HISTORY of sheep- and cattle-raising in Australia's arid regions is essentially a story of the utilisation of marginal land. It is also part of the story of the preliminary assessment and exploitation of a continent newly come to white men's hands. The penetration and conquest, partial though it may still be, of the harsh lands outside the boomerang of eastern Australia's fertile and semi-arid areas and outside the coastal fastness of the south-west corner is an accomplishment made possible only by man's obstinate courage and enduring ambition. It is a story of privation and hardship beyond city imagination, of ingenuity and perseverance, of epic ventures and tragic losses.

The process of occupation began in the early 1850s; the means by which it was carried forward was the extension of pastoral production. By 1914 the boundary of effective white settlement extended as far as it does today. By that time, too, the peak of the region's relative contribution to the support of Australia's pastoral stock had been passed. The numbers of sheep and cattle it carried exceeded, in some years of the twentieth century, those depastured in the 1890s; the performance of those years could not be sustained (Tables 1 and 2). More significantly, the arid zone's share of total sheep and cattle, maintained for many years at about one-third, has fallen to less than one-fifth; it has paid better to increase flock and herd capacity in the non-arid lands. This change serves as a salutary reminder: the crucial feature of arid land occupation was that it depended on making productive use of land that was inferior to other land being used for similar products.

It was (and is) inferior land by reason of rainfall, temperature, soils, natural vegetation, and distance from market outlets and supply centres. It is not, however, uniformly inferior. The arid districts first settled were those in which the combination of climate, fertility, topography, and distance was least disadvantageous when compared with areas of previous settlement. Successively marginal regions were brought into production. Central—and necessary—to this process were changes that made it worth while. (Why blaze new trails on the Georgina if purchasing an established property at Roma will yield as

* Senior Fellow in Economic History, Institute of Advanced Studies, Australian National University.

satisfactory a return?) Rising prices of pastoral products, falling costs of production, or differential movements of cost/price relationships in marginal and intra-marginal districts provide a simplified framework of the economic inducements that created the geographical spread of production. Occupation rarely retreated, and certainly never retreated substantially, when the conditions that led to expansion disappeared. Man is a stubborn animal, especially in time of adversity. But while individuals and the industry might resist pressures under which economic man would wilt, *expansion* of output after occupation was fully established was dependent on the same kinds of market stimuli that originally induced occupation. Moreover, unless output could be increased without further investment to increase capacity, such expansion necessarily involved comparison with the profitability of further production in districts not subject to the arid zone's peculiar disadvantages. The Mitchell grass downs have always been in competition with Victoria's western district; Alice Springs with Goulburn.

In so far as decisions are based on economic criteria, investment—public as well as private—was directed, and will continue in the future to be directed, to those activities and to those areas which, with actual and prospective market conditions and technology, promised the best rate of return.

OCCUPATION

The history of the settlement and progress of the arid region, as a region, has yet to be written. A number of useful, often stimulating accounts of various districts have been written recently by Kelly (1952, 1966), Rose (1954), Bolton (1954, 1963), Bauer (1959, 1962, 1964), Durack (1959), Duncan (1964, 1967), Heathcote (1965), and others. All I can encompass in this paper, which draws heavily on their works, is a quick survey of selected highlights that will identify the realities on which pastoral activity in the arid zone is based.

Three stages of 'occupation' need first to be distinguished: pioneer settlement, essentially tentative, always exploratory, often eccentric; occupation of the bulk of usable land within a district, 'on-the-ground' rather than just on maps, and productive use of that land however light the stocking rate; full utilisation of the land in a district, up to the limits of capacity set by the technology accepted as appropriate to the time and the place.

Explorers opened the way—romantic blundering figures who found pastoral lands where none existed, as well as wily bushmen who knew what they sought and knew how to get back. Settlement and occupation followed them. But neither their timing nor their geographical pattern was related with necessary closeness to prior exploration. Occupation proceeded, in fact, in three well defined spurts.

The first occupation of the area beyond what became Goyder's Line in South Australia, of the central Queensland plains and of the Gulf country, was concentrated in the late 1850s and the first half of the 1860s. It was not by chance that these were the starting points. Agricultural and pastoral pressure on South Australia's narrow non-arid strip induced pastoralists to move to contiguous, arid, land. Queensland's expansive plains, though distant from markets and from the better lands in the south-east corner from which settlers came, are among the most favoured of the arid lands.

Occupation was fingering into the northern Flinders Ranges early in the fifties. Beltana, in the 6-8 inch rainfall belt, was taken up in 1854 and Blanchewater, with an average rainfall of 5 inches, in 1857 (Beltana, 1965). When drought struck the northern runs in 1864-5, the region between Lakes Frome and Torrens and that upward of 150 miles east of the southern Flinders Ranges were fully leased or claimed and there was a scattering of stations south of Lake Gairdner and around Strangeways Spring, Mt Margaret, and Lake Eyre (Northern Runs Commission, 1867). In Queensland settlement moved out from the Darling Downs, fully occupied by the fifties, and from northern ports with narrow coastal hinterlands. By the mid-sixties stations had been established as far west as the Gregory River in the Gulf country (Bauer, 1959); farther south, settlement had passed Aramac, Blackall, and Charleville. By 1866 a fifth of the colony's sheep and cattle were run on arid lands.

The second spurt of settlement fell roughly in the decade 1875 to 1885. It was in this period that the plains and ranges west of the Darling became one of the fastest growing pastoral regions of New South Wales. With the same reckless haste settlement was thickened in the Gulf country and Queensland's central and western districts. In 1880 occupation of the Queensland north and west was concentrated around the headwaters and courses of the main northern rivers and along the line of the western boundary and was interspersed among the channels of the south-west (Bauer, 1959). By 1890 the most westerly pastoral districts were supporting more than two and a half million sheep and nearly as many cattle, while on the Mitchell grass downs and central plains were more than half the colony's seventeen million sheep (Butlin, 1962a: Tables III and IV).

It was at the end of that same decade that the Northern Territory was occupied—twenty years after preliminary exploration and ten years after the completion of the overland telegraph line and the establishment of the first lonely stations. The extension of sheep-raising to Western Australia's arid areas became significant only in the second half of the eighties, at the tail end of this period of expansion, and the settlement of the Kimberleys gathered pace through that decade to the early nineties.

At the end, even the extended end, of that period, occupation was still incomplete in Western Australia. It was rounded out and filled in during the

TABLE 1

ARID AUSTRALIA: ALL CATTLE

Year*	Number† ('000)	Share in Australian Total: All cattle (%)	Share in Australian Total: Beef cattle‡ (%)
1865	253		
1866	267		
1867	190		
1868	234		
1869	248		
1870	255	6	
1871	303		
1872	343		
1873	389		
1874	456		
1875	650		
1876	640		
1877	977		
1878	1,101		
1879	1,374		
1880	1,421	19	
1881-93	Queensland not available		
1894	3,701	31	
1895	4,094	34	
1896	3,958	34	
1897	3,644	35	
1898	3,312	32	
1899	2,968	31	
1900	2,193	26	
1901	2,079	24	
1902	1,926	23	
1903	1,893	27	
1904	1,992	27	
1905	2,161	28	
1906	2,341	27	
1907	2,683	29	
1908	2,918	29	
1909	3,160	30	
1910	3,465	31	
1911	3,471	30	
1912	3,504	30	
1913	3,518	30	
1914	3,529	31	
1915	3,260	29	
1916	3,276	33	
1917	3,656	35	
1918	3,871	33	
1919	3,958	31	
1920	4,110	32	
1921	4,307	32	
1922	4,539	31	
1923	4,446	31	
1924	4,450	33	
1925	4,431	33	
1926	3,856	29	
1927	3,602	30	
1928	3,305	28	
1929	3,015	27	
1930	3,010	27	40
1931	3,106	26	39
1932	3,129	26	39
1933	3,318	26	38
1934	3,386	25	37
1935	3,306	24	37
1936	3,281	24	38
1937	3,261	24	40
1938	3,279	25	41
1939	3,310	26	40
1940-43	Queensland not available		
1944	3,633	26	39
1945	3,675	26	40
1946	3,549	26	40
1947	3,488	26	39
1948	3,703	27	40
1949	3,890	28	40
1950	4,125	28	40
1951	4,034	27	39
1952	3,999	27	38
1953	3,509	22	32
1954	3,639	23	33
1955	3,817	23	33
1956	4,139	24	34
1957	4,161	25	35
1958	3,771	23	33
1959	3,690	22	32
1960	3,676	21	30
1961	3,594	20	28
1962	3,699	20	28
1963	3,789	20	27

first decade of federation. Moreover, while occupation was followed very closely by full—indeed, over-full—utilisation of eastern arid land, in Western Australia and the Northern Territory the two processes were more widely separated in time. Peak cattle numbers in the west were reached before 1920 but full stocking with sheep not until the mid-thirties. Similarly, full use (in terms of the accepted technology) of Northern Territory lands occupied in 1890 was not attained until the mid-twenties.

---

* In the original data the dates are shown variously as at 31 December, or 1 January, 31 March or 30 June of the following year.

† From 1865 to 1879 numbers are for Queensland and South Australia only; no data are available for west of Darling and Northern Territory before 1880; Western Australian data from 1890 only used here.

South Australian figures are not available for 1886-9 and 1894-6.

Queensland figures for 1869, 1872, and 1950 have been interpolated. Coverage for Queensland data changed in 1953; figures from that date are not directly comparable with preceding figures. Trend movements may be extrapolated by comparing 1953-6 figures above with overlapping figures collected on the old basis:

Year	Number	All cattle %	Beef cattle %
1953	4,144	27	38
1954	4,287	27	39
1955	4,468	28	39
1956	4,798	29	39

‡ Statistics not available before 1930.

SOURCES AND GEOGRAPHICAL COVERAGE

Western Australia: includes all of State north and east of 10″ isohyet (note inclusion of high-rainfall Kimberleys); official statistics have been adjusted to conform as closely as possible to the 1963 boundaries of the Statistician's Central, North-western, Pilbara, and Kimberley Divisions plus Kalgoorlie, Coolgardie, Laverton, Leonora and Menzies.

South Australia: roughly, the area beyond 10″ isohyet; official statistics adjusted to conform as closely as possible to the 1963 boundaries of counties Bosanquet, Buxton, Hamley, Hore-Ruthven, Manchester, York and Young plus 'outside the counties'.

New South Wales: west of the Darling; sheep from Butlin (1962a: 300-1 and Table I) and official statistics since 1954; cattle from official statistics adjusted by Butlin's methods before 1893.

Queensland: figures for 1865-1952 are from Butlin (1962a: 282, and Tables III, IV, 304-7, districts I-V) and adhere to his boundaries; from 1953, official statistics for South-western, Far-western and North-western Divisions together with the Local Authority Areas of Aramac, Barcaldine, Blackall, Ilfracombe, Longreach, and Tambo; note the inclusion of Cape area before 1953.

Northern Territory: official statistics for the whole of the Territory; note the inclusion, therefore, of the higher-rainfall areas.

Figures for the individual States have been omitted from this table; they are available from the author on request.

What induced geographical expansion of production during these three periods?

A decade of high wool prices, dating from 1856, underlay the first. Combined with a marked decline from the very high labour costs of the early gold-rush years these prices brought to the sheep industry a prosperity it had not known since the thirties and one more widely shared than that derived from supplying meat to the goldfields. At the same time the cattle industry, which had expanded greatly in the fifties in response to the sudden increase in population, was facing falling prices and declining *per capita* consumption. Both developments served to make sheep more profitable than cattle. Williams (1962) has described how, in the semi-arid Riverina, sheep consequently displaced cattle and how, as pastoralists learned some of the techniques required for successful sheep-raising on the back blocks, there followed a more intensive utilisation of land formerly only partially occupied. Substantially the same pattern was followed in northern New South Wales. One result of this was a rise in the price of pastoral property in the established areas, increasing costs there and reinforcing newcomers' judgment that, even in arid areas, they could produce with profit.

Three factors adequately explain the movement of cattle into arid lands (effectively a redistribution of stock from New South Wales to Queensland) at a time when meat prices were unstable, consumption falling, and total cattle numbers in eastern Australia actually falling. First, cattle could be slaughtered either for meat or for tallow. It was specifically for that latter fate that the Scottish Australian Company, well aware of the meaning of costs and profits, included 7,000 cattle when it first stocked Bowen Downs in 1863 (Macmillan, 1963: 26). Second, cattle could supplement sheep by occupying, and thus securing, back country. It is possible that even at that time some graziers deliberately used cattle as precursors, ecologically as well as financially, of a later concentration on sheep. Third, the expulsion of the cattle industry to more distant or more arid regions created, for a time, the fuel for continued ideas of prosperity. Purchases made by the Scottish Australian Company and by Patsy Durack, who in 1862 set off from Goulburn with £4,500 worth of horses and cattle headed for the Barcoo, provided an additional outlet for stores that kept prices up. And it was at existing prices that new entrants calculated their prospective returns.

The secondary expansionary phase was set off by a short period of boom prices and continued through a longer period of boom expectations that co-existed with falling prices. Butlin (1964: 85-124) has laid bare its ramifications in the sheep industry. In 1872 wool prices reached levels not topped for a half century and from 1871 to 1875 inclusive remained at or above those established in the first half of the sixties. Profits from sheep stations, especially in the geographical heart of the industry—eastern and Riverina New South

Wales and the Queensland Downs—continued to rise, however, right through the seventies. The reason was, simply, that in anticipation of and as a result of the construction of fenced paddocks they were carrying progressively more sheep than they had in the past. On many, income stood two and three times as high as it had in the good years of the sixties.

High profits induced confidence in continued pastoral expansion. Investment in fencing was intensified; the entry of additional producers was encouraged. Land in the established sheep areas was, however, already fully occupied. A rapid rise in the value of pastoral property resulted and, from the sale of going concerns, many who had taken up properties in the preceding two decades reaped very handsome capital gains. But the price of entry to newcomers was high and though some large old estates were subdivided, the number of new properties thus formed was relatively small. The only regions in which a substantial number of new holdings could be established were the arid areas of western New South Wales and Queensland. A decade earlier these regions would have been unattractive, for in the earlier phase of expansion all the better river frontages had been pre-empted. The proven success of paddocking semi-arid lands bred conviction, however, that fences and the excavation of some tanks could open cheap arid lands to sheep and also yield there financial returns comparable with those produced on more valuable land with higher rainfall. Moreover, many of the settlers who moved west of the Darling or into western Queensland were in fact experienced men who had developed and sold improved properties in semi-arid New South Wales and South Australia. Their confidence in the formula was deep-rooted.

Expectations of profit from wool production, which should have fallen as wool prices declined and costs of production began to rise, were retained into the early eighties. Increasingly, however, the natural increase of livestock was seen as the major source of gain. This attitude was sustained by and in turn gave sustenance to the continued multiplication of both sheep population and sheep properties. The thrust of settlement created a boom in inland pastoral real estate. Large properties, well though not fully improved to the standards of the times and carrying large flocks and herds that still fell short of estimated carrying capacity, changed hands for up to £500,000 in 1880 and 1881; others were formed, minimally equipped and stocked, specifically for speculative resale. Demand for stock for such properties, especially in the breeding-deficient areas where many were situated, provided a value for sheep which the prospects for wool alone could not confer.

Fencing made possible the occupation of vast areas. But, particularly on properties distant from rail or river transport, fencing was expensive and carried with it the imperative that investment programs be finalised quickly and that the productive opportunities it offered be utilised fully. It was for this reason, as well as because of a long period of unusually good rainfall, that the spread of occupation and the complete and rapid expansion of sheep numbers coincided so closely.

The implications of the sheep boom extended beyond the sheep industry itself. There was, firstly, a clear and obvious displacement of cattle by sheep on individual properties. In 1879, for example, Cudmore, Swan & Co. amalgamated four cattle runs, aggregating 4,000 square miles, into Milo and Welford Downs. They fenced Milo and built it into a sheep station that carried, at its peak, over half a million sheep (Milo, 1963). The example could be duplicated, in character if not in scale, many times over. It occurred to such an extent that there was, secondly, a wholesale redistribution of cattle between regions and between colonies. The total cattle population of New South Wales was halved between 1876 and 1885; that of Queensland doubled in the same time and increased by another 75 per cent in the next ten years. Half the Queensland increase was accommodated in districts with 20-inch rainfall or more, within 350 miles of the coast. The other half was absorbed, in the second half of the seventies, on the Mitchell grass downs, side by side with sheep; in the eighties in the Gulf and the west.

Only the first stage of this phase of geographical expansion for cattle production was accompanied by a stimulus from the domestic meat market. The decline and slow recovery in total cattle numbers in the sixties, initiated by drought and continued by falling *per capita* consumption, went too far. In the first half of the seventies relatively high cattle prices induced cattlemen to permit their herds to grow. In the context of good seasons their response was taken too far; by 1881 the relation between cattle numbers, population, and consumption per head was even more unfavourable than it had been in 1861. During the eighties, when the Queensland west was filling up and the first great mobs were being driven to the Territory and the Kimberleys, cattle prices were generally falling. Only goldfields' markets afforded any stimulus, and that often shortlived, to regional turn-off or numbers. Total cattle numbers grew, but more slowly than human population. Even though *per capita* beef consumption had risen from the low of the mid-seventies, the industry's turn-off for meat consumption was only 7 to 8 per cent. The tallow market was far less buoyant than it had been in the sixties or even in the seventies.

The incentive to stock the Territory and the Kimberleys came from the prospect of future capital gain. The prospect, hopelessly mistimed in fact, was based on contemporary as well as past experience. In the sixties the Scottish Australian Company was told that Bowen Downs would be very valuable 'if due patience is exercised' (Macmillan, 1963: 28). The forecast was justified. Similarly, Patsy Durack's Thylungra, in the Channel country, repaid patience. After sixteen years, during which 100 cattle had multiplied to nearly 31,000, he sold a controlling interest for £160,000. (That he never received full payment was another story.) Even before he had sold Thylungra, Durack had taken steps to establish his sons in the Kimberleys: had organised a private exploration, secured leases, and overseen the departure of the cattle on their long trek. The new lands satisfied his own land-hunger and his dynamic drive; they offered apparent freedom from the droughts that bleached the Channel country; and, to firstcomers, they held out the same prospect of capital gain that was to be realised by Durack himself in Queensland.

The Duracks took to Western Australia a technology that was essentially the same as that with which the Channel country had been pioneered and as that which had served in the Riverina and arid South Australia in the fifties. Throughout the arid regions cattle ran on the open range; natural water supplies were supplemented, until the eighties, only by occasional excavated tanks or wells; horse paddocks and drafting and branding yards were conveniences whose provision and permanence depended on sophistication and past profits. The use of Aboriginal labour, already widespread though not as quantitatively important as it later became, was a consequence of the same need to minimise working costs in order to offset the additional costs implicit in arid production —and of the scarcity and mobility of white workers.

Arid land sheep-raising had also been conducted on unfenced land in the fifties and sixties, with shepherds caring for flocks of 2,000 sheep and more. Apart from residences, of varying elegance and cost, investment was concentrated primarily on shearing sheds and facilities for washing either wool or, particularly later, sheep (Butlin, 1962b). Sheep-raising was revolutionised, however, by the technology based on fenced paddocks and conserved water. This not only broke through the previous confinement of sheep to the very choicest of arid blocks but, as a once-and-for-all shift, lifted sheep carrying capacity of nearly all arid blocks so greatly that the industry's ability to secure land (and labour) in competition with the cattle industry was sharply and permanently enhanced.

The investment necessitated by the technology was costly. As Butlin (1964: 111-47) has shown, the industry was able to meet it on a scale permitting the actual speed and magnitude of growth only because it was able to borrow heavily. The savings used were provided from Britain; they were channelled into the industry through banks and pastoral finance houses in a stream that was apparently illimitable and at a cost that was, in Australian experience, relatively low. Though individual pastoralists predominated and large individual debts, up to £500,000, were common, lending procedures and the scale of investment brought about a significant change in ownership structure. Partnerships became more important—at least in New South Wales, and probably also in Queensland—because the cost (and risk) of arid land development could be conveniently and profitably shared between urban or established pastoral interests and the developer and because institutional lenders were inclined to be less demanding if arid land assets were not their only security. The formation of operating pastoral companies, such as Queensland Co-operative Pastoral Co. Ltd that bought Durack's property, also represented the need to spread a large investment. They foreshadowed, in addition, the possibility of a direct sale to British investors or of securing additional development finance by the issue of debentures in Britain. Institutional investors of British origin such as the Scottish Australian Co. and the Australian Agricultural Co. were not unknown in the sixties. The Squatting Investment Co. and the Australian Pastoral Co., formed on Australian initiative, the one with Australian and the other with British capital and both with very large holdings, illustrate a change that took place in the seventies and eighties. Finally, the possession of pastoral assets outside the arid regions and a record of successful pastoral experience were assets of more than ordinary value to lenders. Thus C. B. Fisher could finance the stocking of Victoria River Downs on the security of his extensive holdings in the eastern and southern colonies.

REASSESSMENT

Ignorance of the true long-term character of the climate, ecological changes induced by stocking, and (infrequently) altered economic circumstances caused reappraisal of and often profound structural changes in arid land grazing after each phrase of nineteenth-century expansion. An exceptional wet in 1869-70, following on the aftermath of the Queensland financial crisis of 1866, finally convinced people that the coarse grasses of the Gulf were not really suited to sheep raising (Bauer, 1959: 30-1, 33). Drought, burning South Australia's northern runs in 1864-5 after their settlement in a long run of good seasons and striking again in the seventies, sobered pastoralists' ambitions to stock more heavily, taught them that the value even of fences and water improvements was limited on their arid steppes, and turned the attention of many of the more enterprising to other arid regions beyond their colony.

Misjudgment of climate or pasture was probably to be expected: part of the price of pioneering. The major features leading to reassessment were, however, more insidious in their emergence, were more slowly recognised, and were attributable to occupation itself.

The greatest calamity of all was the deterioration of the grazing potential of the New South Wales west and the South Australian north-east. The area west of the Darling attracted a major share of speculative attention in the expansion of the sheep industry in the early eighties. Boom prices were paid for the right to occupy land; vast sums were spent on improvements; heavy debts were incurred; and in 1891 half of all Western Division leases (and probably a higher percentage of leases west of the Darling) were mortgaged to banks or pastoral finance companies (Butlin, 1964: 141). From the beginning, carrying capacity was overestimated. Prime stands of saltbush and mulga and the oaten grasses first seen by new settlers existed because past seasons had been good and because there had been no grazing but the kangaroo's. They were deceptively inviting. But even if pastoralists had realised that their sheep were eating a haystack whose replacement was in the best of times a slow process, it is unlikely, once settlement had taken place, that they would have stocked any less heavily. Wool prices continued falling, with few interruptions; the property boom died. To make their investments pay, later to meet the interest charges or debts, and later still to prevent debts rising, they were driven to make the soil pay. By 1894 there were just under eight million sheep west of the Darling—one-seventh of the whole of the colony's sheep and nearly half as many as there were in all Queensland's vast arid sheep walks. From the late eighties, moreover, the sheep were sharing their feed with uncountable rabbits. In the early nineties the grasses and bushes had been eaten thin and low. Then the long drought of 1895-1902 carried the deterioration to a climax. Starving stock grubbed out grass roots, stripped saltbush and mulga irreparably. Bared of vegetation, surface soils were eroded by winds to create man-made deserts. In only a handful of years since has the region carried as many as half the stock run there in 1894: the restraint imposed on stocking rates by nature was reinforced by government regulation of lessees, imposed in the hope that regeneration could be encouraged.

Denudation of arid South Australian pastoral lands adjoining New South Wales was as devastating, though more drawn out and less spectacular, as that across the border. Badly affected as they were by the same drought, Queensland's arid sheep areas did not undergo the same physical transformation. Although rabbits were widespread in the south and although there were noticeable changes in grass communities, the most apparent and cautionary ecological change came later: the spread of prickly pear and Noogoora burr in the twenty years after the drought broke.

Physical abuse of a different sort became apparent in the Kimberleys by 1910. After nearly thirty years of occupation the river frontages were then displaying gross evidence of erosion caused by overgrazing by sheep and by the increased population of wallabies and kangaroos (Bolton, 1954). Complaints about competition from the euro in north-west Western Australia illustrated how, even in the mid-twentieth century, mismanagement could lead to pasture deterioration and possibly reappraisal (Fennessy, 1962: 236-8).

From the reassessment forced on the industry at the end of the century—by far the most traumatic—came three important consequences. It was realised, in the first place, that throughout the whole of the arid region long droughts were the characteristic, not the exceptional, climatic feature. As a result there was both a renewed emphasis on water improvements and a gradual scaling-down of estimations of stocking capacity. This occurred even in Queensland, where actual pasture deterioration was slightest. The more realistic estimates that emerged have proved, in general, to be those within which the industry can work safely and profitably. In the second place a fairly firm boundary was drawn around the areas best suited for sheep and those for cattle. Although substitution between sheep and cattle has continued, especially in Queensland's south-west, and although marginal regions have been brought into production or abandoned, the pattern of utilisation that emerged from the reappraisal has persisted for more than fifty years.

The third consequence was an increase in absentee, company ownership. This trend had its false starts in the later 1880s. The (eventually repudiated) sale of Victoria River Downs to an English company serves as an example of hopes that individual British investors or syndicates might prove less pessimistic than Australian lenders. Depression and drought, bringing pastoral finance companies into initially reluctant ownership of foreclosed properties in the 1890s, gave the movement its real importance in the sheep industry. Cain (1961), who has analysed the foreclosure process, has also suggested a major reason why some companies, at least, secured good profits from operating those properties: the economies and security derived from the integrated management of multiple holdings in diverse environments. After recovery, they retained some of them until the middle of the century. Other public, and private, companies took advantage of low property prices to do the same. The quarter century up to 1925 probably represents the heyday of the absentee company owner in the sheep industry. A proven potential and low property values provided conditions in the cattle industry, particularly in the north, for a similar trend in ownership. The Kidman interests expanded; Bovril Australian Estates Ltd purchased Victoria River Downs in 1909; Vestey Bros., within two years of their agreement to build the Darwin meat-works, had bought 30,000 square miles.

LONG-TERM TRENDS AND RECENT CHANGE

From about 1910, after drought losses had been made good, the number of stock in arid Australia has been remarkably stable. Sheep numbers have fluctuated between 20 and 30 million; but in only ten of the fifty-three years to 1963 did they exceed the 27 million depastured in 1893 and 1894. Just over 4 million cattle were run in 1895; since 1910 they have risen as high as 4·5 million and never fallen below 3 million. They, also, have exceeded the nineteenth century peak on ten occasions.

Two basic reasons underlie the failure of stock numbers to pierce, in a sustained manner, the ceiling put on capacity in the nineteenth century: an essentially static technology has been used over an essentially static geographic area; and, with one major and one minor exception, market inducements were lacking which would intensify investment permitting increased stocking rates.

Marginal alterations in technology, additional investment, and changes in stock management and breeding practices have been designed to provide protection against seasonal hazard and to increase yield per animal—as wool per sheep, carcass weight, or age of turn-off—rather than to support additional numbers of stock.

Before 1870 the capacity of arid lands to carry sheep was set by the availability of natural water supplies. The expansion of numbers between 1875 and 1894 was made possible by man-made watering points in fenced paddocks. In the twentieth century the limit to capacity is set by the availability of natural pastures. Additional security has been won by increased investment, both private and public, in artesian and sub-artesian bores and wells, in reticulation systems, in improved pumps, and in more and better dams and tanks. Smaller paddocks, more subdivisional fencing, rabbit-proof fences, myxomatosis and 1080, and better pasture management have served to increase available feed. Capacity, whether in the aggregate or in major districts, seems not to have been increased significantly by these developments any more than by closer settlement or the reduction in absentee ownership: 1,300 properties in the Western Division of New South Wales in 1937 ran the same number of sheep as had been run on 865 properties in 1911.

Water supplies and pasture supplies conjointly determined the ceiling on arid land cattle numbers in the nineteenth century. Once stocked to that capacity, as Queensland was in the 1890s, the arid region could support more stock only if water and pasture supplies could be expanded or used more efficiently. Techniques used in the sheep industry—man-made waters and fenced paddocks—were available for adaptation by cattlemen and were in fact used in non-arid fattening districts near metropolitan markets. In the first half of this century boundary and subdivision fencing was commenced or extended, particularly in the better areas on the fringes; more bores were sunk and water facilities in general improved. The total extent and effect of these improvements was, however, slight.

There was, indeed, little incentive for cattlemen to seek increased capacity: cattle prices and the cattle industry were generally depressed throughout the whole period to the end of the 1940s. Even by 1910 it was clear, as it had not been in 1890, that new market outlets for the Australian cattle industry could not be found at paying prices. Commercial export of frozen beef was an established reality, no longer the hope of the future. It stimulated a rise in the national slaughtering rate from the 7 per cent of the early eighties to the 12 and 13 per cent that characterised the first twenty years of this century and probably stimulated some improvement expenditure in Queensland in the five years before 1910. But twenty years' experience had convinced the trade of the limits bounding the market and of the strength of South American competition. The tallow market had collapsed and exports of live cattle to south-east

Asia had been tried and abandoned (Duncan, 1967: 86-102). Subsequent attempts to open new export markets for live cattle, to ship them to domestic markets, to have railways built, or to establish additional freezing works, where successful, were at best limited holding actions of local impact (Kelly, 1952; Rose, 1954). None was capable of stimulating the industry.

Only the Imperial meat purchase during and immediately after the 1914-18 war provoked any positive expansionary response in the industry as a whole. The subsequent collapse of prices in the early twenties initiated a contraction of numbers that, continued by drought, was not reversed until 1930.

In the three years to 1925 rising wool prices offered an inducement to increase flocks. Its effects, masked somewhat by dry conditions in Queensland, were of special importance in Western Australia. Rising profits afforded early settlers with an opportunity to reap capital gains. A boom in pastoral real estate developed, akin in character though not in magnitude to that in the east in the early eighties; pastoralists hurried to install more fences and water improvements so that they could sell or, having bought, could improve their return. Drought meant, however, that their investment was not reflected in increased stock numbers until the later twenties and the first half of the thirties. By then there were 60 per cent more sheep. It was, again, the same story as in the eighties: falling prices and profits per sheep from 1925 led to an expansion of physical output in an effort to bolster income and meet contractual debt obligations. Something of the same pressure was evident in Queensland, where pastoralists were encumbered by debts contracted during the droughts of the twenties.

After the 1939-45 war, the sheep industry experienced a major price stimulus in the late forties and early fifties and the cattle industry the stimulus first of the guaranteed market and prices afforded by the 15-year meat agreement concluded with the U.K. in 1949, then, from the late fifties, of American demand for lean beef and, perhaps of greatest long-term importance, the demand of a richer, much larger domestic population. Yet aggregate sheep numbers in the arid regions still failed to pierce the ceiling established half a century earlier. (Some individual districts, nevertheless, seem to be on a rising trend. The west of Darling is the most conspicuous of them.) Nor, with the exception of one year, has the arid cattle industry been expanded beyond the high point of 1922.

This apparent failure to respond is highlighted by a comparison of the movement of stock numbers in arid and non-arid areas. Taking the long view it is clear (Table 2) that after finding its new post-drought level by 1910, the arid land sheep population retained its percentage share of the national flock, with minor deviations, until the mid-thirties. Since then, it has unmistakably declined. The simple explanation need not be laboured: pasture improvement, the major technological development of the twentieth century affecting potential stocking rates, is for technical reasons still confined to non-arid areas; investors' responses, and the build-up of flocks, have therefore been similarly concentrated.

Arid land cattle numbers behaved less tidily. As a proportion of either all Australian cattle or of all beef cattle (Table 1), they rose to a peak in the mid-twenties, declined to the mid-thirties, and then recovered by the late forties to a plateau lower than pre-1925 levels on which they remained, with deviations, until 1957. It is against that perspective that the recent very rapid and substantial decline in relative importance needs to be viewed.

The earlier decline took place in the context of reductions of herds in both arid and non-arid areas. Drought affected arid areas severely and was essentially responsible for the decline in the twenties. Without it, the arid districts' proportion of cattle may well have increased, for market outlets were closed against them first and most firmly; turn-off in the Northern Territory fell to 2 per cent in 1927 (Kelly, 1952: Appendix F). The decline in relative importance was continued through into the thirties because numbers built up more rapidly in non-arid areas (with better breeding and survival rates) when turn-off slackened very considerably because domestic meat consumption fell during the depression at a time when seasons were good.

Beef cattle numbers expanded so rapidly in non-arid regions in the first five years of the thirties, in fact, that the arid districts were called upon only slightly to help meet the marked rise in the national slaughtering rate in 1936. The national rate at which cattle and calves were slaughtered (slaughtering as a percentage of the previous year's total cattle population) averaged 15 per cent between 1910-11 and 1934-5, fluctuating between 11 and 18 per cent. In 1936-7 it rose to 25 per cent and was maintained at that average until 1952-3. While this episode would repay a good deal more investigation, it seems that it was a response to a lift both in domestic and export markets and that it was seized upon by non-arid-land producers as an opportunity to realise on surplus stock. The number of beef cattle in non-arid areas had fallen by nearly 18 per cent by 1938 and the relative importance of arid land cattle had risen accordingly. By the time the industry settled down to the new norm —at the latest by the mid-forties—it appears that breeding and herd-survival rates in non-arid areas had been adjusted to accommodate most of the increased demands imposed by a 25 per cent slaughter rate and to retain still the capacity to increase total numbers. Some indication of the role that arid production was playing is given by the rise in the Northern Territory's normal turn-off (a slightly different concept) to 11-12 per cent from the mid-forties.

The period of comparative stability, from say 1945 to 1957, when both arid and non-arid cattle numbers moved at roughly the same rate and when exports were tending to grow a little faster than domestic consumption, also embraced another sharp lift in the national slaughtering rate. The rise to 29

TABLE 2

ARID AUSTRALIA: SHEEP POPULATION

Year*	Number ('000)	Share in Australian total (%)	Year*	Number ('000)	Share in Australian total (%)
1865	2,672	9	1907	20,058	23
1866	2,675	8	1908	22,385	25
1867	3,061	8	1909	23,827	25
1868	3,379	8	1910	25,011	26
1869	3,237	8	1911	26,061	27
1870	3,200	8	1912	24,770	28
1871	3,016	8	1913	25,605	29
1872	3,243	8	1914	25,949	31
1873	3,381	7	1915	19,440	27
1874	3,809	8			
1875	4,308	8	1916	20,486	25
			1917	22,590	25
1876	4,418	8	1918	23,870	26
1877	3,746	8	1919	21,208	27
1878	3,998	8	1920	22,099	27
1879	4,759	9	1921	23,451	27
1880	7,420	12	1922	22,858	28
1881	9,538	15	1923	21,977	26
1882	12,601	19	1924	23,846	26
1883	13,711	20	1925	26,525	26
1884	11,182	18			
1885	11,295	17	1926	22,161	22
			1927	21,338	21
1886	11,127	16	1928	22,795	22
1887	14,392	18	1929	23,981	23
1888	15,882	20	1930	27,707	25
1889	17,010	20	1931	28,687	26
1890	22,863	23	1932	28,080	25
1891	26,527	25	1933	26,124	24
1892	26,569	26	1934	28,016	25
1893	27,484	28	1935	23,754	22
1894	25,500	25			
1895	24,550	27	1936	24,366	22
			1937	25,816	23
1896	24,290	27	1938	26,098	24
1897	23,289	28	1939	27,601	23
1898	22,942	29	1940	27,167	22
1899	18,543	26	1941	28,060	23
1900	12,564	18	1942	29,256	23
1901	13,114	18	1943	27,271	22
1902	11,037	21	1944	23,623	22
1903	11,671	21	1945	21,306	22
1904	14,457	18			
1905	15,754	21	1946	19,722	21
			1947	21,167	21
1906	18,348	22	1948	21,016	19

TABLE 2—*continued*

Year*	Number ('000)	Share in Australian total (%)	Year*	Number ('000)	Share in Australian total (%)
1949	22,559	20	1956	24,649	16
1950	22,542	19	1957	23,104	15
1951	21,025	18	1958	22,934	15
1952	21,946	17	1959	24,027	15
1953	19,600	15	1960	22,944	15
1954	21,550	16	1961	23,325	15
1955	25,309	18	1962	24,217	15
			1963	25,792	16

* In the original data the dates are shown variously as at 31 December, or 1 January, 31 March, or 30 June of the following year.

† From 1865 to 1879 numbers are for Queensland and South Australia only; no data are available for west of Darling and Northern Territory before 1880; Western Australian data from 1890 only used here.

South Australian figures are not available for 1886-9 and 1894-6.

Queensland figures for 1868, 1869, 1872, 1893, 1940, 1941, and 1943 have been interpolated. Coverage for Queensland data changed in 1953; figures from that date are not directly comparable with preceding figures. Trend movements may be extrapolated by comparing 1953-6 figures above with overlapping figures collected on the old basis:

Year	Number	%
1953	22,806	18
1954	24,841	19
1955	27,189	20
1956	28,174	19

For 'Sources and Geographical Coverage' see Table 1.

per cent in 1953-4 presaged continued changes that were to take it to over 35 per cent.

Such information as is available suggests that arid lands did not share proportionately in that achieved rise in output per 100 beasts. Northern Territory turn-off percentages established in the late forties have not been exceeded significantly or persistently since then (Kelly, 1952: Appendix F, and similar figures from *Northern Territory Annual Reports*; B.A.E., 1968: 25). Turn-off rates on surveyed Barkly/Gulf properties in 1962-3—1964-5 (Keating, 1967: 198) were scarcely higher than those found by Kelly in the Gulf in the mid-fifties (1959: 62). There had undoubtedly been some investment in these areas, and possibly more in others, as a result of confidence in future prices that was afforded by the Meat Agreement; it had not been sufficient to bring about the necessary improvement in breeding rates or reduction in mortality rates.

The most recent decline in the relative importance of arid herds is characterised by rapidly rising numbers in non-arid areas and an absolute fall followed by stagnant numbers in the arid. Drought, of course, had an important place in this; so had the special attractions of the American market for lean beef which, incidentally, lifted exports to a consistent quarter or more of total output from 1958. The arid land failure to grow has been concentrated in the north: Western Australia, west of Darling, and South Australia until the drought of the late fifties, were increasing their herds quickly.

The impressive feature of that trend, however, is not the north's failure to grow, even though its access to markets has been increased, but the success of non-arid production. It was the non-arid area which produced the lift in national turn-off and which was able to increase its herds very substantially at the same time. The ability to do this, the ability not only to carry more stock but to increase output per beast, is the reason why there was more return from an additional investment in non-arid land than from the same additional investment in arid land.

REFERENCES

Bauer, F. H. (1959). Historical geographic survey of part of Northern Australia Part 1. Introduction and the eastern gulf region. *CSIRO, Land Res., Div. Rep.* No. 59/2.

——— (1962). Sheep-raising in Northern Australia. In A. Barnard (ed.), *The simple fleece: Studies in the Australian wool industry.* Melbourne University Press. Melbourne.

——— (1964). Historical geography of white settlement in part of northern Australia Part 2. The Katherine-Darwin region. *CSIRO Land Res., Div. Rep.* No. 64/1.

Beltana (1965). *A history of the Beltana Pastoral Company Limited.* Privately printed.

Bolton, G. C. (1954). The Kimberley pastoral industry. *Univ. Stud. Hist. Econ.* **2**, ii:7-53.

——— (1963). *A thousand miles away: A history of north Queensland to 1920.* Jacaranda/A.N.U. Brisbane.

Bureau of Agricultural Economics (1968). *The Northern Territory beef cattle industry: an economic survey 1962/63 to 1964/65.* B.A.E. Canberra.

Butlin, N. G. (1962a). Distribution of the sheep population: Preliminary statistical picture, 1860-1957. In A. Barnard (ed.), *The simple fleece: Studies in the Australian wool industry.* M.U.P. Melbourne.

——— (1962b). The growth of rural capital, 1860-1890. In A. Barnard (ed.), *The simple fleece: Studies in the Australian wool industry.* M.U.P. Melbourne.

——— (1964). *Investment in Australian economic development.* C.U.P. Cambridge.

Cain, N. (1961). Companies and squatting in the Western Division of New South Wales 1896-1905. *Econ. Rec.* **37**: 183-206.

Duncan, R. (1964). South Australia's contribution to the development of the Northern Territory cattle industry, 1863-1910. *Hist. Stud. Aust. N.Z.* **43**: 324-42.

——— (1967). *The Northern Territory pastoral industry 1863-1910.* M.U.P./ Monash. Melbourne.

Durack, Mary (1959). *Kings in grass castles.* Constable. London.

Fennessy, B. V. (1962). Competitors with sheep: Mammal and bird pests of the sheep industry. In A. Barnard (ed.), *The simple fleece: Studies in the Australian wool industry.* M.U.P. Melbourne.

Heathcote, R. L. (1965). *Back of Bourke: A study of land appraisal and settlement in semi-arid Australia.* M.U.P. Melbourne.

Keating, C. (1967). The Queensland beef cattle industry: Survey results for the period 1962/63 to 1964/65. *Q. Rev. agric. Econ.* **20**: 192-203.

Kelly, J. H. (1952). *Report on the beef cattle industry in Northern Australia.* B.A.E. Canberra.

——— (1959). *The beef-cattle industry in the Leichhardt-Gilbert region of Queensland: An economic survey.* B.A.E. Canberra.

——— (1966). *The struggle for the north.* Australasian Book Society. Sydney.

Macmillan, D. S. (comp.) (1963). *Bowen Downs 1863-1963.* Privately published.

Milo (1963). *A History of Milo and Ambathala.* Privately published.

Northern Runs Commission (1867). Commission appointed to enquire into the state of runs suffering from drought. Report and minutes of evidence. *South Australian Parliamentary Papers.* Vol. 2.

Rose, F. G. G. (1954). The pastoral industry in the Northern Territory during the period of Commonwealth administration, 1911-53. *Hist. Stud. Aust. N.Z.* **22**: 150-72.

Williams, O. B. (1962). The Riverina and its pastoral industry, 1860-1869. In A. Barnard (ed.), *The simple fleece: Studies in the Australian wool industry.* M.U.P. Melbourne.

14

SOME ECONOMIC ASPECTS OF THE PASTORAL INDUSTRY IN AUSTRALIA

E. J. WARING*

THE INTERACTING physical, social, and political environments provide the working conditions under which farm business management systems evolve. Understanding of those management systems provides valuable datum points for further research and development.

Financial data from a stratified random sample of the population of sheep graziers in Australia have been collected continuously since 1953, as part of the Australian Sheep Industry Survey conducted by the Bureau of Agricultural Economics. Similar information for Australian beef cattle properties is available for the 3-year period from 1 July 1962 to 30 June 1965 (B.A.E., 1965, 1967, 1968 and unpublished; Keating, 1967 and unpublished). This information can be supplemented by the continuous collections of data by the Commonwealth Bureau of Census and Statistics which date back to a much earlier period.

The data are positive in the sense that they relate to what the managers of properties in the low rainfall regions of Australia were doing with the resources at their disposal during the survey period. They do not necessarily indicate the best methods by which these resources might be employed—most of the works of man are in many respects imperfect. Furthermore, the data relate to the existing structure of property sizes and the capital endowments of their operators. Given, say, land reforms, or significant changes in product prices or operator's expectations of future product prices, the economic and financial organisation of the livestock industries in the arid zones might be expected to change somewhat. But as there are few alternatives to livestock production available in these areas, changes might be expected to be less there than those which would occur under comparable stimuli in the areas endowed with higher and more certain rainfall.

* Assistant Director, Bureau of Agricultural Economics, Canberra.

The substantial contribution of members of the Wool and Beef Research Branch of the B.A.E. in the collating of the data presented here is warmly acknowledged. The discussion of the data and the inferences drawn from them are those of the author.

DATA SOURCES

In most of the following discussion the data for the sheep and beef industries will be drawn from the results of the Bureau of Agricultural Economics surveys covering the period 1 July 1962 to 30 June 1965 (B.A.E., 1965, 1967, 1968 and unpublished; Keating, 1967 and unpublished). Unfortunately the surveys differ in the treatment of livestock trading accounts. The published results of the sheep industry survey add to net farm income changes in the value of livestock inventories due to sheep price changes. This is not done in the beef surveys. The results using both methods are shown for purposes of comparison in the tables which follow. In the case of beef cattle the increase in weight of stores with age—the final product of beef cattle—is taken into account in assessing capital value, but no allowance for the value of wool on sheep's backs is made in the source data in the case of sheep. An allowance of $2 per head of adult sheep has been made here in the sheep industry tables to provide an approximate standard basis of comparison. In the data for the sheep industry, the pastoral zone (B.A.E., 1967: 6) is taken to be co-extensive with the semi-arid and arid zone.[1]

In the case of specialist beef producers, the total sample has been divided into an arid and a non-arid component. The original stratified random sampling procedure was designed to provide population estimates within regions some of which do not lie completely within or without the arid zone. However, the classification of properties as 'arid' or 'non-arid' for the present purposes was based on location, approximately following Perry,[2] so that errors—mostly comparatively slight—could be expected in the population estimates obtained by raising the sample estimates drawn on slightly different boundaries.

The years from 1 July 1962 to 30 June 1965 were relatively 'normal' in most areas, although a severe drought began in northern New South Wales and Queensland in the 1964-5 financial year and there was persistent drought in the Alice Springs District of the Northern Territory until 1965. Some attempt to abstract from these drought effects is made in later sections.

[1] In the remainder of this paper the term 'arid' will be used to include 'arid' and 'semi-arid' areas. In general the discussion will relate to the pastoral industries in Australia's rangelands which were defined by Perry (1967). The pastoral zone for the Australian Sheep Industry Survey from 1953 to 1964 was defined as those parts of Australia not included in the wheat-sheep and high rainfall zones. It includes the high rainfall zone of the Northern Territory and coastal Queensland, but there are no sample properties in those regions.

[2] Perry's (1967) definition has not been followed exactly in Queensland and New South Wales. For Queensland, the eastern boundary of the arid zone has been regarded as between the 20″ and 25″ isohyet in southern and central Queensland and almost 30″ in the north. In New South Wales the boundary of the two zones has been regarded as approximately a line drawn from Boggabilla through Moree, Burren Junction, Coonamble, Warren and Condobolin to Hay, and thence due south to the Victorian border. This approximates the 15″ isohyet.

The only omission in the data for the beef industry is in the arid zone of South Australia which, because of administrative difficulties, could not be sampled during the Beef Industry Survey. The industry in that area is, however, comparatively small, and might be expected to resemble in many respects the beef industry in contiguous areas of the Northern Territory. Hence the omission is not especially important for present purposes.

The New South Wales arid zone beef sample is suspect for present purposes. There are few major beef producers in the zone and the sample properties are clustered along the northern sector of the eastern fringe, e.g. near Walgett where for some properties cropping was feasible during the survey years.

THE IMPORTANCE OF SHEEP AND CATTLE IN THE ARID ZONE

The rangelands areas to be discussed cover approximately 75 per cent of the area of Australia. They provide grazing for some 24 per cent of Australia's sheep and 31 per cent of the total cattle population.

Table 1 shows the number of sheep and cattle within and outside the Australian arid zone at 31 March 1957, and also at 31 March 1967. Although for Australia as a whole the numbers of sheep and cattle have increased, as did arid zone cattle numbers immediately prior to 1957, sheep numbers have decreased in the arid zone by 18 per cent and cattle numbers by 16 per cent. This has occurred over a period when beef: sheep products price ratios have been rising. In contrast to the position in the arid zone, sheep and cattle numbers in the remainder of Australia, the non-arid areas, have increased by some 22 per cent and 35 per cent respectively.

To some extent the 1967 statistics reflect the persistent effects of the serious drought in 1965 and 1966 in the eastern pastoral zone of the sheep industry and the persistent drought in the Alice Springs District which terminated about 1965. There is evidence, especially in the performance of the sheep industry in the pastoral zone before the 1965-6 drought began, that other influences have contributed to the decline in the relative proportions of total livestock population depastured in the arid and non-arid areas over the 10-year period.

The considerable development of increased stock-carrying potential through pasture improvement in the non-arid areas, particularly in the high rainfall zone, has certainly been one such influence. Increasingly favourable beef prices, improved transport facilities servicing the arid zone (e.g. beef roads), and the opening of new abattoirs in the north have been others, encouraging more complete musters for slaughter. But breeder numbers have also declined.

TABLE 1

LIVESTOCK IN THE ARID AND NON-ARID ZONES OF AUSTRALIA, 1957-1967 ('000)

State	Arid Zone		Non-Arid Zone		Total	
	1957	1967	1957	1967	1957	1967
			(A) SHEEP			
N.S.W.	18,843	12,740	49,094	51,389	67,937	64,129
Vic.	—	—	25,831	31,239	25,831	31,239
Qld	22,344	18,722	846	583	23,190	19,305
S.A.	2,569	2,330	12,416	15,534	14,985	17,864
W.A.	3,330	4,939	11,556	22,431	14,886	27,370
Tas.	—	—	2,943	4,325	2,943	4,325
N.T.	30	8	—	—	30	8
Total	47,116	38,739	102,686	125,501	149,802	164,240
			(B) BEEF CATTLE			
N.S.W.	145	144	2,284	3,023	2,429	3,167
Vic.	—	—	1,044	1,560	1,044	1,560
Qld	3,143	2,461	2,944	3,559	6,087	6,020
S.A.	182	124	174	340	356	464
W.A.	427	490	306	668	733	1,158
Tas.	—	—	158	269	158	269
N.T.	860	785	316	321	1,176	1,106
Total	4,756	4,004	7,227	9,740	11,983	13,744

Source: Bureau of Census and Statistics and Bureau of Agricultural Economics.
Note: Some totals may not be exact due to rounding.

Wool production has reflected the decline in sheep numbers, decreasing by 16·3 per cent in the arid zone over the 10-year period. Wool production has increased in the non-arid zone by some 22·5 per cent. Accurate estimates of beef production from the two zones are not available, because of cattle movements and the fact that the arid zone is not coextensive with established statistical divisions.

The total number of properties in the arid areas has changed very little in recent years. Table 2 gives a distribution of properties according to herd size by states and zones. Arid zone properties tend to carry larger herds.

TABLE 2

ARID AND NON-ARID ZONE PROPERTIES IN THE SHEEP AND BEEF CATTLE INDUSTRIES, 1962-3 TO 1964-5

(A) SHEEP

	Properties (No.)	Acreage (av.)	Livestock Sheep (av. no.)	Livestock Cattle (av. no.)
Arid Zone				
N.S.W.	3,761	29,628	5,410	116
Qld	2,992	32,675	6,698	183
S.A.	774	58,171	3,501	20
W.A.	353	426,008	9,869	125
Non-arid Zone				
N.S.W.	31,970	2,038	1,892	72
Vic.	26,104	1,383	1,202	43
Qld	1,612	4,005	1,737	106
S.A.	12,307	1,781	1,047	17
W.A.	10,406	2,614	1,558	18
Tas.	2,827	1,738	1,194	55

(B) BEEF CATTLE

	Properties (No.)	Acreage (av.)	Livestock Sheep (av. no.)	Livestock Beef Cattle (av. no.)	Livestock Dairy Cattle (av. no.)
Arid Zone					
N.S.W.	1,318	12,825	7,460	273	—
Qld	786	214,731	884	2,694	4
W.A.	45	922,719	—	10,187	—
N.T.	134	1,134,435	—	8,703	—
Non-arid Zone					
N.S.W.	9,915	2,236	1,169	231	4
Vic.	4,777	1,262	1,161	164	9
Qld	8,236	17,270	340	720	6
S.A.	434	2,270	2,604	154	3
W.A.	1,200	9,486	595	279	16
Tas.	717	2,103	1,762	256	4
N.T.	3	885,200	—	10,589	—

ENTERPRISE COMBINATIONS

Table 3 shows the principal enterprise combinations in the sheep and beef cattle industries.[3] Many properties are engaged in a dual enterprise, particularly in Queensland. However, ecological considerations prevent diversification

[3] For the sheep survey, properties are required to have at least 200 sheep, while the eligibility criteria for the beef survey require that properties have a minimum of 50 beef cattle and that returns from beef cattle are a relatively important proportion of total returns.

in many areas. Where the option to diversify is open to a pastoralist he may do so as a strategy designed to overcome the economic effects of both product price and seasonal fluctuations, and also as a means of reducing the labour force required to produce a given level of income. Although there are certain husbandry difficulties which impede the addition of a cattle enterprise on a sheep property, there is evidence that this is occurring (Reid, 1968: T 2 on p. 48), probably in response to a change in the relative product prices of wool and beef in recent years.

TABLE 3

DISTRIBUTION OF ARID ZONE PROPERTIES: BY ENTERPRISE COMBINATION
(%)

	N.S.W.	QLD	S.A.	W.A.	N.T.
		SHEEP PROPERTIES			
Enterprise Combination					
Sheep only	72·7	77·0	80·0	82·9	—
Sheep/Cattle	14·7	18·7	2·0	17·1	—
Sheep/Cereal	10·4	—	12·4	—	—
Mixed	2·2	4·3	5·6	—	—
Total	100·0	100·0	100·0	100·0	—
		BEEF CATTLE PROPERTIES			
Beef/Sheep	50·0	8·3	—	—	—
Beef only	—	82·9	—	100·0	100·0
Beef/Other	50·0	8·8	—	—	—
Total	100·0	100·0	—	100·0	100·0

In the sheep industry the emphasis is almost exclusively on breeding merinos for wool, despite low lambing percentages. The alternative, running merino wethers for wool, is usually uneconomic as a single enterprise because of the difficulty and cost of obtaining replacements. In the beef industry, the emphasis is mainly on breeding and fattening. Given the relationships between store and fat cattle prices which have prevailed over the immediate past 10-12 years, it can be shown that breeding and fattening is more profitable than breeding stores in arid areas (Rayner, 1968: 153-7). Breeding and fattening is in effect a drought strategy because of its flexibility. Starch (1939) describes flexibility as the ability to 'roll up' and 'unroll'. Graziers can 'roll up' when pasture deteriorates by turning off cattle without reducing the breeding herd, which is difficult to replace in arid areas, or by postponing breeding; when conditions become more favourable graziers can 'unroll' by recommencing their breeding and fattening activities.

Because of the progressive nature of income tax levies in Australia, it is also possible that some pastoralists adopt the policy of breeding and fattening, at least in some seasons, as a means of more nearly equalising incomes between years, even though it may be technically more efficient to produce stores to be fattened elsewhere.

CAPITAL STRUCTURES

The structures of capital values for sheep and beef cattle properties in the arid and non-arid zones are shown in Tables 4 and 5. The values adopted approximate a walk-in, walk-out price averaged over the three survey years. The values adopted for land are based on sales data, and the values assigned to livestock are the averages of opening and closing stock numbers assessed on the basis of prices paid for comparable stock at nearby sales centres. In the case of plant and equipment, values are based on current replacement cost less depreciation appropriate to the age of the item calculated at 'standard basic rates'.[4]

An important feature is the contribution of the value of land to total capital value in all States except the Northern Territory and the Kimberleys region of Western Australia. Land values should logically bear some relationship to the capitalised value of the returns to be expected from the operation of a parcel of land. Not only do net returns per acre and per livestock unit tend to be less the more arid the environment; as a rule investors also seek a higher average rate of return on a risky as compared with a less risky investment. So it is to be expected that the value of land relative to capital investment in livestock should decrease the more variable the rainfall and the lower its annual average. In addition, these climatic factors are correlated with distance from market and hence with cost levels—especially the costs of inwards and outwards freight which are high in the arid and risky areas.

Of the remaining capital items, livestock are generally most important in the arid zone, although in the sheep industry the value of improvements is often more important. Among the structural improvements, the value of water facilities is high; this of course reflects the conditions in the arid zone. The situation in the Alice Springs and Barkly Tableland Districts of the Northern Territory, although not shown separately in this paper, emphasises the importance of water facilities in the capital structure of properties in the arid areas. For the three years 1962-3 to 1964-5 the average capital value of properties in the Alice Springs District, for example, was $172,646 of which $37,274 (22 per cent) was for water facilities. This was to some extent affected by the drought conditions in the Alice Springs District during those years, which reduced the contribution of cattle numbers to total capital value. But the value of water is usually much higher per head of beef cattle in the Alice Springs and Barkly

[4] See *Schedule of Rates of Depreciation—Income Tax Order*, No. 1217, July 1960.

TABLE 4

CAPITAL STRUCTURE FOR PROPERTIES IN THE SHEEP INDUSTRY: FOR ARID AND NON-ARID ZONES

Average per Property, 1962-3 to 1964-5
(%)

	Arid Zone				Non-Arid Zone					
	N.S.W.	Qld	S.A.	W.A.	N.S.W.	Vic.	Qld	S.A.	W.A.	Tas.
Water facilities	7·6	9·0	11·9	13·4	2·9	2·7	3·2	3·9	3·9	2·1
Fencing	7·0	9·9	13·6	17·8	5·7	5·5	5·9	5·2	6·4	7·0
Buildings, fixtures	7·6	9·6	8·4	11·9	6·8	7·4	9·1	7·4	8·0	9·0
Plant and machinery	6·9	7·7	7·8	8·7	10·7	11·2	14·3	15·9	19·1	13·0
Sheep	18·7	22·1	15·7	35·6	12·2	10·1	10·3	9·6	13·4	10·8
Cattle	3·9	5·5	1·0	10·0	4·8	3·7	5·6	1·9	1·8	5·2
Other livestock	0·1	0·3	0·1	0·8	0·1	0·1	0·2	0·3	0·2	0·1
Land	48·1	36·0	41·5	1·9	56·7	59·4	51·6	55·8	47·2	52·9
Total	100·0	100·0	100·0	100·0	100·0	100·0	100·0	100·0	100·0	100·0
Total capital ($)	175,743	155,783	112,150	126,133	95,001	69,895	97,926	55,890	63,982	59,638
Total capital excl. land ($)	91,262	99,745	65,577	123,791	41,134	28,388	47,444	24,720	33,808	28,084
Total capital excl. land and livestock ($)	51,287	56,356	46,791	65,371	24,855	18,690	31,746	18,114	23,899	18,478

TABLE 5

CAPITAL STRUCTURE FOR PROPERTIES IN THE BEEF CATTLE INDUSTRY: FOR ARID AND NON-ARID ZONES

Average per Property, 1962-3 to 1964-5
(%)

	Arid Zone				Non-Arid Zone						
	N.S.W.	Qld	W.A.	N.T.	N.S.W.	Vic.	Qld	S.A.	W.A.	Tas.	N.T.
Water facilities	6·1	9·8	4·4	13·2	2·7	1·6	5·8	2·5	2·0	1·1	1·1
Fencing	5·0	8·1	5·5	8·8	5·4	2·8	7·5	7·5	4·0	7·5	8·8
Buildings, fixtures	8·0	6·1	6·4	7·0	7·8	3·0	4·7	8·7	5·1	6·6	7·5
Plant and machinery	5·0	3·6	3·9	3·7	7·7	2·3	5·5	5·3	7·7	5·6	4·1
Sheep	17·9	1·8	—	—	6·2	3·4	1·9	7·2	3·8	5·3	—
Beef cattle	8·0	60·7	79·8	67·2	15·3	7·0	42·1	8·9	19·3	16·2	78·5
Other livestock	—	0·1	—	—	0·2	0·4	0·4	0·2	1·7	0·4	—
Land	50·0	10·0	—	—	54·7	79·5	32·1	59·7	56·4	57·2	—
Total	100·0	100·0	100·0	100·0	100·0	100·0	100·0	100·0	100·0	100·0	100·0
Total capital ($)	249,500	199,642	376,736	413,954	113,606	204,842	91,366	157,810	94,158	144,181	372,602
Total capital excl. land ($)	124,862	179,816	376,736	413,954	51,552	41,972	62,110	63,600	41,060	61,709	372,602
Total capital excl. land and livestock ($)	60,136	53,928	72,926	135,579	26,755	19,698	21,290	37,830	17,699	28,647	80,210

Tableland Districts than elsewhere in the Northern Territory, as shown in Table 6.

The major reason for these differences is the greater use which is made of natural water facilities in the less arid part of the Territory.

TABLE 6

AVERAGE CAPITAL VALUE OF WATERING FACILITIES PER HEAD OF CATTLE IN THE DISTRICTS OF THE NORTHERN TERRITORY
(3-year average, 1962-3 to 1964-5)

Item	Alice Springs	Barkly Tableland	Victoria River	Darwin and Gulf
Value of water facilities ($)	37,274	100,089	46,596	9,652
Cattle (No.)	2,310	13,503	18,205	10,113
Value of water/head ($)	16·1	7·4	2·6	1·0

GROSS PROPERTY INCOMES

Tables 7 and 8 present data for average returns per property by States and zones for the sheep and beef cattle industries. In both zones in New South Wales, and in the non-arid zone in Victoria, South Australia, and Tasmania, the sheep enterprise contributes considerably more to total revenue than the beef enterprise on properties which are major beef producers. The reverse is true in the non-arid zone in Queensland, Western Australia, and the Northern Territory, and in the arid zone of Queensland, Western Australia (Kimberleys), and the Northern Territory. About 4 per cent of the Australian sheep flock is located north of the Tropic (Cumming and Reid, 1967).

The tables confirm the observation that the generality of properties in the arid zone derive most of their income from the one enterprise—sheep in New South Wales, mid-western Queensland, and the southern part of Western Australia, and cattle elsewhere.

Another interesting point is the lower contribution of sheep trading to total returns in the northern areas, partly a function of distance from market, but mainly because of low reproduction rates and high mortality among young stock in those areas, paralleling the experience in the cattle industry.

The second feature is that total returns for major beef producing properties are broadly similar to those for sheep properties in the arid zone, except in New South Wales and Western Australia, whereas in the non-arid zone the average total returns for beef properties are the higher. This is principally a function of size of enterprise.

TABLE 7

RETURNS FOR PROPERTIES IN THE SHEEP INDUSTRY: FOR ARID AND NON-ARID ZONES

Average per Property, 1962-3 to 1964-5

($)

	Arid Zone				Non-Arid Zone					
	N.S.W.	Qld	S.A.	W.A.	N.S.W.	Vic.	Qld	S.A.	W.A.	Tas.
Wool	27,296	30,651	18,844	43,978	9,543	6,325	9,113	5,235	7,294	6,245
Sheep trading	8,670	1,843	5,086	1,858	2,950	2,350	442	1,756	2,111	2,708
Total sheep enterprise	35,966	32,494	23,930	45,836	12,493	8,675	9,555	6,991	9,405	8,953
Beef cattle enterprise	2,441	3,207	590	3,105	1,805	1,356	2,309	517	530	1,575
Crops	962	145	419	—	5,323	2,829	6,122	4,310	6,222	1,210
Other returns	141	117	137	39	505	627	546	1,192	460	1,500
Total returns	39,510	36,026	25,076	48,978	20,133	13,488	18,532	13,011	16,618	13,238

TABLE 8

RETURNS FOR PROPERTIES IN THE BEEF CATTLE INDUSTRY: FOR ARID AND NON-ARID ZONES

Average per Property, 1962-3 to 1964-5

($)

	Arid Zone				Non-Arid Zone						
	N.S.W.	Qld	W.A.	N.T.	N.S.W.	Vic.	Qld	S.A.	W.A.	Tas.	N.T.
Wool	37,602	3,997	8	—	6,274	6,519	1,834	13,751	2,559	10,340	—
Sheep trading	13,372	−121	—	—	1,875	2,440	355	4,270	799	3,709	—
Inventory change	−1,359	78	—	—	−83	280	21	406	665	371	—
Total sheep enterprise	49,615	3,954	8	—	8,066	9,239	2,210	18,427	4,023	14,420	—
Beef cattle trading	3,911	36,294	67,297	56,887	6,719	5,888	13,010	5,717	8,138	8,764	42,644
Inventory change	1,997	−1,530	−1,712	−5,309	−547	429	−621	−300	−745	−108	−1,401
Total beef cattle enterprise	5,908	34,764	65,585	51,578	6,172	6,317	12,389	5,417	7,393	8,656	41,243
Crops	6,797	2	—	—	4,308	219	1,191	266	203	258	—
Other returns	1,169	504	167	1,622	866	1,754	719	576	4,199	1,855	1,887
Total returns	63,489	39,223	65,760	53,200	19,412	17,530	16,509	24,686	15,818	25,189	43,129

OPERATING COSTS

Tables 9 and 10 show the structure of the operating costs of beef and sheep properties in the arid and non-arid zones.

Properties in the arid zone have far greater total costs than those in the non-arid areas. This might be expected in view of the greater volume of production per property and size of property, on average, in the arid regions.

Average costs per cattle equivalent on beef properties in the arid zone tend to be lower than for the non-arid zone, both generally and on a State-by-State comparative basis.

Generally, the properties in the arid areas of the sheep industry devote some 35 per cent of total costs to labour; the corresponding proportion of costs in the non-arid zone is about 25 per cent. Conversely, the proportion of costs spent on materials is about 24 per cent in the arid areas whereas it is 35 per cent in the non-arid: other major cost groups bear a similar relationship.

The main reason for the differences between labour and materials costs in the two zones of the sheep industry is that properties in the non-arid zone expend considerable sums on fertiliser for both pastures and wheat, and to a less extent on fuel and parasite control. Relatively high expenditure on fuel is incurred in wheat production in the wheat-sheep zone. The lower importance of expenditure on parasite control for sheep in the arid areas is an advantage of that environment.

INCOME AND PRODUCTIVITY MEASURES

Tables 11 and 12 present income measures for properties in the beef cattle industry and in the sheep industry for the arid and non-arid zones in Australia over the survey period.

Within New South Wales, South Australia, and Western Australia, and generally in comparing the results in any of those States with the overall Australian experience, it is apparent that rates of return to total capital including land are generally higher in the arid than the non-arid regions; ranging as high as 11·4 per cent in the sheep dominant enterprises of the beef industry survey in the arid zone of New South Wales. In Queensland the rates of return for the two zones are nearly equal. Arid zone investors probably look for a higher rate of return on total capital expenditure to offset the disadvantages of the zone including, importantly, serious income fluctuation from year to year.

Space limitations prevent a fully detailed discussion of the data presented in the tables. However, in attempting to interpret the figures, certain points deserve to be kept in mind.

TABLE 9

COST STRUCTURE FOR PROPERTIES IN THE SHEEP INDUSTRY: FOR ARID AND NON-ARID ZONES

Average per Property, 1962-3 to 1964-5
(%)

	Arid Zone				Non-Arid Zone					
	N.S.W.	Qld	S.A.	W.A.	N.S.W.	Vic.	Qld	S.A.	W.A.	Tas.
Fuel	4·7	4·8	5·0	6·7	6·5	7·3	7·1	7·0	8·2	5·5
Fertiliser	0·1	—	0·6	—	7·7	8·6	0·4	8·9	14·4	11·9
Maintenance—plant	5·7	4·9	7·1	6·7	7·5	7·9	7·7	9·7	9·8	6·0
improvements	6·5	5·4	9·1	8·7	4·8	3·6	3·5	3·9	2·4	4·5
Other materials	7·5	7·2	3·3	4·2	9·7	7·5	11·5	12·6	8·8	7·2
Total materials	24·5	22·3	25·1	26·3	36·2	34·9	30·2	42·1	43·6	35·1
Freight and cartage	2·2	3·4	1·7	3·4	3·1	1·9	2·0	2·6	1·9	2·7
Marketing and other charges	10·0	9·5	8·2	10·7	6·8	6·8	6·6	5·4	4·7	5·3
Other services	15·8	17·9	9·9	10·3	14·4	12·6	13·3	11·0	9·9	8·8
Total services	28·0	30·8	19·8	24·4	24·3	21·3	21·9	19·0	16·5	16·8
Labour	34·4	35·0	38·1	38·6	25·8	26·9	28·1	21·8	21·6	34·4
Depreciation	13·0	11·8	16·9	10·8	13·9	16·8	19·7	17·2	18·4	13·6
Total	100·0	100·0	100·0	100·0	100·0	100·0	100·0	100·0	100·0	100·0
Average Costs ($)	22,190	25,615	14,575	30,607	11,786	7,474	10,982	7,520	9,541	8,823

TABLE 10

COST STRUCTURE FOR PROPERTIES IN THE BEEF CATTLE INDUSTRY: FOR ARID AND NON-ARID ZONES

Average for Property, 1962-3 to 1964-5
(%)

	Arid Zone				Non-Arid Zone						
	N.S.W.	Qld	W.A.	N.T.	N.S.W.	Vic.	Qld	S.A.	W.A.	Tas.	N.T.
Fuel	3·9	5·1	5·6	5·8	6·7	5·7	7·2	4·9	5·4	5·0	4·8
Fertiliser	—	—	—	—	5·8	7·6	0·1	12·0	11·6	12·0	—
Maintenance—plant	4·9	5·4	5·4 }	10·8	7·5	5·2	7·1	5·6	4·9	6·5 }	10·6
improvements	7·4	5·4	3·9 }		4·4	5·8	4·0	3·7	3·6	3·7 }	
Other materials	6·0	5·2	6·1	3·8	6·2	6·9	10·8	9·7	12·3	7·1	3·3
Total materials	22·2	21·1	21·0	20·4	30·6	31·2	29·2	35·9	37·8	34·3	18·7
Freight and cartage	11·5	10·2	23·7	13·2	6·1	3·3	6·3	4·5	5·3	3·3	6·3
Marketing and other charges	9·2	4·4	4·7	2·9	6·2	9·1	6·9	6·7	8·1	8·8	1·6
Other services	14·7	12·0	6·1	4·8	13·6	17·1	14·7	8·8	8·2	7·2	5·7
Total services	35·4	26·6	34·5	20·9	25·9	29·5	27·9	20·0	21·6	19·3	13·6
Labour	29·3	35·1	30·0	30·8	20·5	21·0	19·5	28·1	21·8	27·1	43·3
Depreciation	13·2	17·1	14·4	28·0	23·2	18·4	23·4	16·2	18·8	19·1	24·2
Total	100·0	100·0	100·0	100·0	100·0	100·0	100·0	100·0	100·0	100·0	100·0
Average Costs ($)	33,490	25,031	45,135	56,134	12,206	10,352	8,950	19,154	10,971	15,751	48,451

R

- Larger properties have a tendency to provide higher rates of return than small: generally arid zone properties and beef properties tend to be large.
- There were drought conditions in the Alice Springs District throughout the survey period, and seasonal conditions were if anything below average in Queensland for the whole period, but New South Wales and Western Australia tended to experience average, or somewhat better, seasons throughout the survey period.
- For Queensland, the comparison between arid and non-arid zones in the sheep industry is virtually a comparison of properties in the dry south-west and by no means fully developed central north, with wheat-sheep properties on the Darling Downs.
- It can also be seen that in the Northern Territory and Western Australia the present title value of leases—the value of land for the comparative purposes of this table—is regarded as zero (14:7).

The rate of return to total capital is not a particularly informative statistic—its principal value is probably for the individual landholder should he contemplate sale of his property and reinvestment of his funds in some alternative enterprise. In fact very few properties change hands in any one year. The factors which operate to set land values are extremely complex, but, in theory at least, include situation characteristics affecting operating costs. Thus, if on one property the provision of water is more expensive than on another, or if one is farther than the other from markets, so that levels of inwards and outwards freight differ, land values should reflect the capitalisation of the resultant differing 'rents to situation', all else being equal. But land values also reflect other factors such as social value, speculative value, and size of property.

At the present time Australia is not short of land as such. It would be possible to crowd up our livestock and cropping enterprises on to a smaller area than encompassed by the existing farmsteads, and have land left over. So in considering the comparative economic efficiency of single farms or aggregates of properties, we may restrict our consideration to those features of land which influence the technical efficiency of production, and these are reflected and measurable in direct running costs.

The value of livestock makes up as much as 80 per cent of the total capital value of some properties in the arid zone. The capital value of livestock is almost directly related to the market price of the product for beef cattle, and for sheep bears a strong relationship to the market prices of sheep, meat, and wool. On specialist cattle properties an increase or decrease in product price is reflected *pari passu* in the value of livestock capital as determined in the survey. Thus the grazier makes an immediate capital gain in his books on the increased value of his store of cattle if the price of beef rises. Much the same

Table 11

INCOME MEASURES (WITHOUT PRICE GAIN) FOR PROPERTIES IN THE SHEEP AND BEEF CATTLE INDUSTRIES: ARID AND NON-ARID ZONES

Average per Property, 1962-3 to 1964-5

	Arid Zone				Non-Arid Zone					
	N.S.W.	Qld	S.A.	W.A.	N.S.W.	Vic.	Qld	S.A.	W.A.	Tas.
(A) SHEEP										
Return to capital and management ($)	11,239	7,318	3,146	3,788	5,459	2,939	4,755	2,377	3,143	1,110
Returns/costs ratio	1·6	1·4	1·3	1·2	1·6	1·6	1·6	1·5	1·5	1·3
Rate of return to capital and management (%)	6·4	4·7	2·8	3·0	5·7	4·2	4·9	4·3	4·9	1·9
with additional $2 per sheep (%)	6·0	4·3	2·6	2·6	5·5	4·1	4·7	4·1	4·7	1·8
excluding land (%)	12·3	7·3	4·8	3·0	13·3	10·4	10·0	9·6	9·3	4·0
excluding land and livestock (%)	22·0	13·0	6·7	5·8	22·0	15·7	15·0	13·1	13·2	6·0
Net farm income per livestock unit ($)	16·4	8·9	10·7	4·1	23·3	24·3	20·1	27·9	23·0	14·0

(B) BEEF CATTLE	N.S.W.	Qld	W.A.	N.T.	N.S.W.	Vic.	Qld	S.A.	W.A.	Tas.	N.T.
Return to capital and management ($)	28,303	12,427	18,873	−4,837	5,511	5,510	5,794	3,780	3,095	7,686	−7,225
Returns/costs ratio	1·9	1·6	1·5	0·9	1·6	1·7	1·8	1·3	1·4	1·6	0·9
Rate of return to capital and management (%)	11·3	6·2	5·0	−1·2	4·9	2·7	6·3	2·4	3·3	5·3	−1·9
with additional $2 per sheep (%)	10·7	6·1	5·0	−1·2	4·6	2·6	6·2	2·2	3·2	5·1	−1·9
excluding land (%)	22·7	6·9	5·0	−1·2	10·7	13·1	9·3	5·9	7·5	12·5	−1·9
excluding land and livestock (%)	47·1	23·0	25·9	−3·6	20·6	28·0	27·2	10·0	17·5	26·8	−9·0
Net farm income per livestock unit ($)	24·9	5·1	2·0	−0·3	18·9	22·6	9·8	11·4	13·1	19·6	−0·5

'Net income' is defined as total returns less total costs. 'Return to capital and management' is net income less an allowance for the operator's labour.

TABLE 12

INCOME MEASURES (WITH PRICE GAIN) FOR PROPERTIES IN THE SHEEP AND BEEF CATTLE INDUSTRIES: ARID AND NON-ARID ZONES

Average per Property, 1962-3 to 1964-5

(A) SHEEP

	Arid Zone				Non-Arid Zone					
	N.S.W.	Qld	S.A.	W.A.	N.S.W.	Vic.	Qld	S.A.	W.A.	Tas.
Return to capital and management ($)	15,567	8,658	8,748	16,618	6,594	4,261	5,797	3,738	5,324	2,662
Returns/costs ratio	1·8	1·4	1·7	1·6	1·7	1·8	1·7	1·7	1·7	1·5
Rate of return to capital and management (%)	8·9	5·6	7·8	13·2	6·9	6·1	5·9	6·7	8·3	4·5
with additional $2 per sheep (%)	6·4	5·1	7·3	11·4	6·7	5·9	5·7	6·4	7·9	4·3
excluding land (%)	17·1	8·7	13·3	13·4	16·0	15·0	12·2	15·1	15·7	9·5
excluding land and livestock (%)	30·4	15·4	18·7	25·4	26·5	22·8	18·3	20·6	22·3	14·4
Net income per livestock unit ($)	21·9	10·2	22·9	13·5	27·0	31·2	23·4	37·1	33·2	21·6

(B) BEEF CATTLE

	N.S.W.	Qld	W.A.	N.T.	N.S.W.	Vic.	Qld	S.A.	W.A.	Tas.	N.T.
Return to capital and management ($)	28,335	16,186	24,874	17,768	5,625	5,833	6,586	5,979	3,309	9,395	25,816
Returns/costs ratio	1·9	1·7	1·6	1·4	1·6	1·7	1·8	1·4	1·5	1·7	1·6
Rate of return to capital and management (%)	11·4	8·1	6·6	4·3	5·0	2·9	7·2	3·8	3·5	6·5	6·9
with additional $2 per sheep (%)	10·7	8·0	6·6	4·3	4·1	2·8	7·2	3·7	3·5	6·4	6·9
excluding land (%)	22·7	9·0	6·6	4·3	10·9	13·9	10·6	9·4	8·1	15·2	6·9
excluding land and livestock (%)	47·1	30·0	34·1	13·1	21·0	29·6	30·9	15·8	18·7	32·8	32·2
Net income per livestock unit ($)	24·9	6·4	2·6	2·3	19·2	23·6	10·9	16·0	13·7	23·2	2·6

is true of sheep properties. Where livestock contribute substantially to capital value and fat and store cattle prices rise, rate of return to capital will rise less than on properties, typically those in the non-arid zone, where cattle contribute less to total capital. But, disregarding possible related concurrent changes in the price of inputs, the landholder's disposable net revenue will increase on all properties, and it is common for established landholders to pay more attention to disposable income than to rates of return determined by accountancy procedures.

Considering the beef industry in the arid zone in recent years, product prices have been increasing more rapidly than the corresponding costs required to produce a given volume of output, so that in the main beef cattle producers have been enjoying rising disposable incomes. Sheep producers have not been faring as well, since prices have been falling while costs rose.

As Tables 11 and 12 show, for the 3-year survey period the rate of return to capital, including livestock but excluding land, varies between the sheep and the beef sector, between States and between zones, in a somewhat different manner from the rate of return to total capital.

At any particular time in recent years surplus forage and not fully employed labour have existed and so has capital improvement capacity to carry more livestock on individual properties and across whole regions, particularly in the non-arid zone. So asset fixities have existed, and the marginal return to running additional livestock on the particular properties would exceed average returns on those properties and on some others. Such a situation could be one factor influencing the disparate rates of growth in livestock numbers in the arid and non-arid zones, but its existence has made livestock a scarce resource in the livestock industries generally. In seeking economic efficiency the aim is to maximise the return to scarce resources.

To the extent that landholders meet in full the cost of services, such as freight, which contribute to the differences in capital values of properties, and are prepared to accept a particular rate of return on the capital asset represented by their livestock store, there are three statistics which are informative in comparing the technical and economic efficiency of livestock production between farms, regions, and zones, and in comparison with other sectors of the economy. These are:

- The rate of return to capital and management excluding land and livestock, representing those items which in the long run at least may theoretically be transferred between districts and between industries: labour and plant can be moved readily, structures can be allowed to depreciate and be replaced elsewhere.
- Net income per livestock unit, which is a measure of value added in a particular industry.

- The returns : costs ratio, which provides some measure of the technical input-output relationships on properties.

These statistics for the sheep and beef industries are set down by States and zones in Tables 11 and 12. Table 11 disregards 'profits' on the increased value of livestock inventories due to price fluctuations to provide measures which are considered to be more indicative of relative efficiencies of resource use between States and zones. Table 12 does take price changes into account.

The influence of differing seasonal conditions is hard to isolate in State-to-State comparisons. But comparing rates of return to capital, excluding land and livestock, by States, zones, and type of livestock in these tables it will be seen that in the sheep industry New South Wales shows the best results in both zones, and the highest result of all in the arid areas.

Table 11 shows that New South Wales ranks highest on every rate of return criterion and the two zones within that State give nearly identical results. Overall the non-arid zone is the significantly better performer on the criteria adopted for the comparisons in this table; the differences between State subzones are most marked in South Australia and Western Australia. The overall ranking of zones is nearly equal if price gain on inventories is taken into account, as in Table 12. Weighting State averages by sheep numbers, the rate of return to total capital and management, regardless of the effect of sheep prices on inventory gains, is higher in the arid than the non-arid zone for Australia as a whole.

Considering the beef industry, rates of return to capital and management excluding land and livestock are not a great deal different in the two zones for Australia, except in New South Wales, where the major arid zone cattle producers could equally be regarded as very large-scale sheep producers. Perhaps the most interesting comparison is the altered ranking and overall great increase in rate of return for the arid and non-arid areas of the Northern Territory when price gain on inventories is taken into account.

Despite the far from favourable conditions which prevailed in the arid zone of Queensland during the survey period, rate of return to capital and management excluding land and livestock in that zone (with or without price gain) compared quite favourably in both the beef and sheep industries with that obtained in the more favoured non-arid zone, including the wheat-sheep sector where production is more intensive due to the inclusion of wheat in the output mix of most farms.

Comparing net income per livestock unit between States, zones, and type of livestock enterprise, striking differences are apparent. There are some differences between performances in different States in the arid zone in the sheep industry over the three survey years, although it is by no means outside the realms of probability that in longer-term averages the arid zone sectors of the four States could show commensurate results. Net income per cattle equivalent, arbitrarily counting eight sheep as equal to one beast, does appear to be quite significantly higher in the non-arid zone than in the arid zone. Ranking in order by industry and zone, there is a general downward progression from non-arid sheep to non-arid beef to arid sheep, and then to arid zone beef at a much lower level.

It is true that arid zone beef properties in New South Wales have a high net income per livestock unit, but it must be recalled that in fact the bulk of their revenue comes from sheep. Judged on the basis of net income per livestock unit it appears that a central planner attempting to maximise returns in the livestock industries with existing technologies, on the basis of the survey results, would rule that the needs of the non-arid areas for livestock of either type should be satisfied in preference to the needs of the arid areas taken in their totality. We can at least say that the cross-sectional data, while not indicating the optimal disposition of resources, indicate that the present disposition is suboptimal since the return to the scarce resource, livestock, is grossly unequal in its various uses—especially regarding the arid and non-arid zones as two 'uses'.

Despite this, on the criterion of returns : costs ratio (see Tables 11 and 12) the arid zone beef industry compares favourably with the sheep industry in both zones and with the cattle industry in the non-arid zone with the exception of the Northern Territory, where on average properties operated at a loss throughout the entire survey period, if the price element of change in inventory values be disregarded. This implies that a lower level of current inputs per head (e.g. labour, veterinary supplies—but a high freight input) was associated with the lower level of outputs per head in the arid compared with the non-arid zone.

TECHNICAL EFFICIENCY IN PRODUCTION

In considering the economic data available from the two surveys for the 3-year period under consideration, the aggregate data for the pastoral zone of New South Wales appear to set it somewhat aside from the rest of the arid zone. It compares favourably in the light of the efficiency factors used with the livestock industries in the non-arid zone generally. The Northern Territory situation is difficult to assess because of the prevalence of drought in the Alice Springs District up to the end of the survey period, but the equally poor results from the high rainfall zone of the Northern Territory, where drought did not prevail, suggest that the livestock industries in the Northern Territory lag in technico-economic efficiency behind the other zones.

With some differences between States, the livestock industries in the remaining sections of the arid zone present a picture of extensive industries producing rates of return to capital investment and returns to current inputs not widely disparate from the livestock industries in the non-arid zone. But output per animal unit in the arid zone is low relative to the non-arid zone. It is notable that at the time of the surveys few of the potential benefits which appear to be offered by the development of tropical legumes in the non-arid zone in northern Australia had begun to be experienced, but even lacking the potential advantages of such new technology, the results disclose a strong economic reason for the observed shift in the relative population of sheep and cattle from the arid to the high rainfall zone since 1957.

The results also suggest that given the probably reasonable assumption that beef prices may if anything improve relative to wool prices in the medium run, while forage supplies in the non-arid zone continue to remain in excess of the needs of a livestock population augmented by natural increase at even the highest observed rates of recent years, this trend might be expected to continue. Given that the net income per beast is higher in the non-arid than the arid zone, some non-arid zone pastoralists could afford to bid higher than arid zone buyers for any breeders offered for sale in the arid zone. In a chain of transactions from west to east, a sale in the arid zone in Western Australia could conceptually result in an increase in breeder numbers on a property in north coastal Queensland. Such sales might be expected to be more prevalent if beef prices steadied at around present levels. Yet even if only because of social and institutional rigidities, it is likely that there will remain substantial numbers of sheep and cattle in the arid zone throughout the foreseeable future.

The statistics of net income per livestock unit highlight the economic consequences of the known characteristics of the livestock industries in the arid zone, in particular the low reproduction and turnoff rates obtained. Table 13 presents five productivity measures by States and zones for the beef cattle industry.

The first, stocking rate expressed as acres per cattle equivalent, measures factors which bear heavily on the higher incidence of costs in the arid zone as compared with the non-arid zone. Although new fencing methods, the motor bicycle, land rover, radio, spotter aircraft and even helicopter have done something to help reduce the tyranny of distance, technological developments have done little to reduce its impact on on-farm costs. Yet at the same time, given the technical relationships existing, arid zone pastoralists appear to have been fairly successful in adopting management techniques which have offset the financial disadvantage which distances impose on them relative to the non-arid zone in the day-to-day working of properties. It is of course relevant that the costs of operations such as fencing or mustering tend to increase approximately as the square root of the number of acres per beast. Despite the lower carrying capacities in the arid zone the labour input per unit of beef turned off is broadly comparable between arid and non-arid zones. But it is the low turnoff rates attained in the arid zone compared to the non-arid—in large measure the outcome of lower branding rates and higher mortality rates—that lead to the low income per head carried which disadvantages the arid zone in comparison with the non-arid.

The B.A.E. has published a study of mortality rates in cattle in northern Queensland (Jenkins and Hirst, 1966) which tends to confirm the commonly held opinions that there are high mortalities in calves between parturition and the time of branding, and that there is a relationship between mortality rates and the incidence of pregnancy in young heifers, which are particularly prone to succumb to the stress of bad seasonal conditions.

TABLE 13

PRODUCTIVITY MEASURES IN THE BEEF CATTLE INDUSTRY: FOR ARID AND NON-ARID ZONES

Average per Property, 1962-3 to 1964-5

Zones	Stocking*	Rates		Equivalent per labour unit	Turnover per labour unit
		Branding†	Turnoff‡		
Arid					
N.S.W.	34·4§	62·9	57·1	365·5	47·2
Qld	76·4	52·9	25·4	606·0	147·6
W.A.	90·6	50·2	13·2	590·2	78·7
N.T.	130·3	31·7	13·7	798·6	109·3
Non-arid					
N.S.W.	5·9	73·2	62·3	187·3	70·7
Vic.	4·0	68·4	57·9	172·5	51·3
Qld	22·4	56·1	34·8	337·5	86·7
S.A.	4·7	87·2	66·2	190·9	40·3
W.A.	25·7	73·5	62·7	192·7	85·4
Tas.	4·4	68·5	67·9	184·2	65·3
N.T.	83·6	38·2	9·4	1,989·2	186·5

*Acres per cattle equivalent.

†No. branded as % of cows and heifers over 1 year.

‡No. turned off as % of beef cattle nos.

§The beef cattle survey contains only four properties in the arid zone of New South Wales as it is a relatively unimportant beef producing area. This statistic is based on the larger sample from the sheep industry survey.

CAN ECONOMIC EFFICIENCY BE INCREASED?

Two factors contribute to an increased turnoff and hence greater gross revenue per beast carried; increased branding rates, and decreased mortality between branding and turnoff. Given a 10 per cent mortality rate, with stock turned off three years after branding, as many as 30 per cent of branded cattle could fail to survive to the time of marketing. However, death rates vary between sexes and age groups. Given a herd containing 14 to 16 per cent heifers which have not yet calved, a 10 per cent death rate overall could reflect a 20 or 30 30 per cent death rate in the young female stock. Under such conditions, both herd growth rates and turnoff rates would be low. If additionally there were a high (and normally unrecorded) death rate among heifers' calves the situation would be the more unfavourable.

Table 14 illustrates the effect on rate of return to capital and management with constant total herd size of increasing branding rates from 45 per cent to 75 per cent under conditions similar to those prevailing in the Alice Springs District. Under the assumptions adopted, a 10 per cent rise in branding rate increases rate of return to capital by over 2 per cent through this range. Herd composition also varies, breeders forming a decreasing percentage of total stock. The budgets presented assume no increase in total cattle carried, which implies that an increasing proportion of females are speyed or culled early with higher branding rates. As an alternative, it is clear that total herd numbers can be expanded more rapidly given high branding rates.

TABLE 14

BRANDING RATE AND RATE OF RETURN TO CAPITAL AND MANAGEMENT

Branding rate (%)	45	55	65	75
Return to capital and management ($)	8,661	16,548	22,915	28,399
Total capital ($)	275,912	271,702	270,535	269,122
Rate of return to capital and management (%)	3·1	6·1	8·5	10·6
Total breeders	1,870	1,716	1,590	1,473
Heifers coming up to first calf	400	397	369	341
Turnoff rate (%)	18·0	21·4	24·1	26·6

Assumptions:

1. Total herd of 3,300 head.
2. Branding rate is the ratio of calves branded to number of females 1½ years and older.
3. Mortality rates assumed 10% total breeders, 5% other branded cattle.
4. 75% of steers turned off at 2½-3½ years, the remainder at 3½-4½ years.

Possible approaches to the problem of decreasing mortality and increasing branding rate generally fall into two groups:

- Improving water and feed supplies,
- Herd control, particularly the control of conception in young females.

It can be calculated from the data in Table 14 that an increase in turnoff percentage of from 10 per cent to 30 per cent absolute would be required to break even on doubled capital, depending on the initial state. Alternatively, a present branding rate of 65 per cent could be increased to 75 per cent at a break-even cost of $5,400.

For the hypothetical 3,300 head beef property of Table 14, $5,400 represents one-fourth of initial cost levels. An increase in branding percentage from 55 up to 75 per cent would cover an increase in costs of some 54 per cent.

Thinking of possible avenues of investment to secure such income increases, the annual cost of a mile of fencing is about $52, and of a water point for 300 head about $820. These estimates include the cost of repairs and maintenance, interest at 6½ per cent, and depreciation.

Accepting that a major problem lies in uncontrolled matings, and specifically in the precocious mating of young heifers which are unable to stand the stresses of bad seasons while in calf, or to rear a calf in a bad year, there is little objective data on which to assess the relative economics of practices which might be adopted. Clearly, however, if the major mortality problem lies with young heifers it would mostly be more economic to apply remedial measures to heifers alone and not the whole herd, nor even to all females indiscriminately.

A simple technique would be to isolate young females from the bulls on the assumption that their mortality would be reduced and their chances of successfully rearing their first calves substantially increased by maintaining them as maidens until two or two and a half years of age. Another possible strategy is to isolate the bulls from the entire female herd until seasonal conditions look favourable for a sufficiently long period ahead to warrant matings. Given the uncertainties of the situation, this may well be a less economic practice than that of isolating only young females from the bulls.

Recent research reports (Robinson and Lamond, 1966) suggest some possibility of developing techniques that might be fairly readily and inexpensively administered to prevent conception in young females. It would not be essential to secure complete musters of young females for this purpose. If only 90 per cent could be mustered and treated, say at branding, in a situation where mortality in young females and their calves would otherwise be high, birth control measures could well prove economic. The potential payoff to such a practice can be indicated by making some, possibly reasonable, assumptions about actual situations.

Assume a herd with 11 per cent maiden heifers, and 20 per cent mortality in heifers on their first calf. Assume further that 25 per cent of heifers' calves die because their mothers are not strong enough to maintain them up to weaning. If such losses could be halved, there would be a 1·6 per cent increase in turnoff rate leading to an increase in gross farm income of $2,912. Assuming the initial branding rate to be 65 per cent as in the budget of Table 14, and assuming all (about 370) maiden heifers treated, the increase in gross revenue per treated heifer would be approximately $7.90.

A sum of $7.90 per heifer is obviously sufficient to cover the cost of administering a moderately priced pharmaceutical product or a significant amount of feed supplement. But if the technique was costless it would add only some $2,900 to present net farm income of about $23,000, a modest payoff to a technical development probably requiring considerable research effort and much increased husbandry standards. Given that death rates in old cows and their calves are less than for heifers, lower returns to a given outlay per head in materials or management effort could be expected from treating the whole female portion.

The foregoing calculations are presented as one suggested method of cost-benefit appraisal of possible research into husbandry practices. They may induce pessimism in some and optimism in others. It is not considered that they should necessarily deter research in these fields. On the other hand, it can be fairly stated that the low gross margins per livestock unit obtained in the arid zone, relative to the more favoured areas, in large measure reflect the harsh conditions under which pastoralists operate, which lead to low conception and branding rates, high mortalities, low rates of liveweight gain, and thus to low rates of increase in breeder population, lack of opportunity for culling of females, and the need to retain stock for long periods before they reach acceptable sale weights. Avenues of research aimed at regulating, rather than manipulating, such conditions, and the development of strains of stock adapted to the regions of the environment should receive emphasis in any overall program for the arid zone.

SOME EFFECTS OF WEATHER UNCERTAINTY

Research into aspects of the economic effects of weather uncertainty in the beef cattle and sheep industries in the arid zone is at present in progress in the B.A.E. Work to date has been mainly concentrated on the cattle industry in the Alice Springs District where seasonal variability is high. Assuming that weather conditions over the last ninety-two years of rainfall recordings are representative of future experience, and assuming also that the conditions in a given year are independent of the conditions in the previous years, about fourteen single dry summers can be expected in 100 years,[5] together with four runs of two, one run of three dry summers. A run of four successive dry summers might be expected about once every 400 years. Some five single non-drought summers can be expected in each 100-year period, four runs of two, three runs of three, two of four, and one of six. About three runs of five good summers can be expected in every 200 years and a single run of seven good summers about once in 125 years. A run of nine good summers can be expected about once in 240 years and a run of ten once every 300 years.

[5] A dry summer was defined as that period during which there are runs of less than four weeks with water in soil store assuming that in any one week, (a) all rain goes into soil store; (b) if soil store is greater than two inches then plants remove water at 0·4 Ew; if soil store is less than two inches (but greater than zero inches) plants remove water at 0·2 Ew (Ew is the evaporation per week from the free water surface of a standard tank evaporimeter).

For the purposes of an analysis of the importance of herd size in determining the economic viability of cattle properties subject to probabilities of good and bad years such as the examples above, a hypothetical model, based on what are believed to be the decision principles adopted by many pastoralists, has been developed.

The model assumes that in a non-drought situation the average carrying capacity would be in the vicinity of five beasts to the square mile, falling to a 'drought grazing capacity' after three successive dry summers averaging about 1·4 beasts to the square mile. The hypothesised strategy of the pastoralist is to destock after one dry summer to a level appropriate in the non-drought situation less one-quarter of the difference between this level and the drought grazing capacity. After the second successive dry summer herd numbers are further reduced by the same amount. After the third dry summer, numbers are reduced to the assessed drought grazing capacity and, should further dry summers occur, these numbers are progressively halved.

If such a destocking policy be followed in drought, it will take the pastoralist, on average, four years to regain non-drought net cash income levels after a single dry summer and around twenty years to recover to non-drought net cash income after four dry summers. Assuming realised beef prices of around $23 per hundred pounds dressed weight, then considering properties of low drought tolerance (i.e. a drought carrying capacity of less than 1·43 beasts per square mile) the expected present value of net returns per head[6] over any 5-year period ranges from minus $3 per head on a property with a total herd size in the range of 1,500-2,000, through minus $7 on a property under similar conditions but carrying between 2,000-3,000 head, to $4 for the 3,000-4,000 total head size group, $18 for the 4,000-7,000 total head size group to $27 per head for the 7,000-13,000 size group.

Given a fall in beef prices of $3 per hundred pounds dressed weight, expected present value of net returns per head over any 5-year period falls by approximately $10 for all size groups ranging from minus $13 for the 1,500-2,000 to plus $17 for the 7,000-13,000 size group. An increase in beef prices of $3 per hundred pounds dressed weight increases the expected present value of net returns per head over the 5-year period by approximately $10 for all size groups.

Although these calculations refer to hypothetical properties, with operational characteristics approximating the means observed in the survey sample, the totality of assumptions made sufficiently approximates actual situations and management strategies to indicate the considerable importance of size of property in determining long-run economic viability. In the period covered

[6] Net returns are defined here as being an amount which is sufficient to pay interest on capital investment at the discount rate of 6½ per cent, replace worn-out assets, and provide a return to management.

by the Bureau's survey, analysis disclosed no statistically significant difference between the performance—measured in terms of income per head and per acre—of large as compared with small properties. There is, however, a clear indication from these calculations of viable property size, that in the long term size of enterprise is an extremely important variable affecting the profitability and long-run economic viability of arid zone beef enterprises.

SUMMARY

The analyses of results from surveys of the beef and sheep industries in the arid zone of Australia presented above, and the economic consequences of certain hypothetical situations which might arise and be consistent with the survey results, indicate that pastoralists in the arid zone have adopted management systems which are a great deal more extensive than in the more favourable high rainfall areas, with relatively low levels of costs, returns, and net incomes per head.

The economic results are consistent with what might be forecast given the harsh environment: low reproduction rates compared with the more favourable environments in which the breeds of livestock used were evolved coupled with slow rates of weight gain and production which are a concomitant of the low and uncertain rainfall of the area, and which are associated with high death rates in stock.

The data provide some support for a hypothesis that the most important measure to undertake to secure acceptable rates of returns to labour and resources employed in the zone is the development of properties of relatively large size, even in comparison with the present quite extensive holdings. It is suggested that many lines of biological research directed to the obvious problems of herd mortality and management would, if successful, require the adoption of management practices of much higher standard and intensity than those at present followed in the zone. The marginal pay off to this additional managerial effort would probably be relatively low.

There also seems reason to believe that although increasing cattle prices, which permit the realisation of capital gains on existing herds, may tend to mask the low gross margins from production realised in the arid zone per livestock unit, nevertheless the forces which have apparently resulted in the movement of livestock from the arid to non-arid zone for some ten years past will no doubt continue to exert a similar pressure under likely price relationships and in response to technical innovations in the high rainfall zone over the next decade or more. Against this, under existing management systems, rate of return on investment and running expenses in the customary accounting sense can be better than on many non-arid zone properties.

REFERENCES

Bureau of Agricultural Economics (1964-5). *The Australian sheep industry survey, 1960-61 to 1962-63.* Canberra.

——— (1967). *The Australian sheep industry survey, 1963-64.* Canberra.

——— (1968). *The Northern Territory beef cattle industry, an economic survey, 1962-63 to 1964-65.* Canberra.

Cumming, J. N. and Reid, G. K. R. (1967). The sheep industry in northern Australia. *Q. Rev. agric. Econ.* **20**: 71-84.

Hansel, W. (1967). Control of the ovarian cycle in cattle—a review. *Aust. vet. J.* **43**: 441-9.

Jenkins, E. L. and Hirst, G. F. (1966). Mortality in beef cattle in north-east Queensland. *Q. Rev. agric. Econ.* **19**: 134-51.

Keating, C. (1967). The Queensland beef cattle industry survey results for the period 1962-63 to 1964-65. *Q. Rev. agric. Econ.* **20**: 192-203.

Perry, R. A. (1967). The need for rangelands research in Australia. *Proc. ecol. Soc. Aust.* **2**: 1-14.

Rayner, I. H. (1968). Comparative profitability of marketing store and fat cattle in north-west Queensland. *Proc. Aust. Soc. anim. Prod.* **7**: 153-7.

Reid, G. K. R. (1968). Economic aspects of the sheep industry in the pastoral zone. *Q. Rev. agric. Econ.* **21**: 36-49.

Robinson, D. J. and Lamond, D. R. (1966). Control of reproduction in sheep and cattle. *Proc. Aust. Soc. anim. Prod.* **6**: 10-18.

Starch, E. A. (1939). Type of farming modifications needed in the Great Plains. *J. Farm Econ.* **20**: 114-20.

15

REQUIREMENTS OF FINANCIAL POLICIES FOR EXTENSIVE RANGE ENTERPRISES[1]

JAMES R. GRAY*

THE SUCCESS of extensive range enterprises on arid lands depends largely on the financial policies that guide these enterprises. The marginality that characterises most of them is such that imaginative financial policy changes are needed in terms of both internal and external capital rationing.

The law of comparative advantage states that a commodity will be produced where it suffers the least disadvantage from competing areas. In most large world areas classified as arid, the law rules that the area will be used for grazing livestock and/or wild game, changed by means of irrigation, or left unused. Additionally, as the alternatives are few and the outcomes relatively uncontrollable, most uses of arid lands fall within the area of uncertainty at worst, and risk at best.

Three major aspects of financial policies are presented. They are management strategies, financial climate in ranching, and policy implications in buying and selling both ranch factors and products. The aspects of management strategies to be dealt with include drought, range improvements or development, and secondary enterprises. Aspects of financial climate are those involving capital rationing of short, intermediate, and long-term capital. In the concluding section, some policy implications are discussed and programs for ranchers are suggested.

MANAGEMENT STRATEGIES

Management strategies are actions taken by entrepreneurs to achieve desired goals. For the purposes here, the goals of most ranchers are assumed to be either profit maximisation or loss minimisation. Although much evidence is being accumulated that these goals may not be valid in some U.S. areas (Martin and Smith, 1968), when coupled with a topic involving financial policies profit maximisation becomes a safe assumption.

[1] Journal Article 321, Agricultural Experiment Station, New Mexico State University, Las Cruces, New Mexico, 88001.

* Professor of Agricultural Economics and Agricultural Business, New Mexico State University, Las Cruces, New Mexico.

The management strategies to be presented deal with three aspects. They are the short-run strategies involved with fluctuating forage supplies because of drought and drought recovery periods, the medium- to long-run strategies dealing with range improvements, and the long-run strategies dealing with diversification by adding other kinds of enterprises to the basic cow-calf or yearling ranch enterprise.

Drought Strategies

In recent years the effects of drought in the United States have been much less severe in reducing ranch production than previously because U.S. ranchers have been much better prepared to cope with drought situations. In the 1890-96 drought, herds were decimated by starvation. In 1929-36, although some livestock starved, the basic breeding herds survived. In 1944-57 very few animals starved, most ranchers survived the drought with intact breeding herds, and production rates were maintained albeit at much higher costs per animal (Boykin *et al.*, 1962).

Drought decision models attempt to organise the situations facing ranchers entering and coming out of a drought period to permit concentration on key points. One model divides the activities or situations into six groups, ranging from resources to outcomes (Gray, 1968). Perhaps the key items are the ranch resources of capital and credit and the decisions available to ranchers in drought situations (Fig. 1). Note how resources are applied to a problem situation, followed by a period of uncertainty, which is alleviated by the rancher's knowledge situation, and eventually leads to decisions or strategies.

Strategies during drought are mainly of six kinds. The first two, that of reducing numbers or ceasing operations, are available to ranchers with restricted capital or credit (Gray and Plath, 1957). Maintaining livestock numbers with reduced production, the third strategy, is a possibility for those with restricted capital (capital rationed externally), while maintaining both livestock numbers and production is a possibility only for those with adequate capital (capital rationed internally). Increased supplemental feeding, the fifth stategy, is a short-run solution which has been successful in the United States but not in Argentina. The differences between these two countries have been a favourable credit system in the former, relatively higher livestock prices, and development of an agricultural industry that can produce and place large volumes of feedstuffs in needed areas. Leasing additional lands, the sixth strategy, is not one widely available. Adjusting cash cost levels is a strategy available to most ranchers at least on a short-run basis. It has been successful in maintaining livestock numbers and production in the southwest United States without materially increasing costs for short droughts. In longer droughts of three years or more, costs increase or herd reductions occur.

Several studies have been made of drought strategies. Most of them have

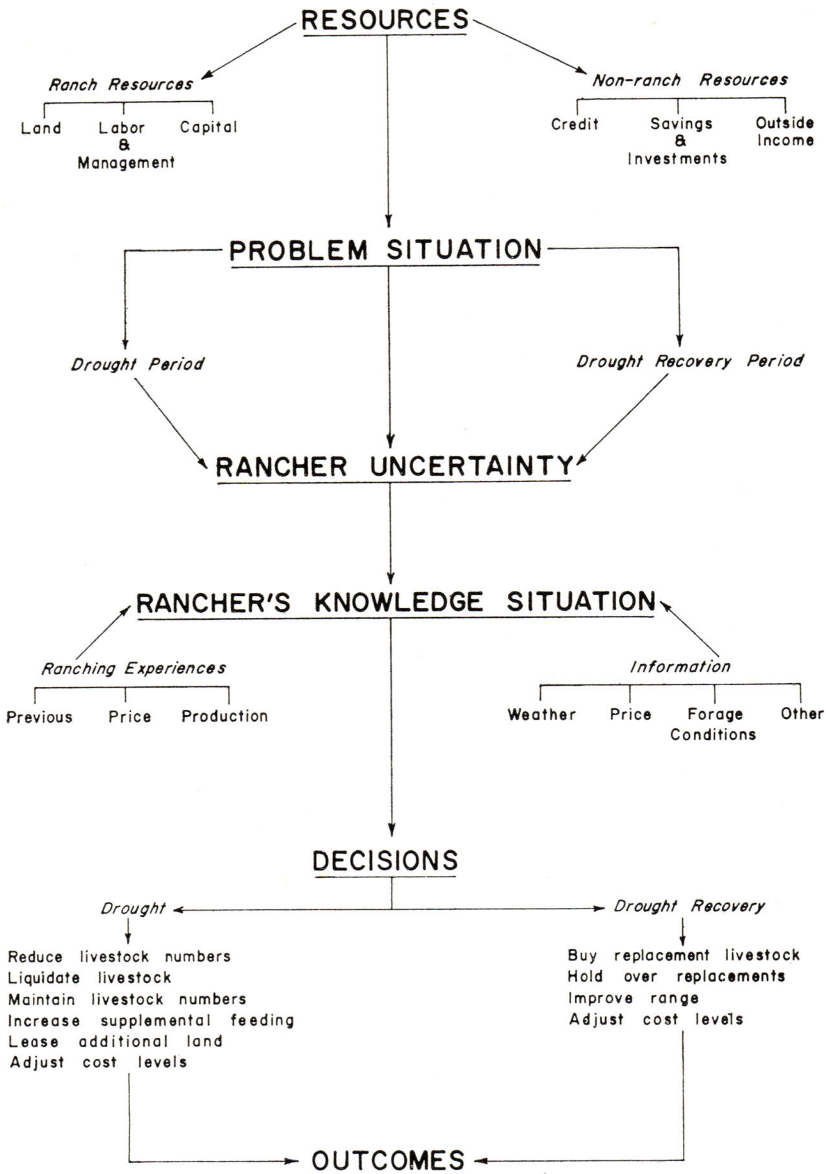

Fig. 1 Drought decision theory
Used by permission from *Ranch Economics* by James R. Gray, copyright 1968 by the Iowa State University Press, Ames, Iowa.

dealt with drought preparation and they have been descriptive (Bonnen and Ward, 1955; Boykin *et al.*, 1962; Nauheim, 1961). The usual recommendations or desirable actions are to store feed, graze ranges lightly even in favourable periods, have a flexible herd which permits cutbacks without damaging the basic breeding herd program, and build cash and credit reserves in favourable periods.

Drought predictions have received considerable attention in the past. In the United States they include a meteorological study (Tannehill, 1947), statistical measures called the coefficient of variability sequence (Clawson, 1943), extreme value theory (Brakensiek and Zingg, 1957), and correlation analysis (Abel *et al.*, 1962). None of these attempts has been particularly successful in predicting drought. Rather, most have presented probabilities on an *ex post* basis, with implications that the results may be useful for extrapolations on an *ex ante* basis.

Some of the more promising research currently being concluded is that using the Bayesian Decision Model, combining probabilities of the states of nature, alternative strategies, and a payoff matrix (Halter, 1963). Research is being completed using the model in widely separated United States range areas (Oregon, Nevada, and New Mexico).[2] Other research efforts attempt linear programming models (including dynamic) in which numerous resource allocative problems other than the special situation of drought are being investigated. Perhaps worthy of special mention is the statistical inventory model being used to determine optimum feed reserves under unstable climatic conditions in Oklahoma (Afzal *et al.*, 1965). Yet to be attempted are various systems analysis techniques, although a few researchers probably are attempting applications of this technique to the drought problem.

Range Improvement Strategies

Before discussing the details of financial strategies dealing with range improvements, some discussion is needed in terms of definitions and unique characteristics of range improvements. Range improvements are management strategies designed to change the ecology of an area, resulting in either improved productivities of rangelands or prevention of undesirable changes in the productivities, both on a long-term basis. Range improvements can be accomplished through a change in management of livestock herds. More commonly, range improvements are accomplished by a capital investment that changes one or more of the natural features of rangelands.

Range improvements assume investment characteristics rather than cost characteristics because almost invariably the returns from range improvements

[2] Minutes, Western Regional Research Technical Committee, W-79, Economic Analysis of Range Livestock Management Decisions on Western Livestock Ranches, Las Vegas, February 7-8, 1968.

extend over more than a single year. (Traditionally, this investment is referred to as a 'cost'.) With the investment in the present (t_o) and benefits in each year of the future ($t_{1 \ldots n}$), difficulties arise in computing net benefits. Usually some form of discounting of future returns is used to permit comparisons with present costs.

Range Improvement Models

A recent publication includes the basics of discounting in terms understandable by the non-economist (Walrath and Gibson, 1968). Included are the elements of discounting and the effects of rates and time on present values.

A slightly more complex model that lends itself well to computing the amounts needed to break even on range improvements is the internal rate of return model (Gardner, 1963). Besides discounting the value of future benefits, the model permits addition of an annual maintenance cost in the calculation. One difficulty with using the model is the assumption of contant amounts of benefits and maintenance costs over the life of the improvement. The model is easily used and it is well within the capabilities of livestock producers for use in making decisions.

A comprehensive treatment of the factors to be considered before adopting the range improvement strategy has been released by the Giannini Foundation in California (McCorkle and Caton, 1962). Questions are answered regarding selecting and evaluating practices and programs, improvements involving time and uncertainty, determining when to reinvest in range improvements, impact of range improvements on the entire ranch operation, range improvements and ranch income relationships, and planning and budgeting. Familiarity with these subjects will permit producers to make meaningful decisions about range improvements, provided they have access to the data needed for the several analyses.

Normally range improvements are not initiated during drought periods because most liquid assets are reserved for feed purchases or unavoidable fixed costs. Several range improvements are, however, amenable as part of the pre-drought strategy. Recently a study was completed of the costs and break-even points for eight kinds of range improvements in the southwest (Gray *et al.*, 1965). Although costs varied widely and in some cases they were not comparable, an average of the ranges in costs reported in the study permitted construction of a crude hierarchy of improvements based on their costs. It is, in U.S. currency and ranging from highest to lowest cost:

1. Artificial reseeding of a square mile ($8,473)
2. Woven wire fencing to enclose a square mile ($6,080)
3. Controlling competitive shrubs on a square mile ($5,139)
4. Barbed wire fencing to enclose a square mile ($2,620)

5. Stock wells ($2,585)
6. Steel rim storage tanks ($2,283)
7. Plastic pipelines of one-and-a-half miles ($1,642)
8. Earthen reservoirs ($1,442).

Two of the three highest cost practices (reseeding and controlling competitive shrubs) affect production directly by changing the species composition of selected range sites. Neither is thought of as primarily a drought strategy. Artificial reseeding in the western United States is restricted to areas with at least twelve inches of annual precipitation. Both practices, if successful, increase production as well as investment costs. The other two higher cost ranking practices are woven and barbed wire fencing. Fencing usually is installed to reduce labour costs or to increase production by forcing animals to utilise ranges more evenly. Fencing usually does not aid directly in drought situations.

The four lowest cost ranking practices are those directly concerned with water. As a rule, water developments obviously are associated closely with areas of limited water supplies in which drought is endemic. Water developments are designed to increase production by permitting animals to utilise ranges more evenly. Although ranchers occasionally are forced to invest in water developments during drought, ranchers in the United States attempt to endure drought by anticipating it beforehand with an adequate number of the more feasible mixes of water developments. In southern New Mexico, ranchers attempt to provide water so that cattle and sheep are not forced to travel more than 1·25 to 1·50 miles from any location on the ranch except during the more lengthy droughts (Gray *et al.*, 1965).

As a rule of thumb, ranchers in the western United States might regard a diversion of less than 5 per cent of net income to payments for range improvements over a period not to exceed fifteen years as a non-critical decision (Gray *et al.*, 1965). But when as much as 10 per cent of income is required, the decision obviously is more serious. If the improvement cost requires more than 20 per cent of the income, improvements costs cannot be paid from ranch income unless the ranch owner has outside sources of income.

Secondary Enterprises

Secondary enterprises are those that have been added to the primary enterprise usually to increase ranch incomes. Often these secondary enterprises are supplementary or complementary to the primary enterprise. The decision to add a secondary enterprise is a critical one on most ranches for several reasons. Usually resources are most adaptable to the primary enterprise and less so to the secondary enterprise, meaning the secondary enterprise will be less profitable. Land resources on ranches are near the extensive margin (Barlowe, 1958) rather than the intensive margin, and returns from the primary enterprise may be high but susceptible to sharp decreases should inputs be increased beyond the extensive margin. Many secondary enterprises are costly, either in terms of additional investments or additional operating expenses. Should the secondary enterprise require public participation, such as recreational enterprises, a geographical problem exists that can be solved only by means of highly costly highway construction. Lastly, cost accounting procedures are needed to determine if the secondary enterprise can be profitable, a technique found mostly only on the largest ranches.

Two types of secondary enterprises were studied recently in New Mexico. The first dealt with a registered herd enterprise (producing registered cattle for sale either to other registered cattle producers or to commercial producers mainly for bull replacements) (Wendland and Gray, 1968). Surprisingly, none of the factors usually associated with success as measured by bull selling price were statistically significant. These factors included values per head of brood cows and herd bulls, years of experience, and amount of care given the herd as measured by the amount of feed fed to the registered herd. The only factor that was statistically significant was the amount spent for advertising. A registered cattle enterprise was less profitable than a combination enterprise, which in turn was less profitable than a commercial enterprise of the same size. Units with high-profit registered cattle enterprises contained more than the usual size of cropping enterprise.

The other study on secondary enterprises is one now coming to fruition. It deals with recreation enterprises on ranches in mountain areas (Bedford, 1968). Five kinds of ranch recreation enterprises were studied, including hunting with and without investment, fish ponds, cabin rentals, and a miscellaneous group including horse rentals and guide services. In general, the capital and labour extensive enterprises were most successful. That is, if ranchers were required to invest much capital or time on the enterprise, it was usually unprofitable for them. The hunting enterprises were most profitable and the cabin rentals were least profitable. As a rule, recreation enterprises should be added only to take advantage of a presently unused ranch resource which can be harvested with little effort or investment on the rancher's part.

FINANCIAL CLIMATE

There are several unique characteristics of ranches that make them different from other kinds of agricultural enterprises and non-agricultural enterprises in terms of financial climate. The ranch is a combination of natural, economic, and personal resources in which the natural resources are overwhelmingly dominant. Any change of the natural resource is extremely costly. With relatively low productivities per acre (the extensive margin), changes must be very selective to be profitable. The product of a ranch is dependent on a renewable resource that, if abused, can shift to the non-renewable category. Abuse can occur either through the actions of man or through uncontrollable actions of nature. Variable costs on ranches are relatively low per acre and per unit of output compared with other kinds of enterprises. Distributions of the capital investment vary widely, depending on the state of development of the particular ranching area or of the nation encompassing the ranching area. Institutional factors exert a strong influence on the distribution of the capital in a ranch firm and these factors are subject to radical changes, particularly in developing nations. For these reasons, financial policies on ranches differ from those of other agricultural and non-agricultural enterprises, they differ from nation to nation, and they differ from time to time.

Land and Long-term Capital Requirements

The sources of capital used in ranching include inheritance, marriage, gifts, savings, renting, and borrowing. Savings, inheritance, and marriage are major sources of capital in many world areas, particularly in the underdeveloped areas. Renting and borrowing are the major sources in developed areas, with inheritance of secondary importance in some areas and for some people. Renting and borrowing imply some form of credit. To balance the quantity and quality of credit and leases in their portfolios, lending and renting institutions attempt to surround the conditions of a transaction so as to minimise their risk. This is accomplished by basing the return of the transaction on the opportunity cost of the capital including risk considerations, the overhead cost, and a profit margin. Uncertainty is avoided unless society is willing to undertake its cost as a subsidy to the lending institution. To reduce costs and rents, credit and rental fees often are packaged using such devices as classifications of loans and rents, security requirements that include real estate, and federal subsidies.

Allocations of capital to land versus other requirements for capital in ranching vary with the stage of development in a nation. In undeveloped areas, very little ranch capital is allocated to land, with almost all capital allocated to livestock. A small portion is used for operating costs. As the nation develops, population increases, and as land begins to feel this population pressure, systems are devised to allocate land to the various needs of the nation. Land values rise, and a larger share of ranch capital is allocated to land purchase or control. A series of interactions occur, with more valuable land being developed for higher production, the developments requiring more capital for operating costs, and more of the net returns being diverted to family living than formerly as the ranch becomes more commercial and less self-sustaining. Eventually, as has happened in the United States and Argentina, as much as 80 per cent of the capital is allocated to land. This heavy allocation creates difficulties, particularly if inheritance laws erect barriers for transferring ranch capital from one generation to the next. Within the generation, national legislation to encourage reallocation of resources to larger proportions of the population creates an uncertainty that inhibits the capital formation needed for successful operations. In the United States, land retirement programs, income tax laws, and monetary policies have inflated land prices well beyond that expected from a gradual rise because of population pressures. It was not foreseen at the time that these programs and policies would affect land prices appreciably. Ranchers have attempted to overcome the effects of these programs and policies with estate planning (before-death disposal of the property to future inheritors to avoid estate taxes) and sell-and-lease-back strategies (shifting capital from real estate to non-real estate items). In the latter strategy, the risk and benefits of land appreciation are shifted to land speculators.

Perhaps the most crucial strategy ranchers can adopt regarding national programs and policies is that of organisation to form political pressure groups that attempt to point out undesirable features of proposed programs. Organisations of ranchers have been successful in this action over the past half century in the United States. Apparently they have had varied success in some Latin American countries, unsuccessful in Mexico and Argentina, and successful in some central and western South American countries.

As the return from ranching is near the extensive margin, long-term capital often is received as credit that has been subsidised. Subsidy imposes a risk that control will pass from the individual to society. If private enterprise is desirable, then the cost of this risk must be balanced against the gain of lower-cost subsidised credit. With long-term capital requirements beyond the capacities of individual ranchers, they must balance their individual needs of a capital subsidy program against individual costs and express their majority opinion as a group voice through their association.

In most cases, ranchers should avoid long-term capital commitments during times of economic stress—drought or depression. Although long-term financial gains are possible in depression periods, the uncertainties are large enough to limit this strategy only to those with large capital reserves.

Improvement and Intermediate-term Capital Requirements

Perhaps the largest deficiency in American agriculture today is the supply of intermediate credit needed for herd and range improvements and purchase of equipment. The major reason is that most credit agencies are geared to provide either short-term or long-term credit. This has occurred partly because of the difficulty in securing adequate security for intermediate-term credit and partly because of its low return.

Usually the credit agencies that do provide intermediate-term credit insert a condition that real estate will perform an additional security function in the loan in event of default. The few agencies that specialise in intermediate-term credit usually are government-sponsored co-operatives. Ranchers attempting to acquire supplies of intermediate-term credit independent of these co-operatives find that they are competing with non-agricultural consumers for the available supply, usually at short-term costs. The strategy that has been most successful in the United States has been the co-operative Production Credit Association movement. The major liability of the movement has been, in some cases, inflexibility and conservative lending policies typical of government-sponsored agencies.

In drought and depression periods, investments in improvements and equipment are often delayed or reduced to a minimum. This is another area that is amenable to the discretion of the rancher. Again, the deflationary gains (investing in deflationary periods) of improvements and equipment purchases must be balanced against the financial reserves of the rancher. Short-term profitability considerations are usually secondary to considerations of long-term survival.

Operating Costs and Short-term Capital Requirements

As stated earlier, adjusting cash cost levels is a strategy available to most ranchers in the short run. During drought and depression periods, once the imperative fixed costs such as property taxes and loan interest charges are paid, capital can be partially diverted from the delayable fixed and variable costs and investments, such as new buildings and equipment, labour, and livestock replacements as well as family living expenses, to the critical feed input category. An approximate ranking of the various categories of costs from those most delayable to these that are least delayable is: (1) ranch buildings and improvements, (2) machinery purchases, (3) miscellaneous costs such as dues, subscriptions, and bookkeeping fees, (4) livestock replacements, (5) labour costs, (6) family living costs, (7) feed costs, (8) grazing fees, (9) machinery repair, (10) livestock medicines, (11) taxes, and (12) loan interest charges. Any of the items before category 7 should be diverted as much as possible to category 7 in the short run. In the long run only the last four or five categories should be maintained.

In a non-drought period, financial strategies should be much different from the above, provided a surplus cash position exists. If data are available, the marginal principle should be used to allocate resources to the various categories of inputs over the chosen planning horizon. Usually, however, even cost accounting is insufficient to reveal the marginal productivities as well as the costs of each increment of input. In this event the rancher may use an informal system of budgeting, with the budgets including expected costs and returns for each planning period of the planning horizon. In general, self-liquidating assets such as livestock replacements should be emphasised if the planning horizon is short—for example, a period of five years or in a drought. If the planning horizon is longer, twenty to forty years, non-self-liquidating assets such as buildings, improvements, and machinery should receive attention. In general, the accumulated long-term gain eventually will be at a higher rate than the short-term gain, although the consequences of a wrong decision—being forced to abandon a long-term horizon because of drought, for a shorter horizon—will be costly. Obviously, the strategy to be selected, and the degree that the strategy is adopted, will depend on both the length of the planning horizon and the risk position of the planner.

POLICY IMPLICATIONS IN SUPPLYING AND SELLING

Range livestock producers in the United States, and possibly in Australia, are parts of mixed capitalistic economies. When purchasing inputs, producers are faced with varying degrees of monopoly prices. Depending on the degree of control exerted by the central government, some inputs have prices based on socialistic considerations. Almost invariably, when the economy is relatively free, producers receive competitive national or world prices for their products. In the current period of world inflation, livestock producers are in a cost-price squeeze that is reducing their net returns to capital (at least in the United States and Argentina) to levels much below most non-agricultural enterprises. There seems to be no solution to this dilemma under the *status quo*. The solution may lie in a trade off of a part of the independence of the individual that is perpetuated in the competitive system for monopoly product prices that can be adjusted to input prices.

It may well be true that a strong bargaining association can achieve market power for agricultural groups. In the United States these associations have not been outstandingly successful because they have been organised on commodity lines. Consumers will substitute freely so long as alternative commodities are available. In the meats sector, pork and poultry substitute for beef, while some of the synthetics are partially successful in substituting for wool. New synthetics are increasing their share of the traditional leather market. New technology in a not-too-distant future may provide a substitute for meat. Presently soybeans are being processed into flakes which are melted into a gel. From the gel a fibre is spun which is pressed into a mass. The product has several advantages over meat in that the ingredients can be predetermined and the product itself is easily storable, capable of mass production and supply control, eligible for copyright and trademark, and of uniform quality.

The import situation in the United States demonstrates another policy available to overseas livestock producers that has not been fully exploited by them. The imposition of the beef import ceiling was a marginal action in that livestock interests in the United States barely mustered enough support for the ceiling. The American consumer has not been organised sufficiently in the past to protect his own interests to any great extent. The success of livestock interests in the United States in having the ceiling imposed was more a result of the failure of overseas producers to reach the American consumer than of any internal program in the United States. Many of the attempts to change or prevent change of consumer tastes by means of advertising by livestock interests in the United States have been wasted by concentrating on media available mainly to producers and their close friends. An overseas supplier must develop a superior product for domestic sale at competitive prices, or a comparable product at lower prices, and sell his product to domestic consumers. Outstandingly successful in this attempt has been one European auto manufacturer, not that his auto necessarily is superior to or comparable with domestic models, but he has attracted domestic consumer interest through his imaginative advertising campaign.

Little can be developed in a discussion about the size of enterprise effect on buying and selling ranch inputs and outputs because ranches in the United States are minimal in size compared to Australian ranches. Some price discounts are available, however, because of volume purchasing in the United States. Although little evidence is available, volume selling permits some adjustment upward in prices received by livestock producers, if for no other reason than that the larger volumes permit quality sorting and price-quality selling practices. Geographic concentrations of production also attract large-scale marketing facilities that are usually more efficient than small-scale or geographically isolated facilities. Transportation facilities especially are attracted to large production centres.

CONCLUSIONS

Financial management of ranches for successful operations or survival during stress periods requires a shifting of ranch resources. The shifts should be preceded by a classification of input categories. Under present technologies, ranchers cannot predict drought; they can only anticipate and endure it.

Solution to the pricing problem on ranches lies in a double approach of organisation to affect the national prices received for ranch products and imaginative advertising programs plus product differentiations to make the products superior on world markets. If consumers are sufficiently informed of situations that are to their advantage, consumer power structures will be erected eventually to force government action to take advantage of these situations.

Range improvements are costly, and because of the extensive margin typical in arid areas, the improvements must be customised and executed only in non-drought periods. Customised means that a balance between costs and gains must be estimated for each improvement in each range site being developed.

As a nation develops, capital in ranching tends to drift from self-liquidating assets toward the non-self-liquidating asset of land. Resources can be reallocated efficiently only when the extreme ranch capital allocative situations as they exist in many United States areas today are avoided by means of estate planning and various selling and leasing arrangements. Federal programs initiated to aid agriculture usually have outcomes unexpected to federal planners. Often the programs are capitalised into land values that hinder ranch expansion and survival. Again, strong organisations of ranchers possibly in association with organisations of producers of competing products are needed to point out and bring pressures to bear on government for long-run desirable programs.

REFERENCES

Abel, H., Rosenfeld, Mrs Lucille and Stephens, W. P. (1962). Range feed forecasting. *Bull. New Mex. agric. Exp. Stn* 462, Las Cruces.

Afzal, M., McCoy, J. H. and Orazem, F. (1965). Development of inventory models to determine feed reserves for beef-cattle production under unstable climatic conditions. *J. Fm Econ.* **47**: 948-62.

Barlowe, R. (1958). *Land resource economics.* Prentice-Hall, Englewood Cliffs.

Bedford, L. D. (1968). Economic analysis of ranch recreation enterprises in New Mexico. M.S. thesis, Dept. of Agri. Econ. and Agri. Business, New Mex. State Univ., Las Cruces.

Bonnen, C. A. and Ward, J. M. (1955). Some economic effects of drought on ranch resources. *Bull. Texas agric. Exp. Stn* 801 (College Station).

Boykin, C. C., Jr., Gray, J. R. and Caton, D. D. (1962). Ranch production adjustments to drought in eastern New Mexico. *Bull. New Mex. agric. Exp. Stn* 470, Las Cruces.

Brakensiek, D. L. and Zingg, A. W. (1957). Application of the extreme value statistical distribution to annual precipitation and crop yields. *U.S. Agr. Res. Serv.* ARS 41-13, Washington.

Clawson, Marion (1943). Economic use and administration of a fluctuating forage supply. Bur. of Agric. Econ., U.S. Dep. Agric. memorandum, Washington.

Gardner, B. D. (1963). The internal rate of return and decisions to improve the range. In Committee on the Economics of Range Use and Development of Western Agricultural Economics Research Council (ed.), *Rept.* 5.

Gray, J. R. (1968). *Ranch economics.* Iowa State University Press, Ames.

Gray, J. R. and Plath, C. V. (1957). Economics of adjusting to drought on eastern Oregon cattle ranches. *Oreg. agric. Exp. Stn. Misc. Pap.* 48, Corvallis.

Gray, J. R., Stubblefield, T. M. and Roberts, N. K. (1965). Economic aspects of range improvements in the southwest. *Bull. New Mex. agric. Exp. Stn* 498, Las Cruces.

Halter, A. N. (1963). A review of decision-making literature with a view of possibilities for research in decision-making processes of western ranchers. In Committee on the Economics of Range Use and Development of Western Agricultural Economics Research Council (ed.), *Rept.* 5.

McCorkle, C. O., Jr. and Caton, D. D. (1962). Economic analysis of range improvement; a guide for western ranchers. *Calif. Agric. exp. Stn Giannini Found. Agr. Econ. Res. Rept.* 255, Davis.

Martin, W. E. and Smith, A. H. (1968). The southwestern range livestock industry as a sector of modern commercial agriculture. In Committee on the Economics of Range Use and Development of Western Agricultural Economics Research Council (ed.), *Rept.* 10.

Nauheim, C. W. (1961). Flexible livestock systems. In Management Strategies in Great Plains Farming. *Great Plains Council Publ.* 19, *Nebr. Agric. Exp. Stn.* MP-7, Lincoln.

Tannehill, I. R. (1947). *Drought: its causes and effects.* Princeton University Press, Princeton.

Walrath, A. J. and Gibson, W. L., Jr. (1968). The evaluation of investment opportunities—tools for decision making in farming and other businesses. U.S. Dep. Agric., *Econ. Res. Serv. Agric. Handbook* 349, Washington.

Wendland, K. H. and Gray, J. R. (1968). Economic aspects of registered cattle enterprises in New Mexico. *Bull. New Mex. agric. Exp. Stn* 538, Las Cruces.

16

RESEARCH, MANAGEMENT, AND ADMINISTRATION OF THE RANGE ECOSYSTEM

C. WAYNE COOK*

A NEW approach to range management concerns the synthesis and use of information dealing with the structure and function of the rangeland ecosystem. Management and administrative decisions can be made more logically if the entire ecosystem is understood. In addition, voids in information are more readily discovered through this approach. Thus, the need for and the role of research become more obvious.

Many flow charts have illustrated the function of the natural and the agricultural ecosystems. The range ecosystem, since it is concerned largely with native vegetation, more nearly resembles the natural ecosystem. However, since livestock are introduced and manipulation procedures are frequently involved, it resembles in part the agricultural ecosystem.

The range ecosystem differs from a natural ecosystem in that it will be purposely manipulated for more effective production of saleable produce (Fig. 1). It differs from the domestic crop ecosystem in that the total standing crop does not represent the exported product.

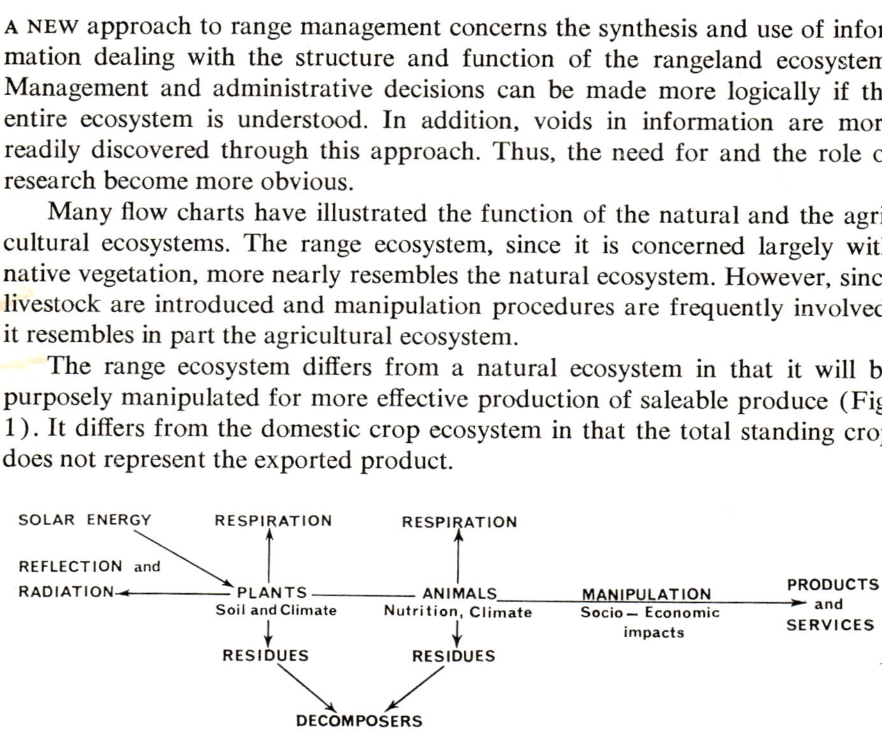

Fig. 1 Energy flow, nutrient cycling, and productivity of the range ecosystem

STRUCTURE OF THE ECOSYSTEM

All physical features along with expressions of flora and fauna characterise the structure of the ecosystem. Structure of the ecosystem is concerned with the whole community complex of plant composition, plant aggregation, life forms, layering, density, crown cover, leaf form, age classes, herbage yield, constancy,

* Department of Range Science, Colorado State University.

fidelity, etc. Measurements describing the structure of the ecosystem involve vegetation patterns, dispersion, heterogeneity, population dynamics: in general, the synecology of each community and the autecology of each species in both above and below ground environments.

Since both plant and soil development on any site are the product of environment, the climate, both micro and macro, along with latitude, altitude, and exposure should be described.

The status of animal life is an important characteristic of the ecosystem. This segment of the biotic community represents the consumers in the food web. Structure of the ecosystem concerns the identification of dominant species of fauna, their population dynamics, dietary habits, and behavioural patterns.

New statistical methodology allows for the evaluation of the influence of each independent factor upon any one of the several dependent factors in the ecosystem. Such programming in ecosystem analysis has been rather meagre to date but in the future such an approach shows great promise (Watt, 1968). By such analysis the effect of environmental factors on yield, frequency, soil nutrients, plant chemical content and many others can be assessed. The effect of each environmental factor can be evaluated when operating separately or when working together with all or any of the other designated environmental factors.

FUNCTION OF THE ECOSYSTEM

The function of the ecosystem is rather complicated and involves many biotic and abiotic interactions. These phenomena are not as well understood as the structure of the ecosystem because of the lack of research in this area. The research concerning functions of the ecosystem has been fragmented and has never adequately explained the operations of the ecosystem.

Soil Organisms

The importance of soil organisms in their role as decomposers and contributors to soil fertility is becoming more evident. The addition of organic matter by fungi is believed to be important to primary productivity on desert biomes. Soil crust algae and other cryptogams may be major nitrogen fixers (Shields and Durrell, 1964). Population levels of soil bacteria have been identified with high soil fertility and optimum structure. Transfer and conversion of soil nutrients for absorption by plants are of paramount importance to nutrient cycling in the ecosystem. However, in arid range soils their role has not been studied or evaluated.

Energy Flow

A primary consideration in the function of an ecosystem is the efficiency of conversion of solar energy to plant material by the primary producers. This is the first step in evaluating efficiency of turnover of energy by the various trophic levels and the efficiency of energy export from the ecosystem.

The conversion of solar energy through photosynthesis by plants is governed largely by light, water, or nutrients. On range ecosystems there is usually unlimited sunlight and the leaf area index is far from complete coverage of the surface area. Therefore, there is little correlation between primary productivity and incident solar radiation. Plant life may be dormant many months of the year because low temperatures and limited soil nutrients and moisture reduce photosynthetic activity. Even at best, less than 2 per cent of the solar energy is converted to chemical energy. However, energy conversion and transfer among different trophic levels in the range ecosystem is an important principle that can serve as a guide to management activities for efficient use of all cycling nutrients in the system.

Duncan (1967) proposed a model for predicting or evaluating efficiency of solar energy conversion by foliage by considering the following canopy architecture: leaf area, leaf angle, leaf position, reflectivity, transmissivity, brightness of sun, and position of the sun. It does not appear that the model would optimise energy conversion in range ecosystems unless it included other limiting factors such as soil nutrients and moisture, either of which should be more important than the leaf characteristics.

Since the function of the ecosystem actually starts with the process of photosynthesis, it is important to understand this physiological phenomenon. Age of leaves is important in the effectiveness of conversion of energy. As would be expected, the more mature leaves are less efficient than the younger ones. Life forms and species differ in the effectiveness of their chlorophyll-bearing tissue in converting radiant energy to chemical energy.

Defoliation by animals, infestation of parasitic insects that suck the life-giving saps from the plants, and diseases that cause malfunction of the plant tissue all decrease the efficiency of food manufactured by plants.

The herbivore population is generally proportional to food produced or food available for consumption. In any event, the entire consumer population is directly or indirectly related to the ability of the plant population to furnish food. Animal population dynamics, of course, depend on more than food supply since diseases, catastrophe, or social behaviour may also be regulatory.

Efficiency of Energy Turnover

Efficiency of solar energy conversion within the ecosystem has received much attention. This appears of paramount importance because of the transfer of solar energy for sustenance of all life in the ecosystem. The existence of the ecosystem itself is dependent upon this phenomenon.

Biological Efficiency

The herbivore population represents the first major turnover in the use of fixed photochemical energy in plants. This involves plant-x-animal and animal-x-animal interactions. The first interaction is obvious since it concerns animal response from consuming plants and plant response from being defoliated by animals. In this case, the stability of plant cover and animal response to consumption of the primary producers is a measure of efficiency of energy turnover.

In evaluating consumer response, it is necessary to consider more than gross energy content of the ingested forage. For instance, according to Cook and Stoddart (1958) maximum digestible energy (TDN $\times$ 2000) per unit area was produced by plants that were harvested only after they completed their annual growth cycle, but maximum digestible protein was produced when harvested some time during mid-growth and again at the end of the growing season. Livestock responses on grassland ranges were satisfactory only when phosphorus and digestible protein were high. This was true even though digestible energy was more than adequate (Cook, 1966).

It would seem that the biological efficiency of the range ecosystem could be based upon (1) the amount of photochemical energy or nutrients produced per unit of land surface area and (2) the amount of this material that is transported from the ecosystem in eatable produce, services, or wealth. The amount of solar energy converted to chemical energy or nutrients and then to the consumers serves as a measure of efficiency of the ecosystem, provided limiting factors are corrected through manipulation consistent with all uses to be made of the area and with cost and return relations.

It is generally acknowledged that as energy passes down the food chain from producer to consumer the efficiency of use of solar energy diminishes. In each case, however, management can increase the efficiency of energy utilisation in the transfer from one trophic level to another. Kozlovsky (1968) points out some interesting considerations in evaluating the efficiency of the ecosystem that applies to both the natural and manipulated system.

Species composition. The total gross energy produced by plants has already been identified as a measure of efficiency of the primary producers, but the transfer of this energy to the consumers would, to a large degree, depend upon the chemical form of the energy-furnishing constituents. For instance, many native browse and forb species produce rather high quantities of resins, waxes, and essential oils that represent chemical energy. However, energy in these forms is not available to the consumers. In studies by Cook *et al.* (1952) it was found that grass species produced approximately 2050 kcal of gross energy per pound of dry matter during early growth and browse species produced about 2322 kcal. The browse yielded only 1205 kcal of metabolisable energy for animal use compared to 1324 kcal of metabolisable energy per pound of dry matter in grasses. Later in the season, when plants were mature, the grasses contained 1859 kcal and browse contained 1922 kcal of gross energy per pound of dry matter. The grasses netted 42·9 per cent usable energy but the browse netted only 33·4 per cent usable energy (Cook and Harris, 1968).

Animal nutrition. Nutritional deficiency among animals is commonly a limiting factor in the efficiency of energy transfer from the primary to the secondary producers. In most range areas of the world the nutritional value of forage can be assessed on the basis of digestible protein, phosphorus, vitamin A, and energy-furnishing constituents. In local areas of higher precipitation and sandy soils, cobalt and copper may be deficient in the forage. Effective conversion or transfer of energy in the food chain is regulated by more than the mere supply of energy in plant material.

It should also be noted that the requirements of range livestock are different for various physiological functions of the animal. For instance, the requirements for lactation are about 25 to 30 per cent higher than for gestation. For gestation they are slightly higher than for maintenance, and for early lactation somewhat higher than for late lactation (Table 1).

TABLE 1

RECOMMENDED NUTRIENT REQUIREMENTS FOR CATTLE UNDER RANGE CONDITIONS DURING GESTATION AND LACTATION ON A DRY-MATTER BASIS*

Phase of production	Nutrient† DP (%)	TDN (%)	DE (kcal/lb)	ME (kcal/lb)	P (%)	C (mg/lb)
Gestation	4·4	46	830	665	0·17	0·6
Lactation						
First 8 weeks	5·4	57	1,120	900	0·22	1·6
Last 12 weeks	4·5	49	880	700	0·20	1·6

*Nutrient requirements are slightly higher for sheep because smaller animals have a somewhat higher metabolic requirement per unit of body weight.

†DP represents digestible protein; TDN, total digestible nutrients (calculated by deducting allowance for high ether extract in browse in the diet); DE, digestible energy (calculated as for TDN); ME, metabolisable energy; P, phosphorus; C, carotene.

Providing proper nutrition through the conjunctive use of seeded range with native range for lactating animals in the Intermountain area of western United States has shown that more pounds of lamb and more pounds of calf at weaning time per breeding female can be obtained (Cook, 1966). In similar fashion, females in gestation that were supplemented to correct protein and phosphorus deficiencies produced more pounds of lamb and calf at weaning time compared with unsupplemented females (Cook and Harris, 1968).

Thus, it is demonstrated that a nutritionally balanced diet can materially increase the efficiency of energy turnover in the ecosystem. This would be true of all animal life in the ecosystem.

Supplements not only furnish the deficient nutrient in the diet but in many cases they enhance the value of the range forage by increasing daily intake and digestibility of ingested range forage. In trials by Cook and Harris (1968), protein supplements consisting of cottonseed meal and soybean meal increased digestibility of protein and cellulose in the range forage on desert ranges used during the winter. Intake of digestible protein and metabolisable energy from the range forage was increased by the protein supplements fed to range sheep at either 0·5 or 0·33 pounds per day. Cottonseed meal supplements in most cases increased daily intake of range forage. Energy supplements consisting of barley or corn reduced the digestibility of protein and cellulose in the range forage and in some cases decreased daily intake of range forage (Table 2).

TABLE 2

AVERAGE DAILY INTAKE DURING WINTER GRAZING AS AFFECTED BY SUPPLEMENTS OF PROTEIN AND ENERGY

Range types and supplement	Daily intake of forage (lb/day)	Digestible Protein: In range forage (%)	Digestible Protein: Total intake from forage (lb/day)	Metabolisable Energy: In range forage (kcal/lb)	Metabolisable Energy: Total intake from forage (kcal/day)
WHEAT GRASSES					
Control	2·31	2·8	·066	458	1058
Corn	2·56	2·5	·064	423	1082
CSM	2·77	3·3	·091	544	1507
SAGEBRUSH-GRASS					
Control	3·21	4·2	·135	318	1021
Corn	3·38	3·4	·115	271	916
CSM	3·69	4·6	·170	366	1351
Barley	2·70	3·8	·103	252	680
SBM	3·01	4·5	·135	358	1078

Proper supplements fed to sheep on winter range during gestation have given as much as a 15 per cent increase in lamb crop, 8 pounds more lamb per ewe, and a pound more wool per animal. Replacement heifers, when supplemented on the winter range, gained 20 pounds more than the controls, and supplemented cows weaned 23 pounds more calf than unsupplemented cows.

Thus, it is shown that a protein supplement, when supplied to grazing animals on dormant desert ranges, actually increases the efficiency of energy turnover in the ecosystem even though protein may not be deficient in the forage.

Energy and productive animal physiology. It is generally acknowledged that sheep are slightly more effective in converting energy into gain and milk than cattle (Kleiber, 1961; Mitchell *et al.*, 1928). There is, however, a rather marked difference in the efficiency of energy conversion from range forage into gain of young weaned lambs and steers compared with gain from lambs and calves receiving milk during lactation. As shown in Table 3, about 40 per cent more gross energy is required per pound gain in sucking calves compared with steers. By knowing the degree of utilisation and the response of the grazing animal, it is possible to calculate the energy production of the ecosystem by the indirect method. Calculations for energy conversion in the range ecosystem (Table 3) with respect to species of animal and physiological function involved the following data: 450 to 500 kcal for heat increment per pound of dry matter consumed (Kleiber, 1961; Morrison, 1951), 33 kcal per 100 lb live weight for walking a mile (Brody, 1945), kcal requirement for steers per pound of gain is about 1900 and for lambs on range forage about 1700 (Garrett *et al.*, 1959), maintenance was calculated by the formula 70 $W_{kg}^{.75}$ (Kleiber, 1961), and 3280 kcal in whole milk was allowed per pound of calf gain and 2650 kcal in whole milk per pound of lamb gain is suggested (Hogue, 1968).

Table 3

COMPARISON OF ENERGY CONVERSION BY LACTATING COWS AND GROWING STEERS ON SPRING AND SUMMER RANGES

	Cow*	Steer
Animal Unit Weight lb/individual	900	650
Acres/individual/month	2·80	2·00
Pounds of oven-dry forage eaten/month	795	564
Gross energy kcal/animal month	1,570,488	1,114,793
Digestible energy kcal/animal month	968,991	687,827
Metab. energy kcal/animal month	772,680	564,780
Energy for maintenance kcal/animal month	193,830	151,230
Remainder after maintenance/animal month	578,850	413,550
kcal for 2 lb steer gain/day for month	——	114,000
kcal in milk production for 2 lb calf gain/day for a month	196,800	——
kcal remaining for heat increment and travel	382,050	299,550
Gross energy kcal/lb gain	26,174	18,580

*Maintenance of a cow during gestation the remaining six months of the year requires an additional 1,170,486 kcal of gross energy per month. Steers would be marketed after six months of grazing.

Common use. Even if we assume that sheep and cattle convert plant nutrients with similar efficiency, we still have marked differences in dietary habits between the two on most range ecosystems. This is a result of the difference between the two in use of plant species and varying topography.

Some range ecosystems are better suited to cattle and others are better suited to sheep. In Table 4 it is shown that a mountain range was slightly better suited to sheep than cattle from the standpoint of herbage consumed but was more effectively used by cattle when comparing the quantity of energy consumed. However, the two using it in common were more effective from the standpoint of both quantity of herbage consumed and efficiency of transferring energy in the ecosystem.

TABLE 4

AVERAGE CONSUMPTION OF ENERGY PER ACRE ON MOUNTAIN RANGE

	Utilisation		
	Cows	Sheep	Common
Grass 500 lb/A	28%	14%	26%
Forbs 400 lb/A	6%	28%	28%
Browse 100 lb/A	5%	20%	18%
Pounds consumed lb/A	169	202	260
Gross energy kcal/lb	1916	1845	1886
Digestible energy kcal/lb	924	702	856
Metabolisable energy kcal/lb	748	555	685
Metabolisable energy eaten kcal/A	126,412	112,110	178,100

MANIPULATION OF THE ECOSYSTEM

In spite of common belief, the natural ecosystem is not always the most efficient in the conversion of solar radiant energy in primary production or the transfer of energy in secondary production. A manipulated system may produce more dry weight of plant material and frequently furnishes a diet more conducive to meeting nutritional requirements of the consumers than the native ecosystem.

Ecosystem Conversion

Transformation of the ecosystem may involve complete conversion of one vegetation type to another by cultivation and seeding. Partial conversion of the ecosystem might be accomplished by burning, by the use of herbicides, or by railing or chaining. As shown in Table 5, the quantity of converted solar energy that can be transferred to livestock may be increased many times under some form of conversion. The efficiency of energy turnover can further be enhanced if the seeded forage more nearly meets the nutritional requirements of the grazing animal than unconverted vegetation does.

Before successful seeding of ranges can be carried out, research must show what sites are suitable for seeding, how best to prepare the seedbed, what species to seed, what season and method to seed, rate of seeding, spacing of seeded rows, and what method of grazing management produces the greatest net economic return over the longest period of time.

The inadequacy of suitable seasonal range for year-long grazing is perhaps the most limiting factor in livestock production in most parts of the world. Research by Cook (1966) indicates that about 80 pounds more beef per cow and 30 pounds more lamb per ewe can be obtained by using improved range for lambing and calving.

Man should recognise the efficiency of natural selection of plant species over time, but he must also investigate the desirabilities and suitabilities of introduced plant species and newly developed varieties. It has been shown that man can improve upon nature in many ways for increased efficiency of the ecosystem. Natural selection during geological time has been serving an entirely different purpose. Man's intelligence and perseverance will find even newer means of increasing the productive efficiency of the ecosystem. This may be far beyond our present expectations, but to be sure, it will far exceed the natural output of the ecosystem.

TABLE 5

TOTAL USABLE ENERGY PRODUCED PER ACRE FOR NATIVE RANGE AND SEEDED RANGE

	Native Range	Seeded Range
Production of usable forage lbs/A	180	952
Degree of utilisation per cent	50	50
Pounds consumed/A	90	476
Gross energy kcal/A eaten	171,000	1,047,000
Digestible energy kcal/A	90,000	571,200
Metabolisable energy kcal/A	63,000	428,400

Controlling unpalatable brush and weeds on range by means of selective herbicides offers great promise. In the future, it may be possible to control any single undesirable species without harming any of the others in the flora. There are presently over 3000 identified herbicides but only a few hundred have been tested and fewer than fifty are currently in use on rangelands. At present, over three-fourths of our poisonous plants can be effectively controlled by selective herbicides.

Research on rangelands of the United States has shown that controlling brush species in seeded and native stands of foothill grass increases the forage yield by as much as 1200 pounds of forage per acre, and control of dense stands of unpalatable forbs on mountain ranges increases the forage as much as 800 pounds per acre. Control of annual weeds and annual grass by selective herbicides prior to seeding and during seedling establishment has made it possible to obtain grass stands where before it was impossible because of weedy competition (Cook and Stoddart, 1964).

Fertilisers

Many studies have shown that the addition of fertilisers on native rangelands has increased the herbage yield many fold (Cooper, 1955; Retzer, 1954; Van Dyne, 1961; Herbel, 1963; Holt and Wilson, 1961; Rogler and Lorenz, 1957; Cook and Jefferies, 1963).

Research in range fertilisation should be concerned with evaluation of the following expected benefits: (1) increased forage yield, (2) increased nutrient yield, (3) increased vigour of plants, (4) increased palatability, and (5) improved distribution of livestock on rough mountain topography.

Fertilising seeded foothill range in western United States has shown that as much as 800 pounds of additional forage per acre can be obtained when 40 pounds of nitrogen are applied (Cook, 1965). On mountain meadow range as much as 1440 pounds of additional forage were obtained when 60 pounds of nitrogen were applied (Table 6). Protein content was increased by one-third through the application of nitrogen, and as much as 300 to 1200 pounds of additional digestible nutrients per acre were produced on seeded foothill and mountain meadow range respectively. Other benefits derived from fertiliser were increased palatability and increased vigour of plants.

TABLE 6

AVERAGE YIELD OF ENERGY PER ACRE OVER A 3-YEAR PERIOD FROM ADDING SIXTY POUNDS OF NITROGEN DURING ONLY THE FIRST YEAR

	Native Range	
	Unfertilised	Fertilised
Production of forage lb/A	620	2,060
Degree of utilisation per cent	50	50
Pounds consumed/A	310	1,030
Gross energy kcal/A eaten	651,000	2,523,500
Digestible energy kcal/A	352,510	1,413,160
Metabolisable energy kcal/A	251,410	1,059,870

Fertilisation of rangelands has not generally been practised by ranchers because of the high cost of fertilisers. However, research by range scientists in various parts of the world has found that the benefits justify the high cost of range fertilisation on many areas.

Mechanical Land Treatments

Contour furrowing, pitting, and water spreading have all contributed to improved moisture absorption and increased forage yield on many range areas in western North America where precipitation is sparse and soil moisture is limiting (Stoddart and Smith, 1955).

In many arid range areas where vegetation is sparse and torrential showers are frequent, effective soil moisture can be increased many fold by land treatments that hold the moisture on the soil and allow it to recharge the soil mantle.

Control of Plant Diseases

Very little research concerning the effect of plant diseases on primary productivity has been carried out on natural ecosystems. Fungi, bacteria, viruses, and nematodes are representative groups of pathogens that attack range plants and reduce productivity and quality of herbage. It is known that these pathogens are ever-present but they almost never reach catastrophic proportions. As a result, plant pathologists have not identified these diseases or assessed their effects upon range plant welfare.

Control of Rangeland Insects

Frequently invertebrates become so numerous that the competitive effect with livestock threatens the stability of the livestock enterprise (Stoddart and Smith, 1955). Among these are: grasshoppers, crickets, army worms, tent caterpillars, harvester ants, termites, snout moths, and many other species of moth larvae and beetles. It is well known that many insects change the species composition of the vegetation completely; yet as managers we generally tolerate the consequences. The rationale for this is known as 'peaks in population dynamics': by ignoring the problem it will disappear over time through a normal decrease in population numbers.

Control of Small Mammals

Perhaps the most notable examples of damage of the range resource from rodents come from Australia and New Zealand. In all parts of the world, peaks in rodent or lagomorph populations have caused serious competition with range livestock. In America, competition exists from such small mammals as the rabbit (*Lepus* spp.), prairie dog (*Cynomys* spp.), kangaroo rats (*Dipodomys* spp.), ground squirrels (*Citellus* spp.), pocket gopher (*Thomomys* spp.), and field mice (*Microtus* spp.). In most cases control methods have not been successful (Stoddart and Smith, 1955). The use of poison baits has been the method most widely used to control rodents. Generally, control methods have merely reduced numbers to a point whereby increased rate of productivity is initiated. Therefore, the relief was only temporary. The reduction would have perhaps occurred eventually as a normal expression of population dynamics. Knowledge of the behavioural patterns and the dietary habits of all small mammals is beneficial in the management of the ecosystem.

MANAGEMENT OTHER THAN ECOSYSTEM MANIPULATION

Range management is more than manipulating the ecosystem for increased quantity and quality of forage. It involves identifying all factors that may be limiting to the function of the ecosystem for maximum productivity.

Identification and evaluation of limiting factors in the ecosystem involve all of the complicated interactions related to the climatic, edaphic, and biotic factors. Management decisions are based primarily upon the knowledge of these factors and their interactions. These include plant-x-plant interactions with climate and soil nutrients, such as plant survival during extreme drought along with low temperatures and with or without adequate soil nutrients. Interactions of paramount importance to management involve plant, animal, and environmental factors. Actually, proper management of the grazing ecosystem is a reflection of both plant and animal responses.

Season and Intensity of Grazing

Research in the western United States (Cook and Stoddart, 1963, 1964) showed that desert ranges have more than twice the carrying capacity when used during the winter compared with use during the spring. In addition, the condition of the range greatly influenced carrying capacity. Desert ranges in poor condition could support only about one-third the numbers of livestock that could be grazed safely on comparable ranges judged to be in good condition. In the Intermountain area of the western United States, most foothill and mountain ranges in poor condition produce only about one-fifth as much forage as good ranges (Cook *et al.*, 1965).

Intensity of grazing profoundly influences the daily intake of forage and the nutrient content of the consumed material (Cook and Harris, 1968). Heavy grazing has been found to decrease daily forage intake by one-third and reduce the digestibility of all nutrients in the diet.

Livestock Distribution

Poor distribution of cattle on rangelands, especially mountainous ranges because of poor salting practices, underdeveloped watering facilities, and insufficient herding away from normal areas of concentration has, in many cases, reduced the grazing capacity as much as 60 per cent (Cook, 1967). Studies indicate that good salting practices can increase cattle production on a range as much as 30 per cent. Water development on several mountainous experimental ranges has increased the capacity by as much as 50 to 100 cow months per development. Driving cattle every week or twice a week from areas where they normally congregate can aid immeasurably in fostering better distribution and range utilisation.

Trail construction through rocks, down timber, and thick brush that hinder livestock travel, offers tremendous possibilities for increasing carrying capacity of the range.

CLIMATE AND MANAGEMENT OF THE ECOSYSTEM

Influence of the vagaries of weather on the function of the ecosystem presents a most formidable problem to range management. Range ecosystems are characteristically located in arid zones throughout the world. The average annual precipitation on native grazing lands of the world lies between about 4 and 20 inches and these areas are subject to intermittent and severe droughts. The effectiveness of the precipitation depends upon soil depth and climatic factors including temperatures, humidity, and wind velocity, all of which affect evaporation and transpiration.

In range ecosystems it is not uncommon for precipitation to be well below normal during any one year or even a series of years. In many localities of the range area of the United States, native vegetation has undergone complete change during drought years, even though grazing was conservative or completely absent. The effects were so profound that recovery has required from five to fifteen years (Stoddart and Smith, 1955; Freschknecht and Harris, 1968).

In many range areas it has been found that forage production is highly correlated with seasonal distribution of precipitation during a particular year. In some cases, this can be used to forecast carrying capacity of the range during subsequent seasons or years (Freschknecht and Harris, 1968; Currie and Peterson, 1966).

Studies by Cook (1969) showed that some plants on desert areas could tolerate only light grazing and during drought years little, if any, grazing. In studies on foothill ranges (Cook, 1966) it was found that range deterioration was a result of drought (60 per cent of normal) and continuous grazing without allowance for recovery following drought. Range deterioration took place during drought years even when grazing was considered conservative (25 to 50 per cent of current year's growth).

Certainly the stability of the range livestock enterprise depends upon the flexibility that can be incorporated into the grazing operations. Generally the more arid the climate the greater the fluctuation in herbage production from year to year. The drier range areas are in a rather delicate balance and once deterioration takes place, it requires relatively long periods for natural restoration (Gardner, 1951; Buffington and Herbel, 1965). Ranges located under more favourable climatic conditions respond more rapidly to alleviation of abuse and as a result they require less flexibility in operation than desert areas. It has been shown on foothill areas of the western United States that vegetation consisting of annual species may be expected to vary as much as 100 per cent from one year to another but perennial vegetation would not be expected to vary more than about 45 per cent (Stoddart and Smith, 1955).

Flexibility in stocking range areas of the United States is employed mainly through conservative grazing and the sale of inferior animals during dry years. Some livestock operations have ranches in two or more areas so that animals can be moved from one to another when drought is severe in some areas but not in others. During the past few years, the U.S. Government has subsidised drought areas by furnishing surplus crop grains at shipping costs to the drought-stricken areas.

PLANT RESPONSES—A KEY TO MANAGEMENT

It is recognised that biological efficiency of the ecosystem is measured by responses of both plants and animals to management. However, once the nutritional requirements of the grazing animal have been provided by various vegetation types, management becomes largely a matter of plant welfare.

Physiological Responses of Plants to Defoliation

The question arises, should the management system allow for a minimum of photosynthetic tissue at all times to optimise the conversion of radiant energy or is it more beneficial to graze rather harshly and obtain more uniform use of the range and allow the plants to regain vigour through deferred-rotation or rest-rotation grazing? Leaves, as they mature, become less effective in their capacity to carry on photosynthesis. Therefore, grazing methods that provide for the removal of apical buds and primordia so that new leaves are formed from stimulated latent or crown buds will generally increase the quality of primary production.

Since grazing prevents the transition from vegetative to reproductive material and since vegetative growth is more nutritious, it is readily understood that secondary productivity of the ecosystem is benefited by grazing during the growing period.

Can the cycle of carbohydrate usage and storage be used as an index to proper season and intensity of grazing? Are plants that draw down on the carbohydrate reserve rather rapidly during initial growth more susceptible to early or heavy use than plants that grow more slowly and draw upon the carbohydrate reserve rather conservatively? Is a plant species that replenishes its reserve rather slowly after initial draw down during early growth more susceptible to continued season-long use than one that replenishes its reserve rather rapidly after initial draw down? To a degree, these questions are related to the growing conditions following defoliation and the ability of the plant to replace its leaf area rapidly.

Ecological Responses of Plants to Grazing

A modern concept of ecosystem management is related to the ecological principle of succession and regression toward and away from the climax. Criteria for identifying the general health or condition of the range with respect to full productive potential has been quantitatively identified. The most effective criterion for identifying range condition (good, fair, or poor) can be accomplished primarily by the per cent plant cover composed of desirable plants, based either upon their climax status or upon their palatability, nutritive value, and soil protective virtues.

The status of stability of the range is referred to as trend in range condition. If the range is deteriorating or changing in the direction of retrogression, it is said to be in a downward trend. If, however, there are signs of plant succession toward climax, the range is in an upward trend. It is reasonable to assume that any range condition (good, fair, or poor) can be managed in a rather stable condition. This is reasonable because the theory of natural succession must accept that stabilising conditions for both soil and plant development exist throughout the various waves of plant expression during serial development in order to arrive ultimately at the climax expression.

Ecosystem management is based upon the ability of range managers to identify range condition and detect trend in it. This is not easy for the inexperienced range technician because objective measures or quantitative criteria have not been clearly identified by research. The basic principles involve the ability of the appraiser to detect change in species composition of the plant cover. In theory, this concerns a change in the occupancy of the perennial species present when dealing with fair and good ranges, but when dealing with a change from fair to poor range the evaluation must consider invading species that require an opening of the community for entry. When range conditions are moving downward there must be a lowering of vigour of the more palatable species, but a range condition that is improving does not necessarily show lowered vigour of the secondary species to be replaced over time by the primary species.

It is generally acknowledged that seed production is necessary if desirable species are to reclaim deteriorated range areas. Seed production, however, may not be required in closed communities judged to be in satisfactory condition, because regeneration through seedlings in a closed climax community is necessary only to the extent of replacing decadent plants.

It is recognised that soil development and soil deterioration are of paramount importance in judging range condition and trend but quantitative information does not allow accurate identification for either situation. It may be that certain hydrological characteristics may be more easily measured and understood than the estimated degree of accelerated soil erosion or actual extent of soil erosion taking place under the present condition.

RESOURCE ALLOCATION

Range is all land producing native forage for animal consumption and land that is revegetated artificially that is managed like native vegetation. Range represents a renewable natural resource that can be harvested annually on a sustained yield basis.

Range management deals with the care and use of rangelands in meeting the current and long-term needs of ranchers, sportsmen, recreationists, water users, foresters, and the meat-eating public in general. Scientific range management stands on the premise that the range resources can be improved and grazed perpetually by domestic livestock and wild game animals, while at the same time producing high-quality watersheds and outdoor recreation.

Therefore, if directed manipulation in the ecosystem is to be made, it should be with the intent of minimising any adverse effect on other uses.

Allocation of multiple use of rangelands will require: (1) establishment of criteria for evaluating suitability of the ecosystem for each use; (2) appraisal of the demand or need for each use; and, (3) decision-making models for allocation of resource uses.

REFERENCES

Blaxter, K. L. (ed.) (1965). *Energy metabolism: Proceedings of European Association for Animal Production.* Pub. 11. Academic Press. London and New York.

Brody, S. (1945). *Bioenergetics and growth.* Reinhold. New York.

Brougham, R. W. (1956). The effect of intensity of defoliation on regrowth of pasture. *Aust. J. agric. Res.* **7**: 377-87.

Buffington, L. C. and Herbel, C. H. (1965). Vegetational changes on a semidesert grassland range. . . . *Ecol. Monogr.* **35**: 136-64.

Cain, Stanley A. and de Aliveira Castro, G. M. (1959). *Manual of vegetation analysis.* Harper. New York.

Cook, C. W. (1965). Plant and livestock responses to fertilized rangelands. *Bull. Utah agric. Exp. Stn* 455.

——— (1966). Development and use of foothill ranges in Utah. *Bull. Utah agric. Exp. Stn* 461.

——— (1967). Increased capacity through better distribution on mountain ranges. *Fm Home Sci.* [Utah Sta.] **28**: 39-42.

——— (1969). Response of desert plants to defoliation. *Bull Utah agric. Exp. Stn.* (in preparation).

——— and Harris, Lorin E. (1968). Nutritive value of seasonal ranges. *Bull Utah agric. Exp. Stn* 472.

——— and Jefferies, N. (1963). Better Distribution of cattle on mountain ranges. *Fm Home Sci.* [Utah Sta.] **24**: 31.

——— and Stoddart, L. A. (1953). Some growth responses of crested wheatgrass following herbage removal. *J. Range Mgmt* **6**: 267-70.

——— and ——— (1963). The effect of intensity and season of use on the vigor of desert range plants. *J. Range Mgmt* **16**: 315-17.

——— and ——— (1964). Range resources, intensive management can meet future demands. *Fm Home Sci.* [Utah Sta.] **25**: 100-1.

———, Harris, L. E., and Stoddart, L. A. (1952). Determining the digestibility and metabolizable energy of winter range plants by sheep. *J. Anim. Sci.* **11**: 578.

———, Kothmann, M. and Harris, L. E. (1965). Effect of range condition and utilization on nutritive intake of sheep on summer ranges. *J. Range Mgmt* **18**: 69-73.

———, Stoddart, L. A. and Kinsinger, F. E. (1958). Responses of crested wheatgrass to various clipping treatments. *Ecol. Monogr.* **28**: 237-72.

Cooper, C. S. (1955). More mountain hay with fertilizer. *Bull. Oregon agric. Exp. Stn* 550.

Cooper, J. P. and Saeed, S. W. (1949). Studies on growth and development in Lolium. I. Relation of the annual habit to head production under various systems of cutting. *J. Ecol.* **37**: 233-59.

Cragg, J. B. (1962-1967). *Advances in ecological research*, Vols. I-IV. Academic Press. London and New York.

Crisp, D. J. (1964). *Grazing in terrestrial and marine environments.* A symposium of the British Ecological Society, Bangor, April 1962. Blackwell Scientific Publications. Oxford.

Currie, Pat O., and Peterson, Geraldine (1966). Using growing-season precipitation to predict crested wheatgrass yields. *J. Range Mgmt* **19**: 284-8.

Duncan, W. G. (1967). *Harvesting the sun.* Model building in photo-synthesis Symposium, Chicago, Ill. 1966. Academic Press. New York and London.

Elmore, C. D., Hesketh, J. D. and Muramoto, D. (1967). A survey of rates of leaf growth, leaf aging and leaf photosynthetic rates among and within species. *J. Ariz. Acad. Sci.* **4**: 215-19.

Freschknecht, N. C., and Harris L. E. (1968). Grazing intensities and systems on crested wheatgrass in central Utah: response of vegetation and cattle. *U.S. Dep. Agric. Tech. Bull.* 1388.

Gardner, J. L. (1951). Vegetation of the creosote bush area of the Rio Grande Valley in New Mexico. *Ecol. Monogr.* **21**: 379-403.

Garrett, W. N., Meyer, J. H. and Lofgreen, G. P. (1959). The comparative energy requirements of sheep and cattle for maintenance and gain. *J. Anim. Sci.* **18**: 528-46.

Herbel, Carlton H. (1963). Fertilizing tobosa on flood plains in semi-desert grassland. *J. Range Mgmt* **16**: 133-8.

Hesketh, J., and Baker, D. (1967). Light and carbon assimilation by plant communities. *Crop Science.* **7**: 285-93.

Hogue, D. E. (1968). Correspondence with respect to data concerning gain of lambs and calves from whole milk. New York State College of Agric. Cornell Univ., Ithaca, New York.

Holt, G. A., and Wilson, D. G. (1961). The effect of commercial fertilizers on forage production and utilization on a desert grassland site. *J. Range Mmgt* **14**: 252-6.

Kleiber, M. (1961). *The fire of life.* Wiley. New York.

Kozlovsky, D. G. (1968). A critical evaluation of the trophic level concept. I. Ecological Efficiencies. *Ecology* **49**: 48-59.

Lewis, T. and Taylor, L. R. (1967). *Introduction to experimental ecology.* Academic Press. London and New York.

Mitchell, H. H., Kammlade, W. G. and Hamilton, T. S. (1928). A technical study of the maintenance and fattening of lambs. *Bull. Illinois agric. Exp. Stn* 314.

Morrison, F. B. (1951). *Feeds and feeding.* Morrison Publishing Co. Ithaca, New York.

National Research Council (1958). Nutrient requirements of domestic animals No. IV. Nutrient requirements of beef cattle revised 1958. Pub. 579.

Pierre, W. H., Kirkham, D., Pesek, J., and Shaw, R. (1965). Plant environment and efficient water use. *Amer. Soc. Agron. and Soil Science of America.*

Retzer, John L. (1954). Fertilization of some range soils in the Rocky Mountains. *J. Range Mgmt* **7**: 69-73.

Rogler, G. A., and Lorenz, R. J. (1957). Nitrogen fertilization of northern Great Plains rangelands. *J. Range Mgmt* **4**: 156-60.

Shields, L. M. and Durrell, L. W. (1964). Algae in relation to soil fertility. *Bot. Rev.* **30**: 92-120.

Southwood, T. R. E. (1966). *Ecological methods.* Methuen. London.

Stoddart, L. A., and Smith, A. D. (1955). *Range management.* McGraw-Hill. New York.

Van Dyne, George M. (1961). Range fertilization studies. Red Bluff research ranch. *Montana Agr. Exp. Sta. Res. Prog. Rept* P.R. 3-5.

Watt, K. E. F. (1968). *Ecology and resource management.* McGraw-Hill. New York.

17

RANGELANDS RESEARCH AND EXTENSION

R. A. PERRY*

ARID AUSTRALIA is a single vast area forming a major part of four states and the Northern Territory. It is of national importance that research and extension services for these lands be improved and expanded, but their low productivity per unit area does not warrant the same expenditure on an area-for-area basis as more humid regions. In the interests of achieving the best use of limited funds, research and extension should be planned for the entire arid zone—regional and state interests should be secondary to the interests of the arid zone as a whole.

In the past, arguments for increased agricultural research and extension in arid Australia have suffered from parochialism. Regional graziers' associations make submissions for laboratories in their district, state governments argue for increased research in their states, and scientists support the establishment of field centres in areas with which they are familiar. All three groups tend to argue against increasing research in arid areas other than the one they are interested in.

Graziers' associations are organised on a state basis. Too often the representatives from the sparsely populated arid areas have their voices drowned in the state association and their proposals do not reach their federal councils. It is time that graziers at Bourke realised that they have more in common with graziers at Longreach, Alice Springs, and Meekatharra than with those at Yass or Tamworth. The situation at political level and with scientists is similar. It is long past time that all people concerned with arid Australia and its future dropped their parochial attitudes and worked instead for the best interest of arid Australia as a whole. We all need to view Australia as it is depicted in Fig. 1.

There is room for many organisations, institutions, and departments, both state and commonwealth, to contribute to rangeland research and extension, and a certain amount of rivalry can be tolerated. However, the rights and jealousies of organisations and states must not be allowed to obscure the common objective. If, for example, a Commonwealth extension service would best serve arid Australia, then one should be established irrespective of the traditional role of the states in this field in more humid areas.

* CSIRO, Rangelands Research Unit, Canberra, A.C.T.

Fig. 1 Arid Australia

In case I am misunderstood I should make it clear that I am not advocating the abolition of state boundaries or state rights in arid Australia. The state governments, as the owners of arid Australia, obviously are vitally concerned in, and important to, its future—it is through them that many of the research results will be applied. What I am advocating is a greater recognition of common interests on the part of all concerned with arid Australia. For example, I doubt whether the seven representatives from arid eloctorates to the Commonwealth Parliament have ever met to discuss their common interests, and I am certain that their state counterparts have not. Similarly, representatives of the various district graziers' associations in the arid zone do not meet; instead they meet with representatives of humid districts at their state branch.

Scientists do have better contact. Since 1960 the 5-yearly meetings of the Arid Zone Technical Conference sponsored by the Standing Committee on Agriculture have encouraged the interchange of ideas and information among scientists, as has the annual Arid Zone Newsletter sponsored by the same body. Both, however, are concerned with the exchange of information rather than the formulation of policies aimed at getting the best value for the arid zone from the limited research funds available.

HISTORY AND SETTLEMENT

For the last twenty to thirty millennia arid Australia has had a sparse Aboriginal population. The Aborigines were hunters who grew no crops and kept no herbivorous animals; they adapted to the environment and had little effect on it. The main herbivores were marsupials. They also existed in equilibrium with the environment, and their populations rose and fell with good and bad seasons.

The first European contact with arid Australia was by British and Dutch sailors who landed on the west coast in the seventeenth century. They were not impressed by the desolate landscapes or by the native inhabitants. The first explorers entered the arid zone from the young settlement of New South Wales in 1829 and by the early 1860s the major exploration phase was over and the general nature of the country was known.

Pastoralists were not far behind the explorers. The earliest applications for leases were filed in New South Wales in 1836 and by the 1860s squatters were scattered over most of western New South Wales, and much of arid South Australia and Queensland were settled sparsely. Settlement gradually spread—from South Australia into central Australia, from Queensland across the Northern Territory to the Kimberley district of Western Australia, and from the south-western corner of Western Australia inland and up the west coast to the Kimberleys. By 1890 pastoralists and their sheep and cattle were scattered the length and breadth of the arid zone, but in South Australia, Western Australia, and the Northern Territory settlement of new land has continued to the present day.

Rabbits (*Oryctolagus cuniculus*) entered arid New South Wales and South Australia in about 1880 and within the next twenty years had spread across the whole of the southern half of the arid zone.

Roads and railways were developed slowly, often to serve mining rather than pastoral industries, and for most of the period of European occupation the lack of good transport facilities has meant that stock could be moved only by droving them along stock routes. This restricted management alternatives, particularly as stock routes were often closed during droughts. In the last two decades the development of better roads and motor transport has had a tremendous effect on the pastoral industries. Now stock can be moved rapidly at any time.

The lands of arid Australia have been grazed by sheep, cattle, and rabbits for periods varying from a few years in the most recently occupied areas to a little over a century in the oldest settled parts, a period short in comparison with most other arid lands of the world. In that short period we have caused far more degeneration of land resources than the Aborigines caused in twenty to thirty millennia. Australia is fortunate indeed that the period has been so short.

REASONS FOR RESEARCH AND EXTENSION

The rangelands support nearly a third of Australia's sheep and beef cattle. That the arid pastoral industries earn over $400 million of export income annually gives some idea of their importance to the national economy and indicates that their continuing productivity is of importance to all Australians, not merely the pastoralists.

In arid Australia the pastoral industries are entirely dependent on native vegetation which, because of the severe stresses imposed by the harsh climate, exists in a delicate balance with the environment. Careful management, based on sound ecologic principles, is needed if the balance is not to be upset.

The pastoral industries are exploitative—they have been mining the vegetation and land resources on which their future and the future of arid Australia depend. Throughout the whole zone land types sensitive to erosion have been degraded severely and even the stable sites have suffered some damage. Here again Australia is fortunate on two counts. First, most of the perennial plant species on which the industries depend in dry periods are relatively unpalatable compared with the short-lived vegetation produced after rains. Thus after rains grazing pressure tends to be concentrated on the short-lived plants, allowing the perennials a natural recovery period. Nevertheless the perennial plant populations have been reduced greatly in many areas. Second, most of the landscapes are relatively stable and not subject to rapid erosion.

The fact that most of the perennials are relatively unpalatable, most of the landscapes relatively stable, and the period of stocking has been short provides no excuse for complacency. They merely explain why arid Australia is not already a dust-bowl. The poor condition of sensitive land types is obvious to the most casual observer, and the steady deterioriation of all land types is shown by the trends in stock populations described in chapter 7.

It is easy to blame the pastoralists for the exploitative nature of the pastoral industries, and certainly they have been the agents of the degeneration which has occurred. The real cause, however, is ignorance—ignorance not only on the part of the pastoralist but also on the part of the governments who own the land, their administrative departments, the financial institutions, and scientists, economists, and politicians. The plain facts are that for productivity of our arid lands to be maintained special management methods, based on the results of scientific and economic research, must be devised and implemented. This requires both research and extension.

An important but neglected aspect of the management of arid lands is that management decisions are generally made not in the context of the biological environment but at the dictates of financial exigencies. The financial and administrative environment has been determined by institutions experienced in humid areas but largely ignorant of the special needs of arid areas. Financial policies for humid areas are based largely on an annual balance sheet, a basis not suited to arid areas where sequences of several good or several bad years are common. Here financial policies need to be based on a longer-term balance, e.g. running 5-year periods. This raises a very important point regarding the extension of improved methods of management. Management needs to be based on biological knowledge but can only be applied by adjusting the financial environment. It is just as important, and possibly more important, for extension to be aimed at financial institutions and government administrators as at the graziers.

PRESENT RESEARCH AND EXTENSION

By comparison with settlement the development of agricultural research and extension services was long delayed and slow. Even now they are hopelessly inadequate for the task confronting them.

Queensland has had stock inspectors and veterinarians stationed at inland centres for several decades and agrostologists for a shorter period. Traditionally most of these people have been concerned with administration and extension and have had little time to spare for research, but in the last decade a small research laboratory has been established at Charleville. Western Australia has one or two graduates stationed in the Murchison and Pilbara districts. In arid New South Wales the State Department of Agriculture has only one graduate (Broken Hill) but the Soil Conservation Service has four at Condobolin, two at Nyngan, and one each at Hay and Broken Hill. The South Australian Department of Agriculture has shown little interest in the arid part of the state and has no staff permanently concerned with it. The Northern Territory Administration established its Animal Industry Branch at Alice Springs just after World War II. Although their headquarters are now in Darwin their Arid Zone Research Institute at Alice Springs is the best equipped and staffed of any laboratory in arid Australia.

From its inception CSIRO has shown a sporadic interest in arid Australia: in its very early years it brought Francis Ratcliffe to Australia to investigate sand drift in the dry areas, and it established Gilruth Plains at Cunnamulla initially for blowfly research and later for pasture and then sheep breeding investigations. In 1953 a small centre was established at Alice Springs; it was virtually closed in 1963, but two graduates were allocated to it in 1968. Also in 1968, Gilruth Plains was closed but the emphasis of the Riverina Laboratory was changed from irrigation to dryland research.

The University of Adelaide is responsible for Koonamore Vegetation Reserve and the University of New South Wales runs Fowlers Gap Arid Zone Research Station near Broken Hill and Mulberrygong Field Station near Hay.

Many individual scientists from Universities, State Departments, and CSIRO, and mostly operating from the large coastal cities, have contributed to arid land research. Some research (e.g. research on textiles and food preservation) is as relevant to the products of the arid zone as to those of the humid areas.

REQUIRED RESEARCH AND EXTENSION

Ecologically, a grazing industry is a climate/land/vegetation/animal ecosystem which must be maintained in some sort of equilibrium if it is to remain viable. The main features of the system are that rain falling on land causes both erosion and soil moisture increments, soil moisture and solar energy produce plant growth, plant production provides both landscape stability and animal food, animals in consuming vegetation reduce the stabilising effect of vegetation on the landscape but produce the only products usable by man from the system. Thus the land and vegetation are intermediate stages in a system concerned with the conversion of rainfall and solar energy into animal products useful to man. Considered in this way, it is clear that the needs of the land and vegetation are more important to the viability of the system than the needs of the animals.

The ecosystem can be likened to a manufacturer's production line in which raw materials are put in one end, go through a series of processes, and emerge at the other end as saleable products. There are some differences—in a production line the steps are a continuous sequence of events each fairly independent of the next, whereas an ecosystem is much more complex with all kinds of 'feedback' mechanisms. Also a manufacturer sets up a production line from a knowledge of all the steps and he can design it to be operated and adjusted to suit his convenience. The main part of the pastoral production line has been set up by nature—we have only a very hazy idea of the processes and cannot control the inputs. Further, if the pastoral production line breaks down it tends to destroy itself irretrievably.

Australia needs to ensure that the pastoral production line is managed so that its productivity is at least maintained and so that, under no circumstances, will it break down. This requires a knowledge of the production line which we do not have and which can only be acquired by research.

A balanced research and extension program for arid Australia needs to include several main components. The existing production line has to be serviced and the operators trained; knowledge is needed of each step in the production line; and the structure and function of the overall production line needs investigating.

The servicing and training component is concerned with keeping the existing industries functioning. It is concerned with such things as animal health, reproduction and nutrition, poison plants, weeds, plant introduction, selection of bore sites, water and soil analyses, soil conservation measures, bush fire control, administrative and regulatory provisions, and with informing and instructing pastoralists and others who affect management decisions. In Australian agriculture, state and territory organisations are responsible traditionally for this component and at present most of their arid zone staff are concerned with it. Queensland and the Northern Territory are the most active, but both require more staff and facilities to carry out their responsibilities adequately. In the other states research and extension services for arid areas are hopelessly inadequate; in South Australia they are non-existent.

The second component—understanding the steps in the production line—is concerned with research on individual factors or individual processes in the ecosystem. This kind of research can be done as short-term projects by individual scientists located in large coastal cities, and probably for this reason most of the past and present arid land research and the bulk of literature deals with it. Many scientists from Universities, state and territory organisations, and various Divisions of CSIRO have contributed.

While most of the past and present research is of this type, the total volume is small and needs to be increased. To a certain extent it will proceed independently of special arid zone research funds because of the intrinsic scientific interest in extreme environments of which the arid environment is an example. It does not require scientists motivated to the welfare of arid Australia, but mission-orientated groups can use many of the results and can encourage an increased volume of work by providing basic facilities (such as office space and ground transport) for visiting scientists. Mission-orientated groups can also influence the nature of the projects by providing limited financial support for selected projects of particular interest to them.

The third type of research—that concerned with the structure and function of the overall production line—has been neglected entirely in Australia and is most in need of strengthening. It requires studies of the whole climate/land/vegetation/animal ecosystem, the processes operating in it, their rates, and their interactions with each other and on the system as a whole. An associated need is for methods of assessing the state or health of the ecosystem at any time, and whether it is improving or deteriorating—in range management terminology we need to develop range condition and trend standards. This third type of research requires integrated studies by interdisciplinary teams of scientists, in close collaboration with economists. It is long-term, mission-orientated research and requires teams of scientists motivated to solving the problems of the arid pastoral industries as well as to extending knowledge. It cannot operate on the basis of intermittent forays into arid areas and requires the allocation of special funds for arid zone research.

No amount of individual research on single factors or single processes will provide an adequate knowledge of the whole interacting system—a knowledge which, together with economic assessments, is absolutely essential for the formulation of intelligent management practices. This type of research has arid-zone-wide relevance and should be organised on a national basis. The recognition of the need for it has led to the recent establishment by CSIRO of a Rangelands Research Unit which plans to locate several interdisciplinary teams in arid Australia. Because of the complete lack of this type of research in arid Australia in the past, the first task of the teams will be to establish the broad outlines of the ecosystems.

The distinction between the three types of research is not sharp. Each type has an important role in a balanced arid land program of research and extension. Each complements and needs the others; they are in no way alternatives.

ORGANISATION OF RESEARCH AND EXTENSION

It is one thing to see the need for, and advocate, increased research and extension for arid Australia; it is a different matter to organise and operate them. Many factors must be considered. Of over-riding importance is the general constraint imposed by the finance available, or likely to be available, and the size and remoteness of the area. Other factors include the diversity of the environment and the pastoral industries, the type of research needed, the requirements for the efficient operation of field centres and laboratories, and the requirements for attracting and keeping staff.

The area is vast and, although the climate, landforms, soils, and vegetation can be described broadly in terms of a small number of major types, there is a multitude of environments. There are two main industries (wool sheep and beef cattle), each with quite a wide range in intensity of operation from place to place. Clearly one research centre cannot cope adequately with the range of environments, but equally clearly a large number of centres is not possible because of the financial constraint. No adequate agroclimatic regionalisation of the arid zone is available, but on amount and seasonality of rainfall four major climatic zones can be distinguished. (For convenience these can be designated as summer rainfall arid, winter rainfall arid, summer rainfall semi-arid, and winter rainfall semi-arid.) At the very minimum research should be undertaken in each of these major climatic zones. More sophisticated agroclimatic analyses are likely to disclose additional zones, e.g. the south-western corner has a higher proportion of winter rain than other parts and there is an area in southern Queensland where, although most of the rain falls in summer, the winter component is more reliable.

Among features desirable for efficient operation of field centres are good land and air transport, availability of normal services (water, sewerage, electricity, mail, telephone), availability of technical services for maintenance of vehicles and equipment, and availability of shops stocking a wide variety of everyday items. These same features are required on a private basis by scientists, support staff, and their families, who also need comfortable housing, a satisfactory cultural and social environment, good medical services, and better-than-average educational facilities. Only the larger Australian country towns come near to fulfilling these requirements and, as far as possible, research centres should be located at such towns.

Another very important aspect is that scientists in isolation are not self-sufficient: they need the constant association of enough other scientists for the whole group to be self-critical and self-stimulating.

Similar arguments apply with extension staff, but in addition they will do their job more capably if they are closely associated with the research workers.

Taken together all these aspects suggest that, within the over-riding financial constraint, the research needs of arid Australia will best be met by concentrating research and extension workers at a small number of field centres located at large towns and having them commute to study sites. Such a policy also has the advantage that adequate clerical and administrative staff can be appointed to support the scientists. The locations almost select themselves—Alice Springs and Broken Hill are the only possibilities in the summer rainfall arid and winter rainfall arid zones, and Charleville is the obvious choice in the summer rainfall semi-arid zone. In the winter rainfall semi-arid zone there are no really suitable places—Deliniquin has an established CSIRO laboratory but it is in the extreme south of the zone. While it is suitable for work on the riverine plain, research on central and northern New South Wales might best be conducted by teams of scientists located in Canberra or possibly Armidale.

The living, cultural, and educational standards desired by scientists and their families are not available at most arid localities. There has been a tendency in the past for both individuals and institutions to consider outback posting as only suitable for young, single graduates. After serving a few years or on marrying, the scientist is rewarded by a posting to a more 'civilised' locality. No formal courses on arid land and its management are available in Australian universities, so the young graduate, trained in humid agriculture or biology, lacks the proper background and orientation for arid zone work. If he has the opportunity and is energetic he will set up some short-term experimental work. By the time he is experienced enough to make a worthwhile contribution to the welfare of arid areas his term is up.

Such an attitude has been detrimental to both research and extension for arid areas and everything possible should be done to change it. Concentration of staff, where possible, at large towns, the provision of first-class laboratories and facilities, attention to such things as housing and living standards, and, where necessary, adjustments to administrative regulations can all help to encourage good quality scientists to make a career of arid land research and extension. Regular association with staff at other centres and at headquarters can help to weld the various groups together and foster a feeling of belonging to an integrated organisation motivated towards worthwhile objectives. The regional groups need to feel that they are part of the whole and need to have a voice in determining its policies.

SPECIAL ASPECTS OF ARID LAND RESEARCH AND EXTENSION

In discussing agricultural research and extension for arid areas the question arises of whether the requirements and philosophy are different from those of humid areas. The answer is not a simple yes or no. In many respects the principles behind any agricultural enterprise apply but there are also many basic differences.

For example, for grazing industries in both humid and arid areas the main outline of the climate/land/plant/animal ecosystem is similar—essentially both are concerned with converting rain and solar energy to animal products. Here the similarity ends. Under humid conditions land, plants, and animals can be manipulated economically—land can be intensively subdivided, ploughed, contour banked, fertilised; plant cover can be changed by sowing more productive and nutritious species; and animals can be rounded-up easily at any time for drenching, hand feeding, and control of breeding. If, at any time, pastures are destroyed by overgrazing, they can be re-established quickly and the cost recovered in a reasonably short period. The situation is quite different under arid conditions. Productivity per unit area is so low that direct manipulation of land and vegetation is not economic and the state and productivity of the whole ecosystem must be controlled by adjustments in the animal factor alone, i.e. opportunities for managing the ecosystem are more restricted under arid conditions. Also a single period of mismanagement which destroys vegetation is disastrous—severe vegetation and land degeneration are virtually irreversible under most conditions and regeneration is slow at any time.

In humid areas the primary aim of agricultural research is to increase the productivity of the enterprise. If for any reason (poor extension services, lack of finance or incentive on the part of the farmer) the research findings are not applied immediately there is a financial loss to farmer and to the national economy but commonly not an irretrievable loss of resources. On the other hand, the first goal of pastoral research in arid areas is to maintain the land resources in a healthy, productive state. Increasing productivity is important but it is a secondary objective. Here delays in application can result in irretrievable destruction of land resources. It is important that extension to graziers be efficient and also that research results are communicated to those responsible for financial and administrative policies. We can lament the ineffectiveness of extension in humid areas; we cannot afford to in arid areas.

The low productivity per unit area and the erratic nature of the rainfall of arid areas also affect some of the technical aspects of research in comparison with humid areas. For example, to run an adequate number of animals in an arid zone grazing trial requires a much larger area per treatment than in a humid area. Even in arid areas it is difficult to get large homogeneous areas and, in any case, a lot of time and effort is required to sample large plots, so much so, that conditions may alter during a sampling period on a large experiment. In addition, to get an adequate sample of the erratic climate, a conventional grazing trial would need to run for a long time (at least twenty and possibly thirty years). Thus a conventional grazing trial would be much more expensive in arid than in humid areas and it is doubtful whether the results can justify the expense. It is more logical for arid pastoral research to concentrate on gaining a fundamental knowledge of the ecosystem, how it operates, and the factors which influence its operation. This can be done by shorter experiments on smaller areas and, provided appropriate environmental factors are monitored, will result in models which can be extrapolated in both space and time. Simple, short-term grazing trials will be needed to validate models.

Much of the agricultural research in humid areas is conducted on experimental farms owned and operated by the research institution. Similar organisation in arid areas would require the purchase and operation of sheep and cattle stations—a very costly undertaking and one which tends to concentrate research on the land types available on the station. For these and other reasons it is likely to be more efficient for rangeland research to be conducted on selected sites on graziers' properties by staff working from field centres or laboratories located in major towns.

These comments should not be construed as a dogmatic assertion that conventional grazing trials or experimental stations have no place in arid land research; rather they are meant to indicate that the special nature of the arid environment and the arid pastoral industries influences the mode of approach.

CONCLUSION

Australia is an urban country with its population concentrated on a narrow coastal fringe. Despite this, it is almost incredible that such an advanced country has almost completely neglected research and extension for its arid areas, which occupy three-quarters of its land surface and contribute substantially to the national coffers. Most of the research which has been done consists of bits and pieces by interested individuals and has proceeded despite the lack of special funds.

The size and remoteness of the arid zone and the special nature of the environment and the pastoral industries pose special problems for the organisation of research and extension but they are not insuperable.

The most pressing need is for investigations of the structure and function of the arid climate/land/vegetation/animal ecosystems (the pastoral production lines). Such investigations can only be carried out satisfactorily by interdisciplinary teams of scientists motivated to solving the problems of our arid pastoral industries and to the future welfare of arid Australia. No amount of individual research will provide the required knowledge of the overall ecosystem. Only when we have such knowledge will we be able to devise the special management practices required for the continued economic utilisation of our arid lands.

The research can best be implemented on a national basis and requires the establishment of several interdisciplinary teams of scientists located in selected areas but working closely together within an integrated organisation. CSIRO has recognised the need and made a modest start but with very limited funds.

An associated need is for high quality efficient extension services fully cognisant of the realities of the whole interlocking biologic/financial/political systems within which management decisions are made. It is pointless to have excellent research without ensuring that the results are applied.

The pressing needs in both research and extension fields require the allocation of special funds. The provision of special funds is an investment for the future of our arid lands which affects the future welfare of all Australians. It is a national as well as a state responsibility, and it is one Australia cannot afford not to accept.

18

EDUCATION AND TRAINING PROGRAMS FOR MANPOWER[1]

THADIS W. BOX*

IT IS APPARENT that as the human population grows there will be increasing demands put upon the arid regions of the world. Authorities may disagree on when world-wide food shortages will occur, but it is obvious that if the present rate of population growth continues pressures will increase for all habitable space on the earth's surface. Optimists believe that the human population will be brought under control before severe world-wide food shortages occur, and much progress has been made. However, it is doubtful that the population can be controlled before it becomes necessary to develop the two remaining large portions of the earth available for increased exploitation—the tropical rain forests and the arid zones.

The problem of arid zone development is acute in that many of the world's fastest growing populations are located in the dry regions of the world. For instance, on the continent of Africa those countries located south of the Sahara have experienced a 2·3 per cent annual increase in population in the 1960-7 period and at the same time have made a 3·7 per cent annual increase in their gross national products. The arid countries north of the Sahara have in the same period had an annual increase in population of 2·6 per cent, the second highest in the world, and an annual growth rate of 2·2 per cent in gross national product, the lowest of all the developing countries of the world (Anon., 1968). Similar figures could be quoted for other desert regions. The populations of the arid nations of the world continue to increase at unprecedented rates, and technology for arid zones has not kept pace. Not only has arid zone technology fallen behind that of the more humid regions of the world, but our attempts to educate people to identify and solve the problems have been largely inadequate.

In the developed nations of the world, insufficient emphasis has been placed on educational or research programs for arid zones. Political pressures from the dry regions have not been great and economic returns from these areas not as certain or as profitable as from the more humid areas. If educa-

[1] This paper is contribution number 65, International Center for Arid and Semi-Arid Land Studies, Lubbock, Texas, U.S.A.

* International Center for Arid and Semi-Arid Land Studies, Texas Technological College, Lubbock, Texas, U.S.A.

tional programs have been lagging in the developed nations of the world, the plight of the developing nations is even more severe. Few programs exist for the education of arid land scientists in developing countries. Far too often an individual is trained in a developed country but fails to return to his homeland, and his expertise becomes available instead to a more developed nation. If he does return he may find himself in a political or administrative position well outside the arid zone, working in an area where the amenities of life are more readily available. Therefore, the developed nations of the world have a responsibility not only to initiate programs for the development of knowledge for their own countries but also to train people to extend knowledge on problems of aridity to the lesser developed countries.

This chapter concerns itself primarily with the problems of education and training of manpower for two developed countries, Australia and the United States. We should always remember, however, that we have the distinct responsibility of developing programs suitable for the solution of world-wide problems.

One question immediately arises in discussing a problem of education and training for arid zones: is there a difference between educational and training programs for work in different ecological areas? I think that there is. It is one of philosophy rather than of practice. Granted, the differences in educational programs for arid and humid areas will be less than for training programs. The major difference between educational programs for personnel working in humid and arid regions will be in the philosophy of resource use. For instance, educational programs for humid regions can well be designed on the philosophy of maximising production. Landscapes can be altered and natural communities replaced, destroyed, and restored within the lifetime of a man. The same is not true in arid regions. Once a landscape has been altered, or a natural community disturbed, it may take generations to restore itself. Educational philosophy for arid regions, then, must be geared more to adapting to environmental changes or conditions than to altering them.

Training programs, or the developing of skills to do specific jobs, obviously will not be the same in arid as in humid areas. Specific techniques, methods, etc. are needed for arid conditions that are not suitable for the more humid areas. Here I confine my remarks primarily to educational programs rather than training for specific jobs.

Sound educational programs are a major key to resource development in arid zones (Thomas and Box, 1968). Staley (1967) has suggested that the four major areas of educational programs necessary for development are:

(1) a general education in which an individual is taught to communicate in his own language, something about his social environment, and elementary ideas of science and computation;

(2) the preoccupational phase in which he is oriented toward the world of work, given testing, guidance, and counselling and given the educational preparation for a family of occupations, e.g. commercial, scientific, medical;

(3) occupational training, in which he is supplied with the specialised tools for a given occupation;

(4) training and re-training throughout the professional career of the individual. The general education in communications, social problems, humanities, sciences, computations, etc. continues throughout the educational program.

In addition to educating for development, there are five specific types of educational programs needed for living and working in arid regions: for general citizenship, for developing scientific manpower, for educating managerial manpower, for training technicians, and for continuing education and retraining.

EDUCATION FOR CITIZENSHIP

Education for citizenship, or general civic responsibility, should be taught at all levels from pre-school throughout the post-graduate period. This phase of education is the responsibility of every teacher in any institution of formal education. In addition, concerned public promotional campaigns should be continually carried out. The goals should be to instil an awareness of the value of natural resources into individual citizens including the unique needs of arid areas.

Perhaps the most important item in the education for citizenship is the instilling of responsibility in the individual for maintaining the quality of the environment. Most developed nations are rapidly changing from affluent societies to effluent societies. The extent to which man's litter and waste products are polluting the available space is alarming indeed. In the United States today the average American deposits 8·5 pounds of solid waste daily in his own ecosystem. The thickly settled areas of the world have been forced to develop complex programs for handling waste products. Unfortunately, the vastness of the arid zone has led many people to a rather careless attitude toward waste disposal. One has only to drive on any outback road in Australia to see beer cans and cigarette packages lining the road and the most desirable camping spots littered with the debris from past occupants. Although the arid regions are remote and large in size, society cannot afford the luxury of leaving its waste unburied and its few precious streams polluted. Increasing emphasis must be placed on aspects of education that will maintain the quality of environment that is related to health, recreation, and aesthetics in the arid zone. To be effective this education for citizenship must begin early and continue throughout the lifetime of all citizens, whether they live in arid or humid areas.

EDUCATIONAL PROGRAMS FOR SCIENTIFIC MANPOWER

There is an increasing world-wide need for scientists trained in arid regions and instilled with the philosophy and educational background necessary for identifying and solving the problems of aridity. In order to fill this need new programs must be developed for the recruitment and education of scientists to work in arid regions.

Attracting Scientists to Arid Zones

There are special problems in attracting people to work in arid zones. Their working areas are isolated and research facilities may be crude. To complicate the situation, most university programs place little emphasis on problems in the arid zone. Educational centres, for the most part, are located in the more populated humid areas and few people actually come into contact with the everyday problems of the arid zone. In some instances counsellors and professors discourage the brighter students from pursuing a career in arid zone development: they will encourage prospective students to stay in the humid areas.

Once a student is trained and employed in an arid zone there are problems in holding him at his location. In general he is isolated from his professional colleagues and may lack the stimulation necessary to continue to be creative. The lack of the social amenities for his family may cause problems at home and the eventual resettlement in another area. Far too often research agencies that conduct work in the arid zone consider the area simply as a training ground for scientists; the investigator who wishes to be promoted must move to a more humid area. Harbison (1968) summarised the problem:

> It is one thing to estimate the needs for manpower of various qualifications, but quite another to induce persons to prepare for and engage in occupations which are most vital to the national growth . . . It is incorrect to assume that relative earnings and status reflect the value of the contributions of individuals to development. Pay and status are more often related to tradition, colonial heritage, and political pressures than to productivity . . . The differentials in compensation between the agricultural officer and the agricultural assistant are too great; and the earnings of scientists and engineers, in comparison with the administrative bureaucrats in government ministries, are too low.

There is no doubt of the need for additional scientific manpower in the arid regions of the world. It is also apparent that these areas present more problems in attracting and holding scientists than do the more populated areas. But regardless of the problems, countries with arid regions must work towards immediate identification of major problems and solutions to those considered to be most pressing.

Types of Scientists Needed

There is a distinct need for two major groups of scientists—one broadly trained at the baccalaureate level and the other specifically trained at the post-graduate level. Regardless of his discipline, the arid zone scientist must be a resource-conscious scientist, not a narrow professionally or vocationally trained individual. For instance, a physiologist working on arid zone plants must be not only a competent physiologist but also able to relate his work to disciplines and the practical problems of the arid world. The Commission on Education in Agricultural and Natural Resources of the U.S. National Academy of Sciences (National Academy of Sciences, 1967) has described the natural resource scientist of the future:

> After they graduate, students . . . will be obliged to face problems having complex biological, sociological and economic attributes and to apply the methodologies of diverse substantive fields. They will have to synthesize information from many areas of knowledge, apply the information to solving problems, and adapt to change.
>
> An understanding of the biotic, physical and human environment is essential to the successful performance of these duties. Obtaining such an understanding through broad, liberal education is clearly more desirable than obtaining it as a remedial measure at the graduate level.
>
> Professional, baccalaureate programmes in renewable natural resource fields will be needed in the future. They should emphasise the natural and social sciences, mathematics, humanities, and communications. A substantial part of the responsibility for 'how-to' training should be shifted to subsequent employers.
>
> There is, above all, a need to develop in the undergraduate a sense of responsibility for the sum total of renewable natural resources, not just for isolated elements.

This broad natural resource training is necessary not only for baccalaureate-trained people but also for people at the post-graduate level. If they have not received it at the baccalaureate level, then they must obtain it somewhere before continuing in post-graduate work.

In addition to general training, students working in arid zones must have a competence in their specific professional field. The entire educational program for an arid zone scientist, then, should include: general education for civic responsibility from the public schools; a general university education in humanities, social sciences, biological sciences, and physical sciences; specific educational philosophy of arid environments and needs; and—specialised education in a given field.

Some people doubt that such educational programs can be developed in a 4-year curriculum. The professional groups such as forestry, range management, and wildlife management in the United States have been debating this problem for years. Recently, engineers in the United States have abolished the

4-year curriculum as a professional degree and have gone to a 5-year curriculum.

I do not think it is necessary to increase the number of years that a student occupies a seat in the classroom. New, imaginative programs are needed that will give the student a sound general education and qualify him for work in a specific field. Far too often in the United States we have not reviewed our programs often enough or dared to throw out obsolete courses and include new ones. The trend has been simply to add an additional course or year on to the student.

It is difficult for a student in forestry, range management, wildlife management, or other natural resource areas to advance professionally with only a baccalaureate degree. Each year we contact employers of our range management graduates from Texas Tech. and ask them how we can improve our product. We have never had a single complaint about the scientific competence of our graduates, but we have been asked many times to improve our students in humanities, communication, and social awareness. Consequently we have added courses in economics, speech, writing, sociology, etc., but we cannot expect our students to continue to have courses added to their 4-year curriculum.

Many governmental agencies are sending their natural resource personnel back to graduate school to earn Masters degrees in administration or social sciences to give them the public awareness needed to do their job. It seems to me that we are inefficient as teachers if we cannot educate a natural resource scientist in four years. But in order to do it, we must take a realistic look at what we are teaching.

Arid Land Scientists in Australia

Australia has one of the largest areas of arid zones in the world. In an attempt to estimate the need for increased manpower in the arid regions, I wrote to fifty-three organisations that had indicated in past arid zone newsletters that they were conducting research in arid Australia. Of these fifty-three organisations, forty-five responded by filling out their questionnaires. These included seventeen Commonwealth agencies, sixteen state agencies, and twelve universities.

Each organisation was asked to list the number of scientists in various categories working wholly or partially in the arid zone of Australia and to estimate the scientific-man-months each contributed to the solution of arid zone problems. In addition, each organisation was asked to estimate the number of scientists and the number of scientific-man-months necessary to complete their mission in 1985. The year 1985 was selected not necessarily to go one year beyond George Orwell's 1984 but because I estimated that if new programs are needed it will take from ten to fifteen years for universities to change curricula, recruit students, and train graduates.

From the forty-five organisations reporting, 434 scientists were employed for approximately 182 scientific-man-years of work in arid regions annually (Table 1). Commonwealth agencies employed more scientists, but universities conducted more scientific-man-years of research in the arid zone.

TABLE 1

SCIENTISTS AND SCIENTIFIC-MAN-MONTHS ENGAGED IN ARID ZONE WORK WITH 45 AGENCIES IN AUSTRALIA, 1969

Type of Agency	Employed 1969		Projected Needs 1985	
	No.	SMM	No.	SMM
Commonwealth (N = 17)	199	805	196	1454
State (N = 16)	69	425	170	1348
University (N = 12)	166	952	289	2763
	434	2184	655	5545

The forty-five institutions indicated an overall increase of 150 per cent in individual scientists needed in the fifteen years. The same institutions estimated a 254 per cent increase in the scientific-man-years required for them to accomplish their mission. These numbers, or the projected numbers, are extremely conservative, since many institutions indicated they could not estimate their needs so far in advance. Others indicated that their actual needs were far greater than the numbers they submitted, but they could not hope to fund their total needs.

The type of training necessary for the future scientists varied with the type of institutions (Table 2). The state agencies indicated a greater need for people trained at the baccalaureate level primarily for research work or soil conservation work, primary industries, etc. Universities were almost unanimous in their need for increased Ph.D. staff. There was a noticeable decline in the estimate of needs for M.Sc. trained people.

TABLE 2

SCIENTISTS EMPLOYED IN AUSTRALIA'S ARID ZONE IN 1969 AND PROJECTED NEEDS FOR 1985

(%)

	B.Sc.	M.Sc.	Ph.D.
1969	47	25	28
1985	53	20	27

TABLE 3

SCIENTISTS, BY DISCIPLINE, AND SCIENTIFIC-MAN-MONTHS EMPLOYED IN 1969 AND NEEDED IN 1985 BY 45 AUSTRALIAN RESEARCH AGENCIES

Discipline	Employed—1969		Needed in 1985	
	No.	SMM	No.	SMM
Agricultural scientists	45	230	79	672
Biological scientists	149	1015	272	2357
Earth scientists	135	266	136	796
Engineers	58	468	67	767
Natural resource scientists	32	151	77	557
Social scientists	15	54	24	196
	434	2184	655	5545

The greatest numbers of individual scientists are now employed in the biological and earth scientist disciplines (Table 3). The greatest increase per cent in projected needs are in natural resource scientists, biological scientists, and agricultural scientists (Table 4). The actual numbers of people needed by the forty-five institutions, listed by educational level and discipline, are included in Table 5. This compilation is conservative in that it considers only those institutions reporting. The real figures could probably be increased by a factor of three or four.

TABLE 4

INCREASE IN SCIENTISTS AND SCIENTIFIC-MAN-YEARS NEEDED BY 45 AUSTRALIAN AGENCIES TO ACCOMPLISH THEIR GOALS IN 1985
(%)

Discipline	Increase in Scientists	Increase in SMYs
Agricultural scientists	175	292
Biological scientists	182	222
Earth scientists	101	299
Engineers	116	162
Natural resource scientists	241	369
Social scientists	160	363

Although the sample is limited, it clearly indicates a need for increased scientific manpower requirements for arid Australia, probably of a thousand to fifteen hundred people with B.Sc. degrees, four hundred to five hundred with M.Sc. degrees, and five hundred to six hundred Ph.D. scientists during the next fifteen years. To put it another way, each university in Australia needs to graduate approximately two hundred people over the next fifteen years for arid zone work.

TABLE 5

ADDITIONAL SCIENTISTS AT EACH EDUCATIONAL LEVEL NEEDED BY 45 AUSTRALIAN AGENCIES BY 1985

(No.)

Discipline	Educational Level		
	B.Sc.	M.Sc.	Ph.D.
Agricultural scientists	53	11	15
Biological scientists	136	41	91
Earth scientists	59	32	44
Engineers	52	23	8
Natural resource scientists	43	13	11
Social scientists	12	6	6
	355	126	173

Supplying Australia's Needs

Certainly the need for arid zone natural resource scientists in Australia is great. Where will these students come from? At the present time there is no adequate natural resource program in Australia. Probably the best equipped Australian student is one trained in an applied science, such as rural science or forestry, with post-graduate training in a specific biological area. In the Australian educational system, however, there are problems in switching fields between undergraduate and post-graduate work; there is little opportunity for re-training and re-orientation. If a geologist, for instance, wants to switch to biological science for post-graduate work, academic roadblocks are almost insurmountable.

There are at least two new programs in natural resources planned in Australian universities. The University of New England has applied for a program to train students in forestry, wildlife management, watershed management, and soil conservation; and the Australian National University is considering a multi-disciplinary approach to training in the natural resources at the post-graduate level. At the present time neither is operational and neither plans to orient its program toward arid zone work.

As we have heard in this symposium, approximately three-fourths of Australia is either arid or semi-arid. If the real potential of this area is to be reached, new programs at both the B.Sc. and Ph.D. levels should be initiated in Australia in range management, wildlife management, watershed management, arid zone forestry, recreation, resource economics, and other resource-oriented fields. These so-called applied sciences should all have a similar basic core of subjects in fundamentals of biology, ecology, mathematics, statistics, general chemistry, physics, meteorology, climatology, geology, soils, economics, resource economics, political science, sociology, philosophy, literature, communications and resource analysis and planning. Courses should be designed to place special emphasis on arid land problems.

In addition to the general educational core, blocks of courses in certain areas of emphasis should be given to students preparing themselves for specific occupations. For instance, people interested in plant sciences should take genetics, plant physiology, entomology, plant pathology and plant taxonomy. Those interested in the animal production areas should take genetics, animal physiology, vertebrate zoology, invertebrate zoology, microbiology, entomology, animal diseases and parasites. Those interested in soil and water sciences should take a block of courses to include geomorphology, hydrology, sedimentation, soil morphology, classification and interpretation. Those interested in social science areas should be required to take a block of courses including advanced economics, resource economics, public administration, social structures, resource policy and administration.

In addition to the general educational core and their area of emphasis, students should take specialised training in their chosen profession. These would normally consist of four or five specialised courses in fields such as range management, forestry, wildlife management, watershed management, and economics.

Students wishing to pursue a career in engineering, architecture, commerce, etc., and/or other fields in the arid zone should take the general educational core of courses and then proceed to their specialised training in engineering, medicine, law, and so on. If elective courses are available students could well take advantage of the chance to increase their biological knowledge.

If the need to train the arid zone resource specialist exists, who will take the lead in developing such a program? Will employers demand this type of individual? Will professional societies outline their needs? I think both employers and professional societies are rather inefficient in designing new educational programs. Far too often employers are concerned with individuals trained for a specific purpose. Professional societies, on the other hand, are generally concerned with perpetuating the particular species of scientists that they themselves are.

I believe that it is the responsibility of the universities to develop new programs. They should anticipate the need and develop the training before industries and societies demand the finished product. It is generally helpful if some group or body sponsors a special study on needs and outlines broad guidelines for the universities to follow. Such was the case when the National Academy of Sciences in the United States sponsored discussion and exchange of ideas on the development of new curricula in renewable natural resources. It seems to me that it is entirely appropriate for the Australian Academy of Science to establish a commission to look into the needs for education of the arid zone scientists and draw broad outlines for the development of curricula.

EDUCATIONAL NEEDS FOR MANAGERIAL MANPOWER

The primary use of most arid lands is the production of animal products from grazing native plants. The pastoral industry based on native plants is no longer a business designed on the mining of a natural resource. In the early days of grazing in America and Australia one had only to declare his intention to become a grazier, have the necessary capital to acquire the land and buy the livestock, and exploit the grazing resource.

Unfortunately, most grazing areas have long been overgrazed, and the ranching business today is a complex enterprise involving decisions in native forage production, animal husbandry, and economics. The rate of return is generally quite low and most livestock producers cannot long survive unless their decisions are based on scientific fact and economic practicability.

The increased complexity of the ranching business in the United States is verified by the fact that there is an increasing demand for scientifically trained managers to run the grazing properties. At our university we are placing an increasing number of our baccalaureate graduates in ranch management on private properties each year. Our Master of Science people are in great demand, particularly if they have been trained in range management at the baccalaureate level and have taken their Masters in economics or animal husbandry or vice versa.

At a recent livestock producers' meeting in West Texas, one rancher told the audience that he knew of at least ten properties in the West Texas area that could benefit from hiring a Ph.D. in range management. Two large West Texas properties have recently hired experienced Ph.D. scientists as general managers. In addition, there is an increasing demand for consultants in the range management area.

There is apparently a high demand for farm managers in Australia. Nelson and Schapper (1966) studied the needs for agricultural manpower in Western Australia and concluded that the training in that state was quantitatively insufficient and qualitatively inadequate. They estimated that there is an annual need of 750-800 new farm managers annually until 1976 and 500-600 thereafter for the next twenty years. In addition, they found that at least 1000 new non-managerial farm technicians were needed per year. They reported that only 185 certificate- and diploma-holders in agriculture were graduated per year from all agricultural high schools and colleges. Practically no university degree students went into farm management.

The result was that in Western Australia practically all farm management is self-taught and farmers are training the non-managerial workers. Only a small minority of the agricultural personnel receive formal agricultural schooling at any level. Although no comparative figures are available for other states in Australia and none for the arid zone *per se*, I suspect that the need for managerial personnel in the arid zone is at least as great on a percentage basis as the entire state of Western Australia. Managerial personnel usually are taught through the jackaroo system.

The increasing complexity of the pastoral industry dictates that people with better education are needed to manage the properties in the arid zone. Agriculture is no longer just farming; the agricultural enterprises of the arid zone are particularly difficult to manage since they are based almost entirely on the ecological potential of the land and the economics of the country in which the arid land occurs. I have no accurate estimate of the extent to which university-trained range managers would be accepted in Australia. I think it is significant that our university has had calls from three Australian firms during the last year for range management graduates to assume responsibility of the management of Australian properties. I have also had several property owners ask me about the feasibility of sending their sons to the United States for a degree in range management.

If the properties located in the arid zone of Australia are to be managed on a scientific basis, new programs in training pastoral managers must be initiated. There are several possibilities. University programs involving the integration of ecology, animal husbandry, and economics could be initiated, similar to the program outlined by the 19-member Range Management Education Council of the United States. In Australia, this would probably be a joint program between faculties of rural science, economics, and biology.

Some combination of university education and jackaroo training seems to me to be the most desirable method of educating and training a scientifically oriented yet practical range manager. For instance, a prospective student could spend two years at a university obtaining the core of general education courses and some orientation in agriculture. He would then be assigned to a property for one to two years of jackaroo training, returning to the university for his final two years of specialised training. Such a program would not only educate and train badly needed managers for arid zone properties but would also bring the university professors into direct contact with the practical problems faced on the properties in the arid zone.

A third possibility would be the combination of a diploma from an agricultural high school or college with a university degree in rural science. Regardless of the system used there appears to be a real opportunity for some university to establish a program especially to educate managerial personnel for the arid zone properties. Unless one of the universities takes this challenge, ranch management will continue to be largely a self-taught perpetuation of the *status quo*.

TECHNICIAN TRAINING

The demand for scientific and managerial personnel will increase the pressures on universities to supply the manpower needs. However, in addition, increasing numbers of technician-trained individuals will be required to carry on the scientific investigations and intensive management generated by the new scientists and university trained managers, and these people will need vocational training. One apparent source of this training is to increase the number of agricultural high schools and colleges. However, in many cases the people handling the technician jobs may have been out of high school for several years. Some may have a leaving certificate; others may have dropped out years ago. There is an apparent need for short courses to train technicians to do specific jobs in the arid zones. These short courses can probably best be handled as continuing education programs through established universities. In some cases these programs may be handled by extension people actually working within the arid zone.

There may also be a case for special vocational schools in arid zones. Many local inhabitants, such as Australian Aborigines, may be unemployable as technicians with their current level of education. With proper adult education and specialised 'skill' training they may become valuable additions to the manpower supply of the arid regions.

These vocational schools are obviously outside the scope of the standard educational institution. Special curricula would have to be developed to give adult students, with varying and minimal formal education, opportunities to gain competence in skilled jobs associated with research and industry.

By training local people for skilled jobs, the problem of attracting and holding competent personnel in remote areas is largely overcome. Serious consideration should be given to developing programs to train local 'unemployables' for specific skills.

EDUCATIONAL PROGRAMS FOR PEOPLE NOW EMPLOYED IN THE ARID ZONE

There is an immediate and growing need for continuing educational programs at all levels for people now working in the arid zones.

A dean at a leading university recently said that all university diplomas should be printed in disappearing ink because within ten years the education students had received would be obsolete. There is no doubt that, with the knowledge explosion and the new techniques and methods continuously being developed, scientists must be retrained periodically. The problem in arid regions is even more critical in that many people working in the arid areas were originally trained in the humid regions.

By the same token farm managers, investors, and others working in the arid regions need to be continually brought up on the newer developments in arid zone technology, and technicians must be continuously retrained to operate new and more sophisticated equipment.

Whose responsibility, then, is this retraining program? Is it the agencies that employ the people; is it the state departments that have responsibility for the land within the arid regions, or is it the universities?

My personal bias is toward the universities. Knowledge is their business. They are charged with the development of knowledge and must continuously be at the forefront of all fields. Not only must they lead in the development of knowledge, but they must identify areas for further research and train the people to do the work. They also have the responsibility of disseminating knowledge and making it available to society. Indeed, they should lead in the application of new knowledge, in relating new developments and new technology to real world problems, regardless of their unpopularity. Of all the organisations only the university is free to develop research into controversial areas. Industry-oriented research must work toward the solution of problems specifically related to a given industry. Governmental agencies are directly responsible to voters or to politicians who must answer to voters. The academic freedom enjoyed by most universities is necessary to allow free inquiry and the development of knowledge. However, if the university fails to relate this knowledge to the problems of the community it serves, then it is slack in its duties. It is the responsibility of the university, then, to see that the knowledge it develops is successfully applied. This responsibility may mean that extensive retraining programs need to be developed by the university.

THE FUTURE NEEDS FOR EDUCATION IN ARID REGIONS

I have tried to say here that arid areas are an important portion of the world's surface and they will become increasingly more important. There is a need for more people equipped to identify and solve the problems relating to aridity. Present programs, for the most part, are inadequate to educate and train these people. There is a need for creative new approaches to education of arid zone personnel.

Political pressure is not great for new programs for arid zones because of their sparsity of population and their present relative unimportance economically. Therefore, if new programs are to be developed, the burden for the development will largely fall on the universities. Universities must continuously be looking to the future, anticipating needs and attempting to educate people for the development of the earth's resources.

New social pressures and increasing technology will change the requirements of the scientists of tomorrow. Industry recognises this fact and attempts to forecast manpower needs as much as ten years ahead. The university then, in order to be ahead of industry, must look at least fifteen years in the future. If we wait for fifteen years to start educating people to work in arid zones then we may be too late to save this valuable resource.

I would suggest that one of the important developments that could come out of this symposium would be the appointing of a commission from the Australian Academy of Science to investigate more fully the manpower needs for developing Australia's arid zones and to suggest new curricula and programs for meeting the need. The natural resources of the arid zone are important and the most important resource is the human resource. Society cannot afford to let the human resource go undeveloped.

REFERENCES

Anon. (1968). How much progress are developing countries making? *OECD Observer* **36**: 16-19.

Harbison, F. (1968). A systems analysis approach to human resource development planning. In *Manpower Aspects of Educational Planning*, UNESCO, Paris, 57-78.

National Academy of Sciences (1967). *Undergraduate education in renewable natural resources— an assessment.* Publ. 1537. Natl. Acad. Sci., Washington, D.C.

Nelson, J. S. and Schapper, H. P. (1966). *Manpower training for agriculture in Western Australia.* University of Western Australia Press. Perth.

Staley, E. (1967). Planning occupational education for training and development. *Manpower Journal* **3**: 7-14.

Thomas, G. W. and Box, T. W. (1968). A formula for resource use in arid lands. In *Proceedings of a Symposium for Increasing Food Production in Arid Lands.* ICASALS Publ. 3. In press.

INDEX

Text set in 10 pt Linotype Times Roman 2 pt leaded
and printed on 85 gsm English Finish paper by Halstead Press Pty Ltd, Sydney

Designed by John Pitson